Plant–Insect Interactions—2nd edition

Plant–Insect Interactions—2nd edition

Guest Editor

Francisco Rubén Badenes-Pérez

MDPI

Basel • Beijing • Wuhan • Barcelona • Belgrade • Novi Sad • Cluj • Manchester

Guest Editor
Francisco Rubén
Badenes-Pérez
Plant Protection
Spanish National Research
Council (CSIC)
Madrid
Spain

Editorial Office
MDPI AG
Grosspeteranlage 5
4052 Basel, Switzerland

This is a reprint of the Special Issue, published open access by the journal *Plants* (ISSN 2223-7747), freely accessible at: https://www.mdpi.com/journal/plants/special_issues/0E15YME735.

For citation purposes, cite each article independently as indicated on the article page online and as indicated below:

Lastname, A.A.; Lastname, B.B. Article Title. *Journal Name* **Year**, *Volume Number*, Page Range.

ISBN 978-3-7258-4573-6 (Hbk)
ISBN 978-3-7258-4574-3 (PDF)
https://doi.org/10.3390/books978-3-7258-4574-3

Cover image courtesy of Francisco Rubén Badenes-Pérez
Drone fly *Eristalis tenax* L. visiting flowers

Contents

plants

MDPI

Editorial

Plant–Insect Interactions: Host Plant Resistance, Biological Control, and Pollination

Francisco Rubén Badenes-Pérez

Instituto de Ciencias Agrarias, Consejo Superior de Investigaciones Científicas, 28006 Madrid, Spain; fr.badenes@csic.es

1. Introduction

The evolving field of plant–insect interactions impacts basic and applied fields of plant sciences, entomology, and agronomy. Within this field, pest management and plant pollination receive a lot of attention because of their implications for maximizing crop yield [1–4]. Host plant resistance and biological control are important aspects of pest management that can bring alternatives to the problems arising with insecticide use [5,6]. This Special Issue presents a collection of papers addressing the interactions of insects and mites with plants, while highlighting the importance that pest management and pollination have in this field.

2. Overview of Published Articles

1. Zhu et al. (Contribution 1) investigated gene expression response to mechanical damage and feeding by *Phthorimaea operculella* Zeller (Lepidoptera: Gelechiidae) at different time points. Compared to mechanical damage, feeding by *P. operculella* induced moregenes and resulted in a stronger gene expression.
2. Zhang et al. (Contribution 2) studied the adaptive responses of *S. frugiperda* Smith (Lepidoptera: Noctuidae) to nutritional and enzymatic variations in different maize cultivars. This study highlights the different adaptations of *S. frugiperda*'s digestive and detoxification systems.
3. Volp et al. (Contribution 3) investigated the avoidance of and preference for pigeonpea flowers and pods of first-, second-, third-, and fourth-instar larvae of *Helicoverpa armigera* Hübner (Lepidoptera: Noctuidae). Early instars were shown to prefer flowers, while older instars preferred to feed on pods. Boring through pigeonpea pod walls imposed a physiological cost for third- but not for fourth-instar larvae.
4. Rodríguez-Leyva et al. (Contribution 4) analyzed the volatile organic compounds (VOCs) released by the cactus pear *Opuntia ficus-indica* (L.) Miller as a result of herbivory by the cochineals *Dactylopius coccus* Costa and *D. opuntiae* Cockerell. Cactus pear produced different VOCs depending on the *Dactylopius* species affecting the plant. Among the VOCs identified, methyl salicylate, terpenes, and *p*-vinylguaiacol were suggested as being likely to play a defense role in *O. ficus-indica*.
5. Abbes et al. Al-Azzazy and Alhewairini (Contribution 5) studied the suitability of three solanaceous crops—eggplant, potato, and tomato—to cotton mealybug *Phenacoccus solenopsis* Tinsley (Hemiptera: Pseudococcidae). Using age-stage two-sex life tables, the study showed that tomato, followed by potato plants, were more suitable hosts for *P. solenopsis* than eggplant.

 https://doi.org/10.3390/plants14101488

6. Al-Azzazy and Alhewairini (Contribution 6) tested the use of the predatory mites *Phytoseius plumifer* Canestrini and Fanzago and *Euseius scutalis* Athias-Henriot (Phytoseiidae) as biological control agents against the grape erineum mite, *Colomerus vitis* Pagenstecher (Eriophyidae).
7. George et al. (Contribution 7) investigated the potential of different fungal endophytes, specifically *Beauveria bassiana* strains, in colonizing cotton plants and their efficacy against tarnished plant bug *Lygus lineolaris* Palisot de Beauvois (Hemiptera: Miridae). These endophytes can colonize different plant parts, affecting cotton plant growth as well as the development and mortality of *L. lineolaris* adults and nymphs. The new *B. bassiana* strain JG-1 affected the olfactory response of *L. lineolaris* adults and caused significant mortality in bioassays.
8. Benvenuti (Contribution 8) reviewed the mutualistic interactions between pollinators and weeds, discussing them in terms of food reward and attractiveness, and analyzed the specialization of these interactions.
9. De Brito Machado et al. (Contribution 9) investigated how the chemodiversity of leaves and reproductive organs affect pollinator visitation in *Piper mollicomum* Kunth (Piperaceae). They showed the importance of understanding the complex interactions between plant chemistry, environmental factors, and plant–insect interactions in *P. mollicomum*.
10. Pimkornburee et al. (Contribution 10) investigated the settling preferences and feeding behavior of the silverleaf whitefly *Bemisia tabaci* Gennadius (Hemiptera: Aleyrodidae) on six cassava cultivars using electrical penetration graph techniques. They suggest that cultivars with large trichomes and a low trichome density are more resistant to whitefly infestation and the subsequent transmission of Sri Lankan cassava mosaic virus than those with smaller trichomes and a lower trichome density.
11. Jansen-González et al. (Contribution 11) investigated the parasitic relationship between *Arastichus gallicola* Ferrière (Hymenoptera: Eulophidae), an ovary-galling wasp, and the inflorescences of *Thaumatophyllum bipinnatifidum* (Schott ex Endl.) Sakur., Calazans & Mayo (Araceae). Although ovule fertilization is not required for gall formation by *A. gallicola*, pollination substantially enhanced gall retention by reducing *T. bipinnatifidum* inflorescence abscission.

3. Key Messages

The impact that insects have as pollinators in angiosperms is one of the most prominent aspects of studying plant–insect interactions [7]. With the decline in pollinators [8,9], their conservation is very important. Weeds can be an important part of pollinator conservation [10,11]. The process of insect galling by eulophid wasps can interact with pollination and fruit retention in trees [12,13]. Plant chemistry also plays an important role in plant–insect interactions [14]. Besides being a key part of the integrated pest management of insect pests, host plant resistance can prevent damage caused by insect-transmitted diseases [15,16]. Studies on host plant resistance and biological control are important in both generalist species that cause important economic losses in many crops, like the generalists *B. tabaci*, *H. armigera*, *L. lineolaris*, *P. solenopsis*, and *S. frugiperda* [17–21], and in specialists like *P. operculella*, *D. coccus*, and *D. opuntiae* that damage only a few crop species [22,23]. Endophytic entomopathogenic fungi can be used as part of the integrated pest management of some of the insect pests included in this Special Issue [24,25]. Some of these insect pests have also developed resistance to many insecticides and insecticide use presents the risk of disrupting the biological control of insect and mite pests [18,26–30].

4. Future Directions

Plant–insect interactions encompass many relationships deserving further research [31–33]. With climate change and habitat loss being such pressing issues [34,35], the effects that these can have on plant–insect interactions should be further studied. Studies conducted in geographical locations that are biodiversity hotspots should offer many opportunities for the study of plant–insect interactions and conservation [36]. Alternatives to insecticides should also be investigated further, especially in insect pest species that can easily develop resistance to insecticides. The compatibility of different biological control agents for improving pest management should be further investigated [37–39]. In terms of host–plant resistance, plants like *Barbarea vulgaris* W. T. Aiton (Brassicaceae), which can provide resistance to several types of pests, are of particular interest for further research [40–43].

Funding: This research received no external funding.

Acknowledgments: I thank the authors who contributed to this Special Issue. I also thank Rene Lv, the Managing Editor of this Special Issue of *Plants*, for all his help.

Conflicts of Interest: The author declares no conflicts of interest.

List of Contributions:

1. Zhu, C.; Yi, X.; Yang, M.; Liu, Y.; Yao, Y.; Zi, S.; Chen, B.; Xiao, G. Comparative Transcriptome Analysis of Defense Response of Potato to *Phthorimaea operculella* Infestation. *Plants* **2023**, *12*, 3092. https://doi.org/10.3390/plants12173092.
2. Zhang, Q.; Zhang, Y.; Zhang, K.; Liu, H.; Gou, Y.; Li, C.; Haq, I.U.; Quandahor, P.; Liu, C. Molecular Characterization Analysis and Adaptive Responses of *Spodoptera frugiperda* (Lepidoptera: Noctuidae) to Nutritional and Enzymatic Variabilities in Various Maize Cultivars. *Plants* **2024**, *13*, 597. https://doi.org/10.3390/plants13050597.
3. Volp, T.M.; Zalucki, M.P.; Furlong, M.J. Ontogenetic Changes in the Feeding Behaviour of *Helicoverpa armigera* Larvae on Pigeonpea (*Cajanus cajan*) Flowers and Pods. *Plants* **2024**, *13*, 696. https://doi.org/10.3390/plants13050696.
4. Rodríguez-Leyva, E.; García-Pascual, E.; González-Chávez, M.M.; Méndez-Gallegos, S.d.J.; Morales-Rueda, J.A.; Posadas-Hurtado, J.C.; Bravo-Vinaja, Á.; Franco-Vega, A. Interactions of *Opuntia ficus-indica* with *Dactylopius coccus* and *D. opuntiae* (Hemiptera: Dactylopiidae) through the Study of Their Volatile Compounds. *Plants* **2024**, *13*, 963. https://doi.org/10.3390/plants13070963.
5. Abbes, K.; Harbi, A.; Guerrieri, E.; Chermiti, B. Using Age-Stage Two-Sex Life Tables to Assess the Suitability of Three Solanaceous Host Plants for the Invasive Cotton Mealybug *Phenacoccus solenopsis* Tinsley. *Plants* **2024**, *13*, 1381. https://doi.org/10.3390/plants13101381.
6. Al-Azzazy, M.M.; Alhewairini, S.S. The Potential of Two Phytoseiid Mites as Predators of the Grape Erineum Mite, *Colomerus vitis*. *Plants* **2024**, *13*, 1953. https://doi.org/10.3390/plants13141953.
7. George, J.; Glover, J.P.; Perera, O.P.; Reddy, G.V.P. Role of Endophytic Entomopathogenic Fungi in Mediating Host Selection, Biology, Behavior, and Management of Tarnished Plant Bug, *Lygus lineolaris* (Hemiptera: Miridae). *Plants* **2024**, *13*, 2012. https://doi.org/10.3390/plants13152012.
8. Benvenuti, S. Weed Role for Pollinator in the Agroecosystem: Plant–Insect Interactions and Agronomic Strategies for Biodiversity Conservation. *Plants* **2024**, *13*, 2249. https://doi.org/10.3390/plants13162249.
9. De Brito Machado, D.; Felisberto, J.S.; Queiroz, G.A.d.; Guimarães, E.F.; Ramos, Y.J.; Moreira, D.d.L. From Leaves to Reproductive Organs: Chemodiversity and Chemophenetics of Essential Oils as Important Tools to Evaluate *Piper mollicomum* Kunth Chemical Ecology Relevance in the Neotropics. *Plants* **2024**, *13*, 2497. https://doi.org/10.3390/plants13172497.
10. Pimkornburee, S.; Pombud, S.; Buensanteai, K.; Namanusart, W.; Aiamla-or, S.; Roddee, J. Impact of Cassava Cultivars on Stylet Penetration Behavior and Settling of *Bemisia tabaci*

Gennadius (Hemiptera: Aleyrodidae). *Plants* **2024**, *13*, 3218. https://doi.org/10.3390/plants13223218.

11. Jansen-González, S.; Teixeira, S.P.; Pereira, R.A.S. Conflicting Dynamics of Galling and Pollination: *Arastichus gallicola* (Hymenoptera, Eulophidae), a Specialized Parasitic Galler in Pistillate Flowers of *Thaumatophyllum bipinnatifidum* (Araceae). *Plants* **2024**, *13*, 3520. https://doi.org/10.3390/plants13243520.

References

1. Waterfield, G.; Zilberman, D. Pest Management in food Systems: An economic perspective. *Annu. Rev. Environ. Resour.* **2012**, *37*, 223–245. [CrossRef]
2. Klein, A.M.; Vaissière, B.E.; Cane, J.H.; Steffan-Dewenter, I.; Cunningham, S.A.; Kremen, C.; Tscharntke, T. Importance of pollinators in changing landscapes for world crops. *Proc. R. Soc. B Biol. Sci.* **2007**, *274*, 303–313. [CrossRef] [PubMed]
3. Zalucki, M.P.; Shabbir, A.; Silva, R.; Adamson, D.; Shu-Sheng, L.; Furlong, M.J. Estimating the economic cost of one of the world's major insect pests, *Plutella xylostella* (Lepidoptera: Plutellidae): Just how long is a piece of string? *J. Econ. Entomol.* **2012**, *105*, 1115–1129. [CrossRef]
4. Badenes-Pérez, F.R. Benefits of insect pollination in Brassicaceae: A meta-analysis of self-compatible and self-incompatible crop species. *Agriculture* **2022**, *12*, 446. [CrossRef]
5. Stout, M.J.; Bernaola, L.; Acevedo, F. Recent history and future trends in host–plant resistance. *Ann. Entomol. Soc. Am.* **2024**, *117*, 139–149. [CrossRef]
6. Naranjo, S.E.; Ellsworth, P.C.; Frisvold, G.B. Economic value of biological control in integrated pest management of managed plant systems. *Annu. Rev. Entomol.* **2015**, *60*, 621–645. [CrossRef]
7. Peris, D.; Ollerton, J.; Sauquet, H.; Hidalgo, O.; Peñalver, E.; Magrach, A.; Álvarez-Parra, S.; Peña-Kairath, C.; Condamine, F.L.; Delclòs, X.; et al. Evolutionary implications of a deep-time perspective on insect pollination. *Biol. Rev.* **2025**. [CrossRef]
8. Potts, S.G.; Biesmeijer, J.C.; Kremen, C.; Neumann, P.; Schweiger, O.; Kunin, W.E. Global pollinator declines: Trends, impacts and drivers. *Trends Ecol. Evol.* **2010**, *25*, 345–353. [CrossRef]
9. Dicks, L.V.; Breeze, T.D.; Ngo, H.T.; Senapathi, D.; An, J.; Aizen, M.A.; Basu, P.; Buchori, D.; Galetto, L.; Garibaldi, L.A.; et al. A global-scale expert assessment of drivers and risks associated with pollinator decline. *Nat. Ecol. Evol.* **2021**, *5*, 1453–1461. [CrossRef]
10. Rollin, O.; Benelli, G.; Benvenuti, S.; Decourtye, A.; Wratten, S.D.; Canale, A.; Desneux, N. Weed-insect pollinator networks as bio-indicators of ecological sustainability in agriculture. A review. *Agron. Sustain. Dev.* **2016**, *36*, 1–22. [CrossRef]
11. Benvenuti, S. Wildflower strips in the agroecosystem for pollinator biodiversity restoration: Which plant species are capable of self-seeding? *Ecol. Eng.* **2025**, *212*, 107486. [CrossRef]
12. Badenes-Pérez, F.R.; Johnson, M.T. Ecology and impact of *Allorhogas* sp. (Hymenoptera: Braconidae) and *Apion* sp. (Coleoptera: Curculionoidea) on fruits of *Miconia calvescens* DC (Melastomataceae) in Brazil. *Biol. Control* **2007**, *43*, 317–322. [CrossRef]
13. Jansen-González, S.; de Padua Teixeira, S.; Kjellberg, F.; Pereira, R.A.S. Same but different: Larval development and gall-inducing process of a non-pollinating fig wasp compared to that of pollinating fig-wasps. *Acta Oecologica* **2014**, *57*, 44–50. [CrossRef]
14. de Brito Machado, D.; Ramos, Y.J.; de Queiroz, G.A.; e Defaveri, A.C.A.; Gobatto, A.A.; de Lima Moreira, D. Study of volatile chemical constituents and insect-plant interaction in *Piper mollicomum* Kunth (Piperaceae) from Tijuca Forest, Rio de Janeiro—RJ, Brazil. *Rev. Virtual Quim.* **2021**, *13*, 1216–1225. [CrossRef]
15. Roddee, J.; Kobori, Y.; Hanboonsong, Y. Characteristics of sugarcane white leaf phytoplasma transmission by the leafhopper *Matsumuratettix hiroglyphicus*. *Entomol. Exp. Appl.* **2019**, *167*, 108–117. [CrossRef]
16. Heck, M. Insect transmission of plant pathogens: A systems biology perspective. *mSystems* **2018**, *3*, e00168-17. [CrossRef]
17. Bellotti, A.C.; Arias, B. Host plant resistance to whiteflies with emphasis on cassava as a case study. *Crop Prot.* **2001**, *20*, 813–823. [CrossRef]
18. George, J.; Glover, J.P.; Gore, J.; Crow, W.D.; Reddy, G.V.P. Biology, ecology, and pest management of the tarnished plant bug, *Lygus lineolaris* (Palisot de Beauvois) in Southern Row Crops. *Insects* **2021**, *12*, 807. [CrossRef]
19. Waqas, M.S.; Shi, Z.; Yi, T.-C.; Xiao, R.; Shoaib, A.A.; Elabasy, A.S.; Jin, D.-C. Biology, ecology, and management of cotton mealybug *Phenacoccus solenopsis* Tinsley (Hemiptera: Pseudococcidae). *Pest Manag. Sci.* **2021**, *77*, 5321–5333. [CrossRef]
20. Downes, S.; Kriticos, D.; Parry, H.; Paull, C.; Schellhorn, N.; Zalucki, M.P. A Perspective on management of *Helicoverpa armigera*: Transgenic Bt cotton, IPM, and landscapes. *Pest Manag. Sci.* **2017**, *73*, 485–492. [CrossRef]
21. Prasanna, B.M.; Bruce, A.; Beyene, Y.; Makumbi, D.; Gowda, M.; Asim, M.; Martinelli, S.; Head, G.P.; Parimi, S. Host plant resistance for fall armyworm management in maize: Relevance, status and prospects in Africa and Asia. *Theor. Appl. Genet.* **2022**, *135*, 3897–3916. [CrossRef] [PubMed]

22. Kebede, T.G.; Birhane, E.; Ayimut, K.-M.; Egziabher, Y.G.; Belay, T. Arbuscular mycorrhizal fungi-induced resistance in *Opuntia ficus-indica* (L.) Miller plant against cochineal insects under various soil water levels. *Arthropod-Plant Interact.* **2024**, *18*, 253–264. [CrossRef]
23. Rondon, S.I. The potato tuberworm: A literature review of its biology, ecology, and control. *Am. J. Potato Res.* **2010**, *87*, 149–166. [CrossRef]
24. Panwar, N.; Szczepaniec, A. Endophytic entomopathogenic fungi as biological control agents of insect pests. *Pest Manag. Sci.* **2024**, *80*, 6033–6040. [CrossRef]
25. Russo, M.L.; Jaber, L.R.; Scorsetti, A.C.; Vianna, F.; Cabello, M.N.; Pelizza, S.A. Effect of entomopathogenic fungi introduced as corn endophytes on the development, reproduction, and food preference of the invasive fall armyworm *Spodoptera frugiperda*. *J. Pest Sci.* **2021**, *94*, 859–870. [CrossRef]
26. Zalucki, M.P.; Adamson, D.; Furlong, M.J. The future of IPM: Whither or wither? *Aust. J. Entomol.* **2009**, *48*, 85–96. [CrossRef]
27. Van den Berg, J.; du Plessis, H. Chemical control and insecticide resistance in *Spodoptera frugiperda* (Lepidoptera: Noctuidae). *J. Econ. Entomol.* **2022**, *115*, 1761–1771. [CrossRef]
28. Wang, Q.; Luo, C.; Wang, R. Insecticide resistance and its management in two invasive cryptic species of *Bemisia tabaci* in China. *Int. J. Mol. Sci.* **2023**, *24*, 6048. [CrossRef]
29. Walsh, T.K.; Heckel, D.G.; Wu, Y.; Downes, S.; Gordon, K.H.J.; Oakeshott, J.G. Determinants of insecticide resistance evolution: Comparative analysis among heliothines. *Annu. Rev. Entomol.* **2022**, *67*, 387–406. [CrossRef]
30. Alhewairini, S.S.; Al-Azzazy, M.M. Side effects of abamectin and hexythiazox on seven predatory mites. *Braz. J. Biol.* **2023**, *83*, e251442. [CrossRef]
31. Giron, D.; Dubreuil, G.; Bennett, A.; Dedeine, F.; Dicke, M.; Dyer, L.A.; Erb, M.; Harris, M.O.; Huguet, E.; Kaloshian, I.; et al. Promises and challenges in insect–plant interactions. *Entomol. Exp. Appl.* **2018**, *166*, 319–343. [CrossRef]
32. Sharma, G.; Malthankar, P.A.; Mathur, V. Insect-plant interactions: A multilayered relationship. *Ann. Entomol. Soc. Am.* **2021**, *114*, 1–16. [CrossRef]
33. Badenes-Pérez, F.R. Plant-insect interactions. *Plants* **2022**, *11*, 1140. [CrossRef] [PubMed]
34. Harvey, J.A.; Tougeron, K.; Gols, R.; Heinen, R.; Abarca, M.; Abram, P.K.; Basset, Y.; Berg, M.; Boggs, C.; Brodeur, J.; et al. Scientists' warning on climate change and insects. *Ecol. Monogr.* **2023**, *93*, e1553. [CrossRef]
35. Hanski, I. Habitat loss, the dynamics of biodiversity, and a perspective on conservation. *Ambio* **2011**, *40*, 248–255. [CrossRef]
36. Fonseca, C.R. The silent mass extinction of insect herbivores in biodiversity hotspots. *Conserv. Biol.* **2009**, *23*, 1507–1515. [CrossRef]
37. Al-Azzazy, M.M.; Alhewairini, S.S. Life history of the two predacious mites species, *Amblyseius swirskii*, and *Neoseiulus cucumeris* (Acari: Phytoseiidae), as biological control agents of the date palm mite, *Oligonychus afrasiaticus* (Acari: Tetranychidae). *Braz. J. Biol.* **2024**, *84*, e283484. [CrossRef]
38. Hernández-Valencia, V.; Santillán-Galicia, M.T.; Guzmán-Franco, A.W.; Rodríguez-Leyva, E.; Santillán-Ortega, C. Combined application of entomopathogenic fungi and predatory mites for biological control of *Tetranychus urticae* on Chrysanthemum. *Pest Manag. Sci.* **2024**, *80*, 4199–4206. [CrossRef]
39. Badenes-Pérez, F.R.; Márquez, B.P.; Petitpierre, E. Can flowering *Barbarea* spp. (Brassicaceae) be used simultaneously as a trap crop and in conservation biological control? *J. Pest Sci.* **2017**, *90*, 623–633. [CrossRef]
40. Agerbirk, N.; Olsen, C.E.; Bibby, B.M.; Frandsen, H.O.; Brown, L.D.; Nielsen, J.K.; Renwick, J.A.A. A saponin correlated with variable resistance of *Barbarea vulgaris* to the diamondback moth *Plutella xylostella*. *J. Chem. Ecol.* **2003**, *29*, 1417–1433. [CrossRef]
41. Shinoda, T.; Nagao, T.; Nakayama, M.; Serizawa, H.; Koshioka, M.; Okabe, H.; Kawai, A. Identification of a triterpenoid saponin from a crucifer, *Barbarea vulgaris*, as a feeding deterrent to the diamondback moth, *Plutella xylostella*. *J. Chem. Ecol.* **2002**, *28*, 587–599. [CrossRef] [PubMed]
42. Christensen, S.; Enge, S.; Jensen, K.R.; Müller, C.; Kiær, L.P.; Agerbirk, N.; Heimes, C.; Hauser, T.P. Different herbivore responses to two co-occurring chemotypes of the wild crucifer *Barbarea vulgaris*. *Arthropod-Plant Interact.* **2018**, *13*, 19–30. [CrossRef]
43. Badenes-Pérez, F.R.; López-Pérez, J.A. Resistance and susceptibility to powdery mildew, root-knot nematode and western flower thrips in two types of winter cress (Brassicaceae). *Crop Prot.* **2018**, *110*, 41–47. [CrossRef]

plants

MDPI

Article

Comparative Transcriptome Analysis of Defense Response of Potato to *Phthorimaea operculella* Infestation

Chunyue Zhu [1], Xiaocui Yi [1], Miao Yang [1], Yiyi Liu [1], Yao Yao [1], Shengjiang Zi [1], Bin Chen [2,*] and Guanli Xiao [1,*]

[1] College of Agronomy and Biotechnology, Yunnan Agricultural University, Kunming 650201, China; zcynzb@163.com (C.Z.); 18287196028@163.com (X.Y.); m15350739150_2@163.com (M.Y.); 18313033732@163.com (S.Z.); booya611@163.com (Y.L.); yaoyao990926@163.com (Y.Y.)

[2] College of Plant Protection, Yunnan Agricultural University, Kunming 650201, China

* Correspondence: chbins@163.com (B.C.); glxiao9@163.com (G.X.); Tel.: +86-13708876067 (B.C.); +86-13888217416 (G.X.)

Abstract: The potato tuber moth (PTM), *Phthorimaea operculella* Zeller (Lepidoptera: Gelechiidae), is one of the most destructive pests of potato crops worldwide. Although it has been reported how potatoes integrate the early responses to various PTM herbivory stimuli by accumulatively adding the components, the broad-scale defense signaling network of potato to single stimuli at multiple time points are unclear. Therefore, we compared three potato transcriptional profiles of undamaged plants, mechanically damaged plants and PTM-feeding plants at 3 h, 48 h, and 96 h, and further analyzed the gene expression patterns of a multitude of insect resistance-related signaling pathways, including phytohormones, reactive oxygen species, secondary metabolites, transcription factors, MAPK cascades, plant–pathogen interactions, protease inhibitors, chitinase, and lectins, etc. in the potato under mechanical damage and PTM infestation. Our results suggested that the potato transcriptome showed significant responses to mechanical damage and potato tuber moth infestation, respectively. The potato transcriptome responses modulated over time and were higher at 96 than at 48 h, so transcriptional changes in later stages of PTM infestation may underlie the potato recovery response. Although the transcriptional profiles of mechanically damaged and PTM-infested plants overlap extensively in multiple signaling pathways, some genes are uniquely induced or repressed. True herbivore feeding induced more and stronger gene expression compared to mechanical damage. In addition, we identified 2976, 1499, and 117 genes that only appeared in M-vs-P comparison groups by comparing the transcriptomes of PTM-damaged and mechanically damaged potatoes at 3 h, 48 h, and 96 h, respectively, and these genes deserve further study in the future. This transcriptomic dataset further enhances the understanding of the interactions between potato and potato tuber moth, enriches the molecular resources in this research area and paves the way for breeding insect-resistant potatoes.

Keywords: plant–insect interactions; insect resistance-related pathways; mechanical damage; potato tuber moth herbivory stimulation

1. Introduction

During long-term plant–insect interactions, plants have evolved complex defense strategies, including constitutive and inducible defense mechanisms to defend themselves against phytophagous insects [1,2]. Plants' own inherent morphological structures or biochemical properties that resist insect feeding or pathogen invasion, such as waxy cuticles, glandular hairs, trichomes, spines, cell walls and lignin layers [3,4], as well as natural compounds in the body that sense and inhibit insects [5], all of which are categorized as constitutive defenses. Induced defenses are mostly activated only when the plant is exposed to external stimuli and play an important role in both direct and indirect ways for a short period of time [1,6]. In inducible defense, plants achieve direct defense against insects by producing toxic secondary metabolites (e.g., terpenoids, phenolic compounds, nitrogenous compounds such as alkaloids) and defense proteins (e.g., protease inhibitors PI, polyphenol oxidase, lectins, etc.), and or by reducing their own nutrient levels [1,7–9]. Indirect defense mainly refers to the release of pest-induced volatile chemicals (HIPVs) from plants that serve to attract predators of those herbivores to withstand insect infestation [10].

In general, after being subjected to insect feeding, the following steps occur in plants: (i) plant recognition of damage-associated and herbivore-derived molecular patterns (DAMPs and HAMPs) [11]; (ii) early signaling responses in plants; (iii) hormonal signaling responses; and (iv) reconfiguration of the transcriptome, metabolome and proteome [12]. Specifically, early signal events occurring before herbivore attack-related gene expression in plants mainly include ion imbalances, depolarization of membrane potential (Vm), variations of cytosolic Ca^{2+} concentrations, reactive oxygen species (ROS) burst and mitogen-activated protein kinases (MAPKs) signaling cascades [13]. Up to now, many HAMPs such as volicitin (termed 17-hydroxylinolenoyl-L-Gln, a fatty acid–amino acid conjugate (FAC) elicitor), califerins (the sulfooxy fatty acids), inceptin (a peptide fragment from chloroplast ATP synthase of cowpea plants digested by *Spodoptera frugiperda* larvae), a mucin-like protein (NlMLP) and glucose oxidase (GOX) have been isolated from insect oral secretions OS (e.g., regurgitant, saliva, insect-associated microorganisms including oral secretion bacteria) and interact with plant pattern recognition receptors PRRs (e.g., G-type lectin receptor kinases and leucine-rich repeat LRR-RK, etc.) [11,12]. The calcium ion is considered a second messenger in plant signaling pathways. A study has shown that rapid and highly-localized calcium elevations in Arabidopsis around the feeding sites of green peach aphid (*Myzus persicae*) [14]. Herbivorous insect Spodoptera littoralis feeding induced not only Ca^{2+} elevation in Arabidopsis and lima bean (*Phaseolus lunatus*) but also H_2O_2 production in lima bean leaves. Spodoptera littoralis infestation leaves have higher H_2O_2 concentrations when compared to mechanically damaged leaves [15,16]. Kandoth et al. demonstrated that the tomato MAPKs LeMPK1, LeMPK2, and LeMPK3 are important components of the systemin signaling pathway and are also required for successful defense against herbivorous insects in tomatoes [17]. There are close interactions between early defense signals related to plant–herbivore interactions [12,18]. Although some signal-transduction mechanisms are also evident in plant–pathogen interactions [13], further studies are required on the trade-offs between resistance to pathogens and herbivores in plants. Early signaling events are followed by changes in the signaling networks mediated by multiple hormones (such as JA, SA, ABA, and ETH), which in turn regulate downstream defense genes and the production of metabolites. Numerous studies have shown that JA plays a central role in regulating plant responses to phytophagous insects, while other hormones act in concert or antagonize the JA signaling pathway [1,12]. The selective activation of downstream-related pathways or signals by plants depends on the insect and plant species, and time differences in insect infestation can also lead to dramatic changes in plant defense responses [18–20]. Transcription factors play a key regulatory role in the global defense deployment of plants [12]. Rice transcription factor WRKY70 is required for protease inhibitor activation and resistance to herbivore *Chilo suppressalis* by positively regulating jasmonic acid (JA) biosynthesis, and WRKY70 enhances rice susceptibility to *Nilaparvata lugens* [21]. TFs regulate plant stress responses in an interactive manner by mediating secondary metabolism, such as the MYB-bHLH-WDR complex in Arabidopsis that regulates flavonoid biosynthesis [22].

As we mentioned earlier, early signaling events are rapidly activated thanks to the fact that plants can precisely perceive insect-derived elicitors (e.g., physical damage, insect oral secretions OS, ovipositional fluids, frass, etc.) [11,13]. Thus, at the plant–insect interface, mechanical damage and insect oral secretion stimuli OS are two main herbivory-associated stimuli [19,20]. Mechanical damage is often used as a control in experiments to investigate plant defense responses against herbivory. The results of microarray experiments in several plant species showed a huge overlap in gene expression patterns between mechanical damage and herbivore feeding, yet there are also transcriptional responses that are specific to both stimuli [1]. However, there are also examples that have demonstrated that gene expression and transcriptional profiles differ significantly between wounding alone and insect-induced activation, and saliva was identified as the agent responsible for the difference in the responses [23,24].

The potato, *Solanum tuberosum* L. (Solanales: Solanaceae), is one of the four major food crops, along with rice, wheat and maize. The potato is not only an essential vegetable crop and fodder crop but also an important raw material for the food industry, and its production is important for sustaining livelihoods and enhancing nutritional balance, especially in developing countries [25]. Potato tuber moth (PTM), *Phthorimaea operculella* (Zeller) (Lepidoptera: Gele-chiidae), which originated in Latin America, is an important pest of Solanaceae crops and is especially devastating to potatoes. Potato tuber moth larvae mine into the leaves and stems of potatoes or excavate tunnels into the tuber to feed, while PTM adults lay eggs in cracks on the surface of potato tubers, thus affecting potato growth and storage [26]. Due to the high reproductive potential of PTM, research on chemical and biological control has received much attention [27,28]. However, there is still a lack of green and effective control technology for this insect, such as improving potato resistance to this insect with the help of transgenic technology [29,30]. In this context, it is crucial to understand potato–moth interactions. Currently, there is a distinct lack of reported research on the molecular mechanisms of potato tuber moth feeding induced by potatoes at home and abroad. Therefore, in this study, we analyzed key signaling events occurring in potatoes in response to mechanical damage and PTM herbivory stress at multiple time points in the broad context of overall plant defense and compared the differences between the two stimuli, thus attempting to summarize the molecular mechanisms of potato under potato tuber moth feeding stress.

2. Results

2.1. Overview of Illumina Sequencing within Different Treatment Samples

To investigate the transcriptomic differences between mechanical damage and potato tuber moth (PTM) feeding at different time points, we performed RNA-Seq analyses on PTM-infested (P), mechanically damaged (M) and blank control (CK) potato plants.

The number of raw data obtained from 27 libraries (9 undamaged control for 3 h, 48 h and 96 h samples: CK3-1/CK3-2/CK3-3/CK48-1/CK48-2/CK48-3/CK96-1/CK96-2/CK96-3; 9 mechanical damage for 3 h, 48 h and 96 h samples: M3-1/M3-2/M3-3/M48-1/M48-2/M48-3/M96-1/M96-2/M96-3; 9 PTM infestation for 3 h, 48 h and 96 h samples: P3-1/P3-2/P3-3/P48-1/P48-2/P48-3/P48-1/P96-1/P96-2/P96-3) ranged from 37,271,202 to 51,880,920, and the number of clean data ranged from 37,224,532 (99.87%)~51,836,050 (99.91%). More than 81.9% of the clean data were successfully mapped to the reference genome of the potato, and the unique_mapped reads ranged from 14,048,153 (79.21%) to 42,974,467 (85.96%). The GC ratio ranged from 44.37% to 48.15%, and the Q30 values of these unique genes ranged from 91.08% to 94.09% (Table S1).

Principal component analysis (PCA) revealed that the gene expression profiles of PTM-infested plants were clearly separated from control plants, whereas the gene expression profiles of mechanically damaged plants partially overlapped with those of PTM-infested and control plants, which suggests principal component analysis (PCA) effectively separated the different transcriptome samples by treatments and PTM herbivory attack triggered stronger responses relative to mechanical damage (Figure 1A). PC1 accounting for 69.5% of the variation indicated a significant separation among PTM infestation for 3 h, 48 h and 96 h plants, whereas there was a weaker separation among mechanically damaged plants for 3 h, 48 h and 96 h along the PC2 axis (14% of the variance), and there was only a partially separation among control plants for 3 h, 48 h and 96 h, which suggests changes in transcriptomes could be linked to the developmental changes and stress response according to the time course within each treatment (Figure 1A). The correlation coefficients between different samples are shown in Figure 1B.

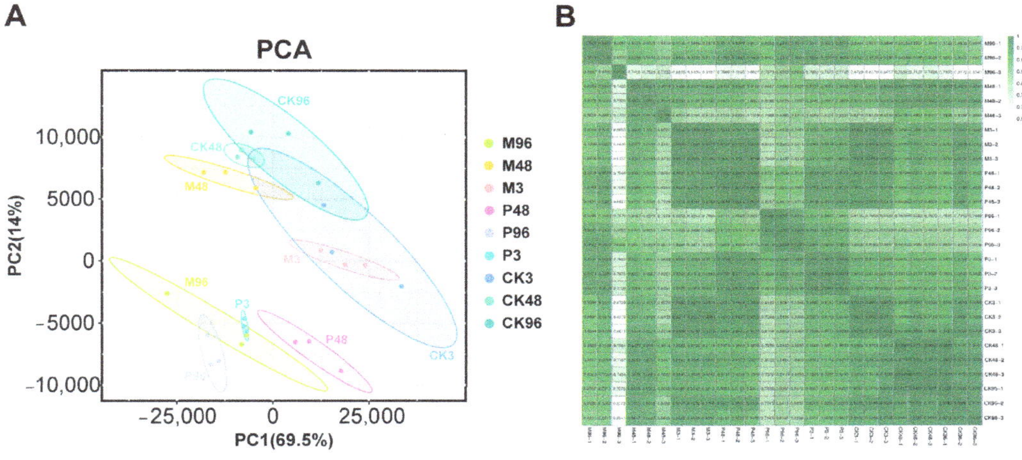

Figure 1. Global changes of transcriptome data for all treatment samples. (**A**) Sample principal component analysis diagram. The grouping patterns of samples were with respect to the first two principal components. A 69.5% variance is explained in PC1, and 14% variance is explained in PC2; the ellipse represents the 95% confidence interval of each treatment group. (**B**) Pearson correlation heatmap between different samples.

2.2. Identification and Functional Annotation of Differentially Expressed Genes (DEGs)

2.2.1. Transcriptomic Changes in Response to Mechanical Damage (CK-vs-M)

To understand the transcriptional changes in potato leaves after mechanical damage, transcription profiles of the mechanical wounding and control plants for 3 h, 48 h and 96 h were compared. Padj < 0.05 and an absolute value of Log_2FoldChange (relative to the control) ≥ 1 was used as the standard for screening DEGs. A total of 1531 (764 up and 767 down) and 5766 (3280 up and 2486 down) DEGs were found for the comparisons: CK48-vs-M48 and CK96-vs-M96, respectively, whereas 21 (20 up and 1 down) DEGs induced only within CK3-vs-M3 comparison (Figure 2A). The total number of up-regulated genes (4064 DEGs) was more than that of down-regulated genes (3254 DEGs). The number of DEGs within intersections of CK3-vs-M3 (U/D) compared with other comparisons was less than 10 because of a small number (21 DEGs) of genes (Figure 2B). Within up-regulated genes, 466 and 2986 DEGs were uniquely expressed at 48 h and 96 h, respectively, and 116 DEGs were shared in common between two time points. A total of 396 and 2108 DEGs were uniquely expressed respectively within down-regulated genes after 48 h and 96 h, while 196 genes were common DEGs between the two time points (Figure 2B). The results indicated mechanical damage induced a significant transcriptome change in potato leaves through the time course.

GO term enrichment and KEGG pathway enrichment were performed on the total number of genes (21 DEGs for 3 h, 1531 DEGs for 48 h and 5766 DEGs for 96 h) after mechanical damage to predict their functions. The top 20 GO terms with the most significant enrichment including cellular components (CC), biological processes (BP) and molecular functions (MF), and pathways were presented in Supplementary Tables S2 and S3. Of the GO enrichment, the DEGs functions were related mainly to membrane (CC), negative regulation of biological process (BP), response to stimulus (BP), hydrolase activity (MF) and ion binding (MF) for 3 h, 48 h and 96 h. KEGG pathways were mainly associated with biosynthesis of secondary metabolites, protein processing in the endoplasmic reticulum and plant hormone signal transduction from 3 h to 96 h (Tables S2 and S3).

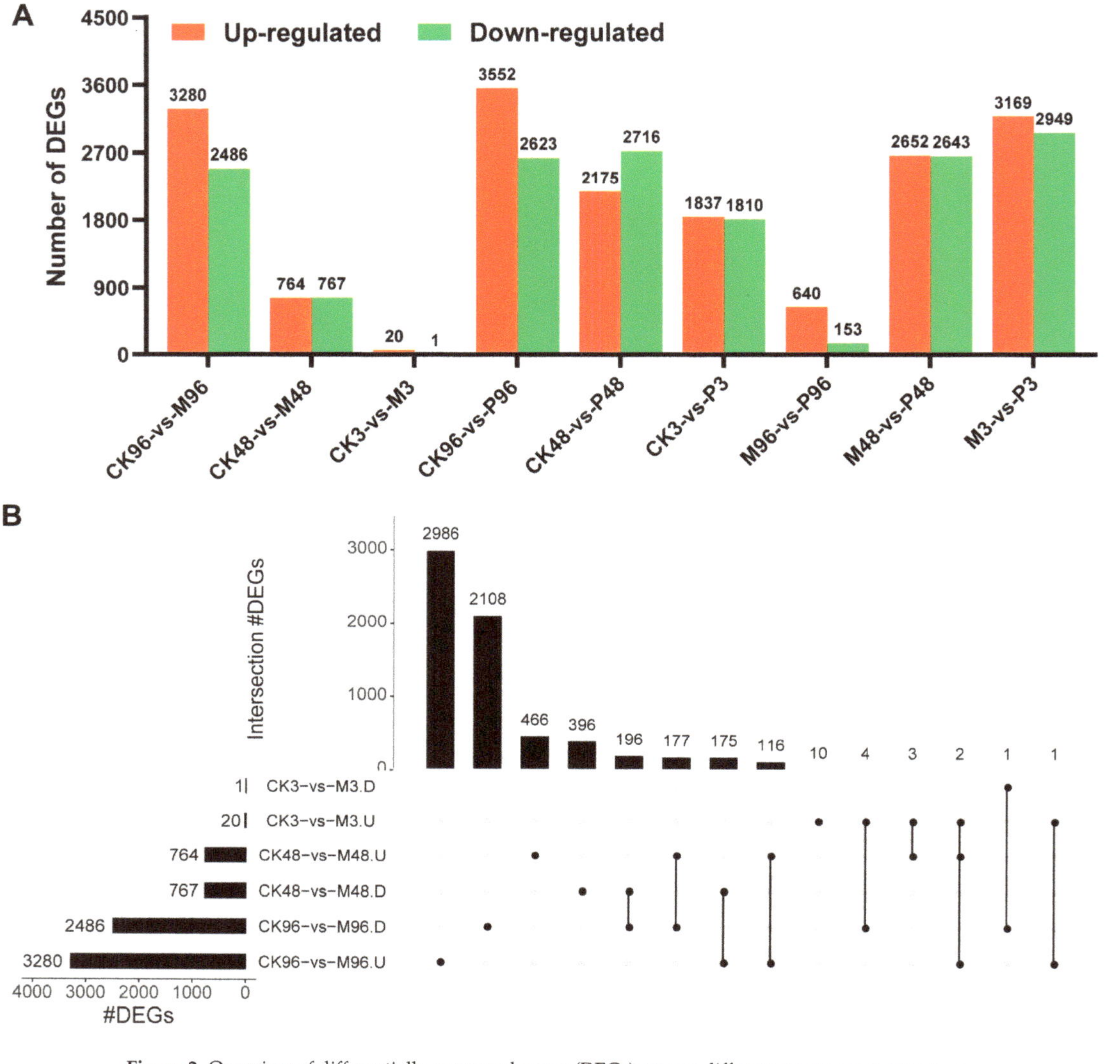

Figure 2. Overview of differentially expressed genes (DEGs) among different treatment comparisons for the same time points. (**A**) Number of up- and down-regulated differentially expressed genes (DEGs) among different comparisons at each time point. (**B**) UpSet intersection diagram illustrating the unique and shared DEGs between mechanical damage plants and control plants for 3 h, 48 h, 96 h. The intersection of shared (unique) DEGs are represented by a line connected to more plots (one plot), and the number of DEGs are shown on the top of the vertical bars. The black dots represent different comparison groups. U—up-regulated genes; D—down-regulated genes.

In addition, KEGG enrichment analysis was also performed for up- and down-regulated genes at 3 h, 48 h and 96 h after mechanical damage. Up-regulated genes were mainly concentrated in protein processing in the endoplasmic reticulum (48 h) and plant hormone signal transduction (3 h and 96 h) pathways, while down-regulated genes were mainly involved in DNA replication (48 h) and biosynthesis of secondary metabolites (96 h) Table S4).

2.2.2. Transcriptomic Changes in Response to PTM Infestation (CK-vs-P)

The transcriptomes of PTM infestation and the undamaged control were compared to study the transcriptional changes induced by potato tuber moth feeding. The data exhibited 3647 (1837 up and 1810 down), 4891 (2175 up and 2716 down) and 6175 (3552 up and 2623 down) DEGs after 3 h, 48 h and 96 h, respectively. The number of DEGs induced by PTM infestation was more than that by mechanical damage at the same time, which suggested that PTM infestation triggered stronger responses than mechanical damage (Figure 2A). A total of 309 and 495 DEGs were continuously up-regulated and down-regulated for 3 h, 48 h and 96 h after PTM feeding. A total of 916, 814 and 2079 DEGs were uniquely up-regulated genes at 3 h, 48 h, and 96 h, which represent 49.9%, 37.4% and 58.5% of the total number of up-regulated DEGs at 3 h (n = 1837 DEGs), 48 h (n = 2175 DEGs) and 96 h (n = 3552 DEGs). 600, 1131 and 1064 genes were down-regulated DEGs uniquely for 3 h, 48 h and 96 h after insect herbivory (Figure 3).

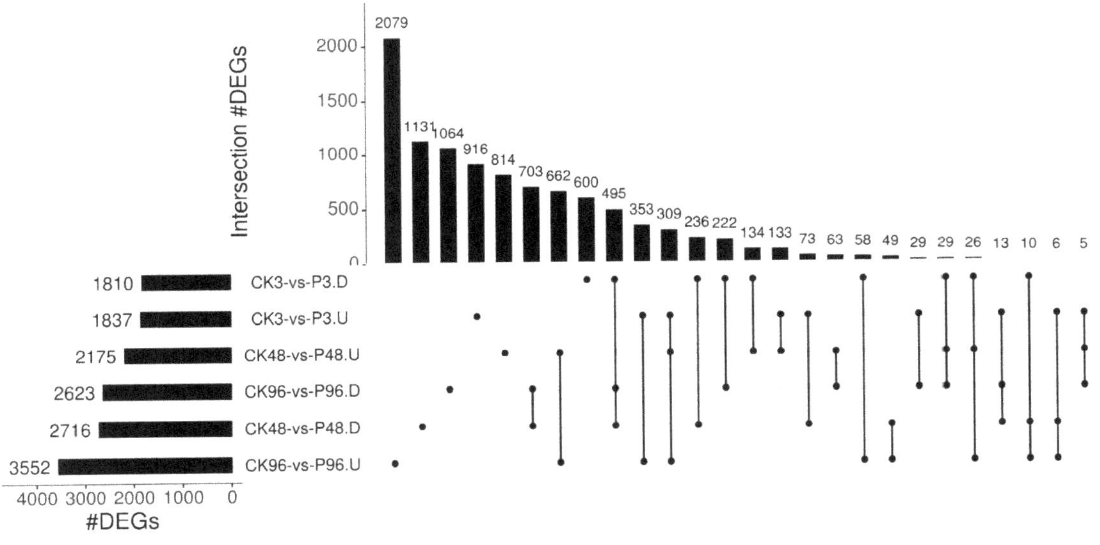

Figure 3. UpSet intersection plot of DEGs induced and suppressed by PTM infestation relative to the control of the same time. The intersections are shown with a line connected to one or more plots, and the number of common and unique DEGs is displayed above the vertical bars. The black dots represent different comparison groups. U—up-regulated genes; D—down-regulated genes.

The GO enrichment analysis of these DEGs characterized functions as: cell and cell part (CC), single-organism process (BP), transferase activity (MF). Ribosome, phenylpropanoid biosynthesis and carbon metabolism were the dominant KEGG enrichment pathways (metabolic pathways, biosynthesis of secondary metabolites and plant hormone signal transduction were excluded) (Tables S2 and S3).

After PTM infestation, the up-regulated DEGs functions were related mainly to ribosome (3 h), phagosome (48 h) and plant hormone signal transduction (96 h). Down-regulated DEGs were mainly associated with plant–pathogen interaction (3 h), ribosome (48 h), and protein processing in the endoplasmic reticulum (96 h) (Table S4).

2.2.3. Transcriptomic Comparison Analysis between the Two Treatments (M-vs-P)

A total number of 12,206 DEGs were altered, of which 6118 (3169 up and 2949 down), 5295 (2652 up and 2643 down) and 793 (640 up and 153 down) DEGs were detected respectively for 3 h, 48 h and 96 h by comparing the transcription profiles of PTM-infested and mechanically damaged plants (Figure 2A).

Interestingly, the least amount of DEGs was found at 96 h and the largest amount of DEGs was detected at 3 h within M-vs-P, which was the opposite of the case for CK-vs-M and CK-vs-P comparison groups, where the largest number of DEGs was found at 96 h, and the least number of DEGs was observed at three h. It is likely due to the potato leaves the internal state of insect feeding for 96 h is similar to that of mechanical damage for 96 h, which is consistent with the results of PCA data. 2231 DEGs, 1554 DEGs and 196 DEGs occurred uniquely, respectively, for the comparisons: M3-vs-P3. U, M48-vs-P48. U and M96-vs-P96. U, while 60 DEGs were common genes among three comparisons (Figure 4A). 1725, 1587 and 54 DEGs were uniquely expressed respectively within three comparison groups: M3-vs-P3. D, M48-vs-P48. D and M96-vs-P96. D, whereas only 15 DEGs were shared in common among three comparisons (Figure 4A).

GO enrichment analysis showed genes functions related mainly to plastid (CC), single-organism metabolic process (BP), oxidoreductase activity (MF) and catalytic activity (MF). KEGG analysis of genes comprising M-vs-P groups were enriched in pathways related to MAPK signaling pathway-plant (3 h), biosynthesis of amino acids (48 h), and biosynthesis of secondary metabolites (96 h). The detailed information of KEGG enrichment terms for up- and down-regulated genes in the M-vs-P comparison groups were presented in Supplemental Table S10. Functional analysis results of the M-vs-P were similar to the CK-vs-P comparisons groups (Tables S2 and S3).

In addition, the distribution of the abundances of transcripts among CK, M and P plant libraries were also compared strategically according to the method from Wang Dan et al. [19]. Venn diagrams revealed that 2976, 1499 and 117 DEGs were expressed only in M-vs-P comparisons for 3 h, 48 h and 96 h, respectively (Figure 4B–D). These DEGs may be related to other potato tuber moth herbivorous stimuli (e.g., insect oral secretions OS, walking, frass depositions, ovipositional fluids, etc.), which requires further validation through experiments.

2.3. *Expression Analysis of DEGs Involved in Plant Hormone Biosynthesis*

When plants are subjected to external stimuli (mechanical damage, herbivores attack, pathogen infection and abiotic stress), phytohormones can accumulate as signaling molecules, and induce the expression of downstream defense genes to activate the immune system against environmental stresses. In terms of JA, SA, ET and ABA biosynthesis pathways related to plant resistance, potato plants had different phytohormonal responses to external stress across the time course [18]. The expression level and functional annotations of related genes were presented in Supplementary Table S5.

α-linolenic acid (18:3) is a precursor of jasmonic acid (JA) synthesized from the octadecanoid pathway occurring in the chloroplast, peroxisome and cytoplasm sequentially in plants [31]. In the potato, LOX enzymes are divided into 9-LOX in cytosol and 13-LOX in chloroplast depending on the oxidation position of α-linolenic acid [32]. In terms of α-linolenic acid metabolism pathway, 9-LOX genes leading to the production of 10-oxo-11,15-phytodienoic acid (10-OPDA) were not found, whereas six 13-LOX genes related to the formation of 12-oxo-phytodienoic acid (12-OPDA) were up-regulated within all comparisons, except for the down-regulated gene encoding LOX2.1 (PGSC0003DMG400032207) in M3-vs-P3 (Figure 5, Table S4). The LOX product (9Z,11E,15Z)-(13S)-13-Hydroperoxyoctadeca-9,11,15-trienoic acid (13(S)-HpOTrE) was converted progressively by AOS and AOC to the final product 12-OPDA of the plastid-located part of JA biosynthesis. Subsequently, 12-OPDA is catalyzed gradually by OPDA reductase (OPR3) and OPC-8:0 CoA ligase 1 (OPCL1), resulting in the JA production ultimately according to β-oxidative steps in peroxisome [33]. Most of the genes encoding AOS, AOC, OPR3 and OPCL1 were up-regulated between all comparisons, whereas one AOS gene was down-regulated in 3 h or 48 h P/M comparisons, and one AOC gene was also down-regulated at 3 h P/M comparison. Among genes related to the enzymes of fatty acid β-oxidation, ACX1(PGSC0003DMG400004827) in CK48-vs-M48 and AIM1 (PGSC0003DMG400005498) in M48-vs-P48 had a low expression, respectively. MFP (PGSC0003DMG400003906) was down-regulated at 48 h both in P/CK and P/M comparisons, and PED1(PGSC0003DMG400015808) was also down-regulated in M3-vs-P3 comparisons (Figure 5, Table S5).

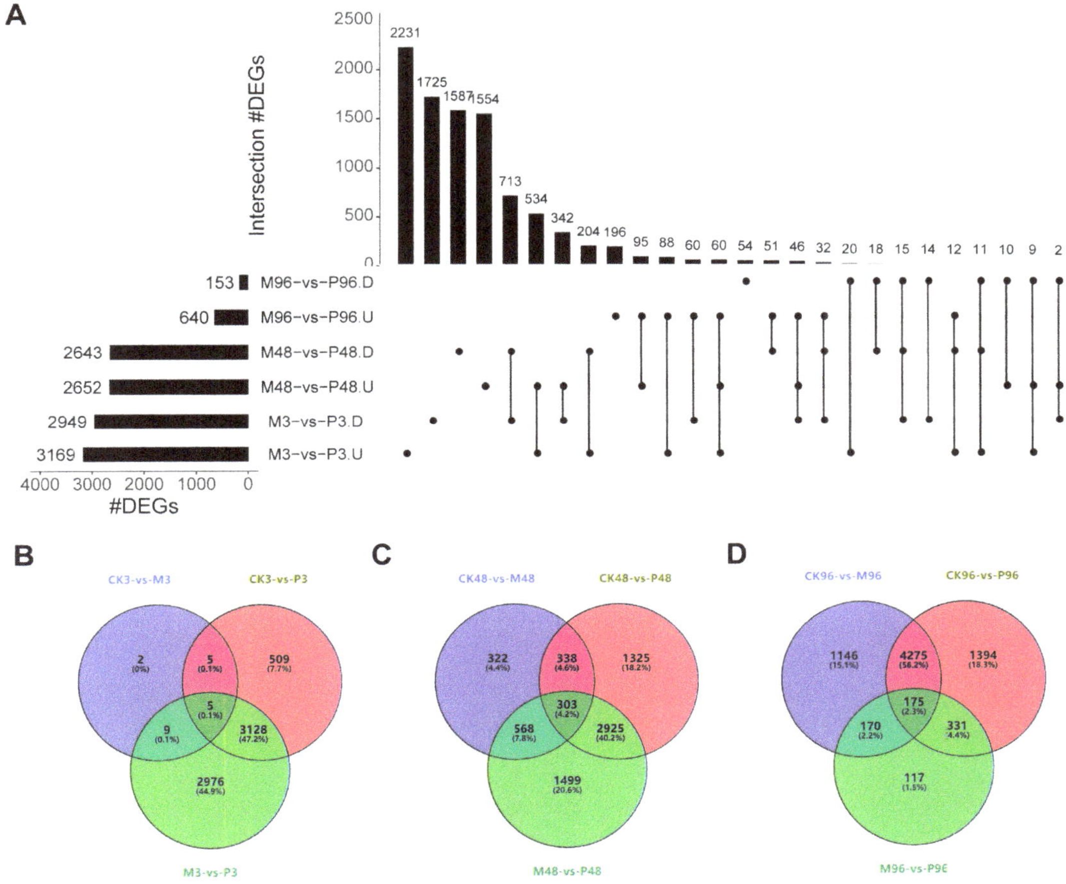

Figure 4. Comparative analysis of differentially expressed genes (DEGs) response to PTM infestation and mechanical wounding. (**A**) UpSet intersection diagram of up- and down-regulated DEGs from PTM-infested plants compared with mechanically-damaged plants at the same time. The black dots represent different comparison groups. U—up-regulated genes; D—down-regulated genes. (**B–D**) Venn diagram of differentially expressed genes (DEGs) that were common and unique among the CK, M, and P plant libraries after (**B**) 3 h, (**C**) 48 h, and (**D**) 96 h.

In the cytoplasm, JA is metabolized into different structures by various chemical reactions, such as methyl jasmonate (MeJA), jasmonoyl-isoleucine (JA–Ile), cis-jasmone (CJ) and 12-hydroxyjasmonic acid (12-OH-JA). Among all genes, JMT (PGSC0003DMG400003743) had the lowest FC value in CK96-vs-M96, and remained down-regulated in CK48-vs-P48, CK96-vs-P96 and M48-vs-P48 comparisons, while the other JMT gene had a weak induction at 96 h M/CK and P/CK. Except for the CYP94C1(PGSC0003DMG400019471), all genes encoding JAR, CYP94B3 and CYP94C1 were up-regulated. The above results indicated that the DEGs related to MeJA production were inhibited, whereas JA–Ile synthesis was strongly induced (Figure 5, Table S5).

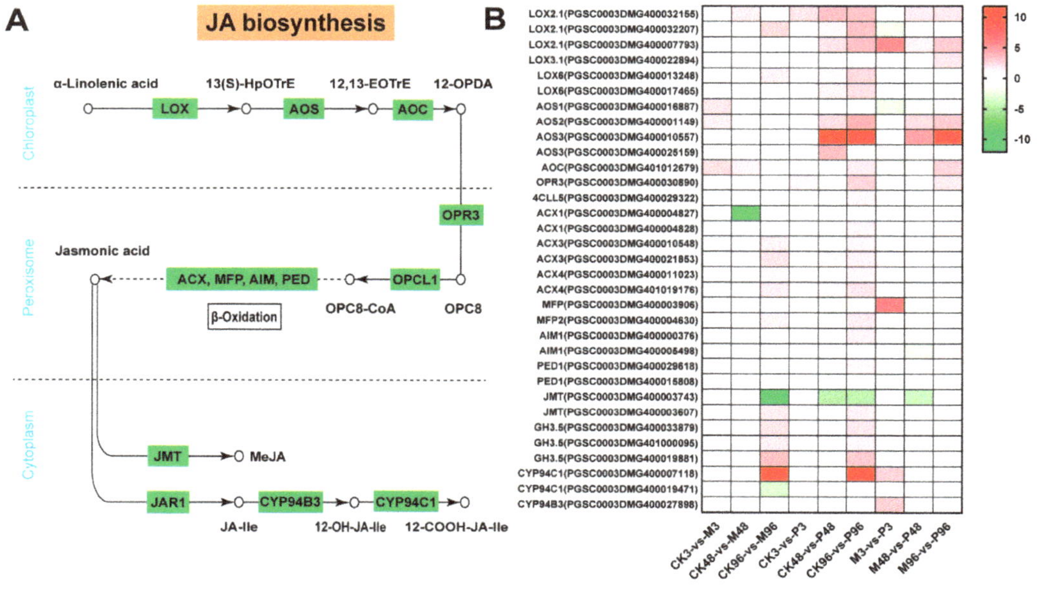

Figure 5. Analysis of jasmonic acid (JA) pathway gene expression. (**A**) Schematic diagram of the JA biosynthesis pathway. The circles represent the production, the green rectangles represent the genes that encoded for biosynthetic enzymes. LOX—lipoxygenase; AOS—allene oxide synthase; AOC—allene oxide cyclase; OPR3—12-oxophytodienoate reductase; OPCL—4-coumarate-CoA ligase; ACX—acyl-coenzyme A oxidase; MFP—enoyl-CoA hydratase/3-hydroxyacyl-CoA dehydrogenase; AIM—Fatty acid beta-oxidation multifunctional protein; PED—3-ketoacyl-CoA thiolase; JMT—JA carboxyl methyltransferase; JAR1—jasmonate resistant 1; CYP—cytochrome P450; 13(S)-HpOTrE— (9Z,11E,15Z)-(13S)-13-Hydroperoxyoctadeca-9,11,15-trienoic acid; 12,13-EOTrE—12,13-Epoxyoctadecatrienoic acid; 12-OPDA—12-oxo-phytodienoic acid; OPC8—3-oxo-2-(2′(Z)-pentenyl)-cyclopentane-1 octanoic acid; OPC8-CoA—3-oxo-2-(2′(Z)-pentenyl)-cyclopentane-1 octanoic acid coenzyme A; MeJA—methyl jasmonate; JA–Ile—jasmonoyl-isoleucine. (**B**) Heatmap of JA biosynthesis-related gene expression. Red represents up-regulated genes and green represents down-regulated genes. Color coding represents the range of log2 (foldchange relative to control). GH3.5—indole-3-acetic acid-amido synthetase GH3.5.

Overall, among the 33 JA synthetic pathway genes, there were 3 (3 up and 0 down), 3 (2 up and 1 down) and 14 (12 up and 2 down) DEGs detected in the potato at 3 h, 48 h and 96 h after mechanical damage, respectively. A total of 2 (2 up and 0 down), 12 (10 up and 2 down) and 25 (24 up and 1 down) DEGs after PTM infestation were observed for three time points, respectively. In P/M comparisons, there were 9 (5 up and 4 down) DEGs for 3 h, 8 (4 up and 4 down) DEGs for 48 h and 10 up-regulated genes for 96 h, respectively (Figure 5, Table S5). Results suggest that mechanical injury and PTM feeding lead to the up-regulation of most JA biosynthesis-related genes, and more DEGs are detected according to the time course.

Plants synthesize salicylic acid through two pathways: the isochorismate (ICS) pathway and the phenylalanine ammonia-lyase (PAL) pathway [34]. Chorismate is required as the common precursor in both pathways. In the ICS pathway, chorismate is catalyzed by ICS enzymes to produce isochorismate, which are transported from plastids to cytosol through EDS localized on chloroplast envelope and subsequently was conjugated with glutamate via the PBS3 to produce isochorismate-9-glutamate (IC-9-Glu). Ultimately salicylic acid was produced via spontaneous decay of IC-9-Glu or catalytic reaction of IC-9-Glu by

EPS1. In the other pathway, chorismate was catalyzed to produce phenylalanine in chloroplasts, which entered the cytoplasm and was converted by PALs to produce trans-cinnamic acid. After entering the peroxisome, trans-cinnamic acid is catalyzed to be benzoic acid, resulting in the production of salicylic acid finally in cytosol [34,35].

In terms of the DEGs encoding phenylalanine ammonia lyase (PAL), five DEGs were down-regulated among the first seven comparisons, whereas the opposite was the case for the TPA1 gene where the gene was up-regulated among these comparisons. Notably, all the three PAL genes detected in M96-vs-P96 comparison were up-regulated. The expression of the gene encoding chorismate mutase (CM) was similar to the gene encoding PAL (Figure 6, Supplemental Table S5). Among the genes encoding ICS, EDS, PBS and EPS, only EPS1 (PGSC0003DMG4000025842) was up-regulated in CK96-vs-P96 and M3-vs-P3 comparisons; other genes were down-regulated. Among the total of 11 up-regulated genes in 9 comparisons groups, there were 5 genes with high fold change (FC > 5), 1 gene with FC > 4, 3 genes with FC > 3, 2 genes with FC > 2, which suggested only several genes involved in SA biosynthesis pathway were highly induced in response to two treatments (Supplemental Table S5).

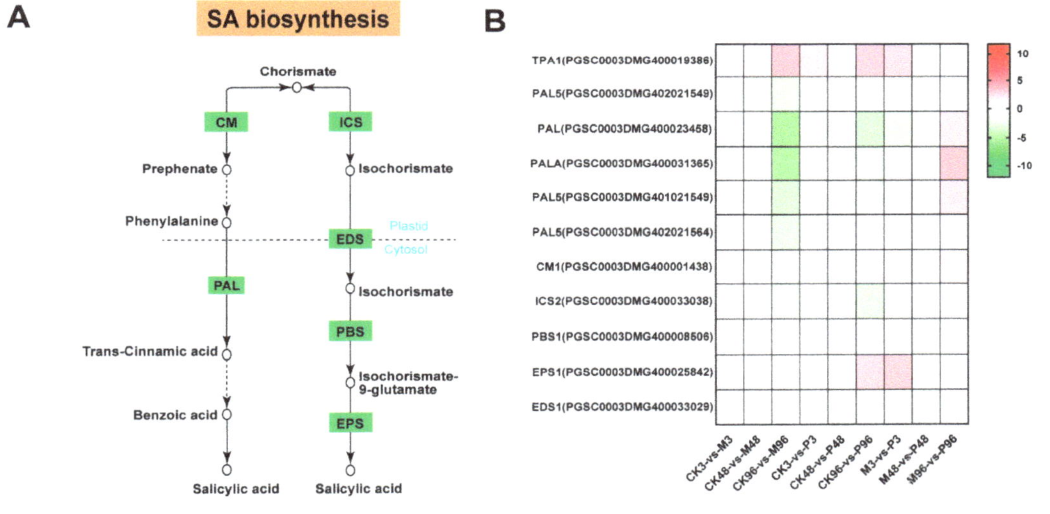

Figure 6. Analysis of salicylic acid (SA) pathway gene expression. (**A**) Schematic diagram of the SA biosynthesis pathway. The circles represent the production, the green rectangles represent the genes that encoded for biosynthetic enzymes. CM—chorismate mutase; PAL— phenylalanine ammonia lyase; ICS—isochorismate synthase; EDS—enhanced disease susceptibility; isochorismate is transported by the multidrug and toxin extrusion (MATE) transporter EDS from plastid to the cytosol. PBS—avr PphB susceptible; EPS—enhanced pseudomonas susceptibility. (**B**) Heatmap of JA biosynthesis-related gene expression. Red represents up-regulated genes and green represents down-regulated genes. Color coding represents the range of log2 (foldchange relative to control).

The currently accepted ABA biosynthesis pathway in plants is from the oxidative cleavage of carotenoids. β-carotene is utilized via various enzymes to produce trans-isomer zeaxanthin [36]. The conversion of zeaxanthin to violaxanthin catalyzed by zeaxanthin epoxidase (ZEP) within a two-step epoxidation via antheraxanthin is the formal first step of ABA biosynthesis [37]. Subsequently, violaxanthin was converted by neoxanthin synthase (NSY) and an unknown isomerase according to the two pathways to generate 9′-cis-neoxanthin and 9′-cis-violaxanthin [38]. The committed step of ABA biosynthesis is that 9′-cis-epoxycarotenoid dioxygenase (NCED) enzyme family split the cis-xanthophylls (9′-cis-violaxanthin and 9′-cis-neoxanthin) to synthetize C15 xanthoxin [39]. The above steps are carried out in plastids, and the next steps of the conversion from xanthoxin to ABA take place in the cytosol. The xanthoxin is converted into abscisic aldehyde (ABAld) utilizing a short-chain alcohol dehydrogenase (SDR), which is eventually oxidized to be abscisic acid (ABA) by abscisic-aldehyde oxidase (AAO) [40,41]. In addition, xanthoxin can also first be catalyzed to xanthoxic acid by AO isoform(s) and then xanthoxic acid is converted to ABA. The minor shunt pathway involving abscisic aldehyde as an intermediate is also an important source of ABA [42]. Overall, most of the genes involved in the ABA biosynthesis pathway in all comparisons were clearly down-regulated. Only one gene encoding ZSD1(PGSC0003DMG400023193) with FC > 5 was highly expressed, while two gene encoding AAO3 (PGSC0003DMG402018708) with FC > 2 were low expressed both in CK96-vs-M96 and CK96-vs-P96 comparisons (Figure 7 and Table S5).

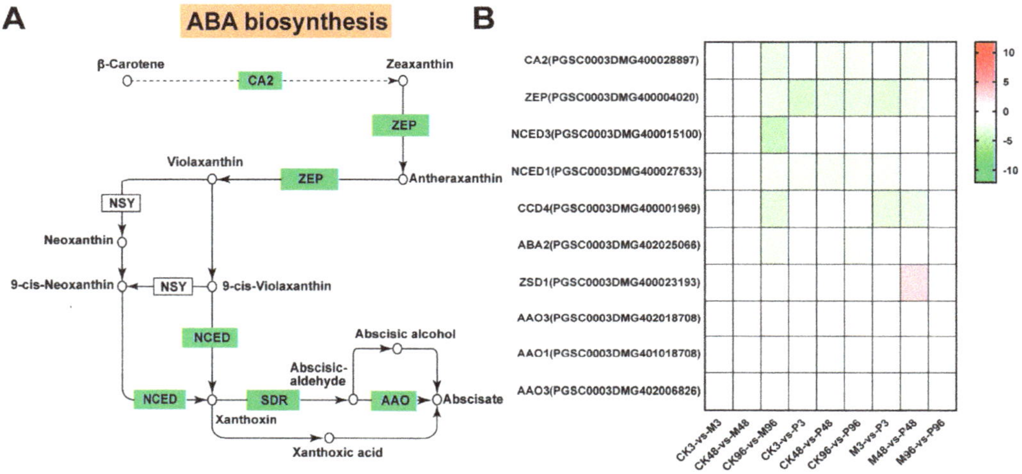

Figure 7. Analysis of abscisic acid (ABA) pathway gene expression. (**A**) Schematic diagram of the ABA biosynthesis pathway. The circles represent the production, the green rectangles represent the genes that encoded for biosynthetic enzymes. CA2—beta-carotene 2-hydroxylase; ZEP—zeaxanthin epoxidase; NSY—neoxanthin synthase; NCED—9-cis-epoxycarotenoid dioxygenase; SDR—short-chain dehydrogenase/reductase; AAO—abscisic-aldehyde oxidase. (**B**) Heatmap of ABA biosynthesis-related gene expression. Red represents up-regulated genes, and green represents down-regulated genes. Color coding represents the range of log2(foldchange relative to control). CCD4—carotenoid cleavage dioxygenase; ABA2—xanthoxin dehydrogenase; ZSD1—secoisolariciresinol dehydrogenase.

Compared with other plant hormones, the biosynthesis of ethylene is relatively simple. The first step is that L-methionine as the substrate was converted by S-adenosylmethionine synthase (S-AdoMet or SAM synthase) to generate S-adenosyl methionine (S-AdoMet or SAM) [43]. The subsequent step of S-AdoMet converted to 1-aminocyclopropane-1-carboxylate (ACC) utilizing ACC synthase (ACS) is the rate-limiting step [44]. Eventually, ACC was catalyzed by 1-aminocyclopropane-1-carboxylate oxidase (ACO) to produce ethylene [45]. Among the 13 genes of ethylene biosynthesis pathway, mechanical damage induced a total of 3 up-regulated gene expressions and PTM infestation induced a total of 5 up-regulated gene expressions in the corresponding treatment groups. A total of six up-regulated genes appeared in M-vs-P comparisons. Among the 11 up-regulated genes, there were a total of 2 genes with FC > 2, 5 genes with FC > 3, 1 gene with FC > 4, 1 gene with FC > 5, 2 genes with FC > 10, 2 genes with FC > 40 and 1 gene with FC > 500, which suggested ethylene pathway were highly induced locally (Figure 8 and Table S5).

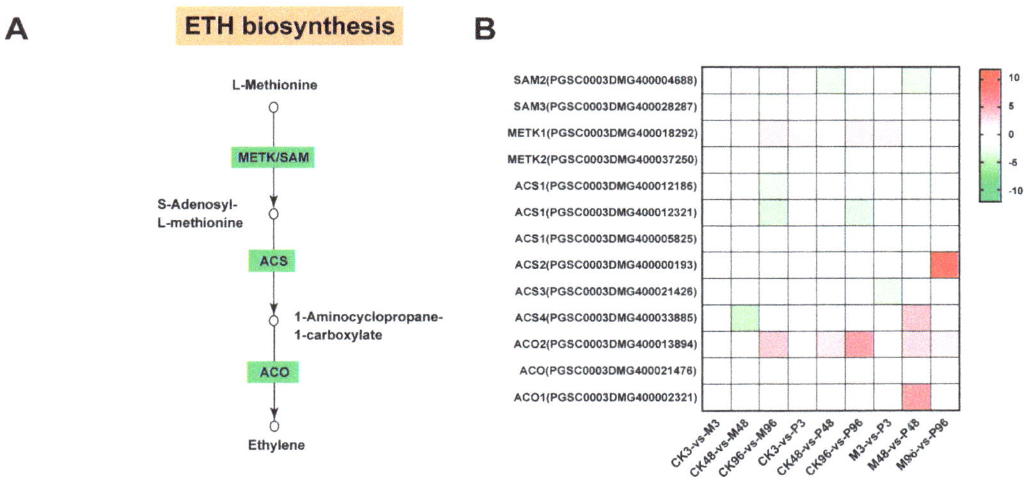

Figure 8. Analysis of ethylene (ET) pathway gene expression. (**A**) Schematic diagram of the ET biosynthesis pathway. The circles represent the production, the green rectangles represent the genes that encoded for biosynthetic enzymes. METK/SAM—s-adenosylmethionine synthase; ACS—1-aminocyclopropane-1-carboxylate synthase; ACO—1-aminocyclopropane-1-carboxylate oxidase. (**B**) Heatmap of ET biosynthesis-related gene expression. Red represents up-regulated genes and green represents down-regulated genes. Color coding represents the range of log2 (foldchange relative to control).

2.4. Expression Analysis of DEGs Involved in Reactive Oxygen Species (ROS) Signaling

Reactive oxygen species (ROS) mediate the normal metabolic and developmental processes in plants as key signaling molecules and enable plants to respond quickly to exogenous abiotic and biotic (including different pathogens and pests) stress [46]. We investigated genes encoding the key enzymes involved in metabolism reactions related to ROS production and scavenging in different comparisons. The gene expression of each comparison is shown in Supplemental Table S6.

A total of 13 DEGs (9 up and 4 down) and 20 DEGs (8 up and 20 down) related to ROS generation were observed at 48 h and 96 h post mechanical damage, respectively, whereas there was no DEGs occurring at 3 h in potato leaves. PTM infestation induced 21 DEGs (11 up and 10 down) for 3 h, 25 DEGs (12 up and 13 down) for 48 h and 27 DEGs (13 up and 14 down) for 96 h associated with ROS production, respectively. Compared with mechanical damage, PTM infestation triggered more up- and down-regulated DEGs at the same time point. Within comparisons of PTM infestation relative to mechanical damage, 30 DEGs (16 up and 14 down), 38 DEGs (18 up and 20 down) and 10 DEGs (8 up and 2 down) were found at 3 h, 48 h and 96 h, respectively (Table 1, Supplemental Table S6). The genes encoding acyl-CoA oxidase (ACX) had the largest group of the up-regulated genes within M/CK and P/CK comparisons, respectively, while the number of amine oxidase (AO) genes and gibberellin dioxygenase (GAOX) genes within down-regulated genes was the largest group in two comparisons.

Table 1. DEGs associated with reactive oxygen species (ROS) generating.

Category		AO	GOX	HPPD	GAOX	ND	COX	RBOH	SOD	FIN4	QSOX	ACX	AAO	ERO	FMO	Total
CK3-vs-M3	up	0	0	0	0	0	0	0	0	0	0	0	0	0	0	0
	dw	0	0	0	0	0	0	0	0	0	0	0	0	0	0	0
CK48-vs-M48	up	1	0	0	2	1	1	0	1	1	0	0	0	2	0	9
	dw	0	1	0	1	0	0	1	0	0	0	1	0	0	0	4
CK96-vs-M96	up	0	0	0	1	0	0	2	0	0	0	4	1	0	0	8
	dw	4	0	1	3	0	2	1	0	1	0	0	0	0	0	12
M/CK Total	up	1	0	0	3	1	1	2	1	1	0	4	1	2	0	17
	dw	4	1	1	4	0	2	2	0	1	0	1	0	0	0	16
CK3-vs-P3	up	0	0	0	2	4	0	0	1	0	2	0	1	0	1	11
	dw	4	1	1	1	0	0	2	0	0	0	0	0	0	1	10
CK48-vs-P48	up	3	0	0	1	0	0	3	1	0	0	2	1	0	1	12
	dw	2	0	1	4	1	2	2	0	0	0	0	1	0	0	13
CK96-vs-P96	up	1	0	1	3	0	0	1	0	0	0	5	2	0	0	13
	dw	3	0	1	3	1	2	2	1	1	0	0	0	0	0	14
P/CK Total	up	4	0	1	6	4	0	4	2	0	2	7	4	0	2	36
	dw	9	1	3	8	2	4	6	1	1	0	0	1	0	1	37
M3-vs-P3	up	1	1	0	4	3	0	0	1	1	1	1	1	0	2	16
	dw	7	1	1	0	1	0	2	0	0	0	0	0	0	2	14
M48-vs-P48	up	5	1	0	1	2	0	4	3	0	0	0	1	0	1	18
	dw	5	0	1	3	1	5	1	1	1	0	0	1	1	0	20
M96-vs-P96	up	4	0	0	2	0	0	0	0	0	0	1	1	0	0	8
	dw	1	0	0	0	0	0	0	1	0	0	0	0	0	0	2
P/M Total	up	10	2	0	7	5	0	4	4	1	1	2	3	0	3	42
	dw	13	1	2	3	2	5	3	2	1	0	0	1	1	2	36

Up—up-regulated genes; dw—down-regulated genes; AO—amine oxidase; GOX—glycolate oxidase; HPPD—4-hydroxypheny lpyruvate dioxygenase; GAOX—gibberellin dioxygenase; ND—NADH dehydrogenase; COX—cytochrome c oxidase subunit; RBOH—respiratory burst oxidase homolog; SOD—superoxide dismutase; FIN4—L-aspartate oxidase; QSOX—sulfhydryl oxidase; ACX—acyl-CoA oxidase; AAO—aldehyde oxidase; ERO—ER oxidoreductin; FMO—flavin-containing monooxygenase.

In terms of the genes encoding ROS scavenging, CK-vs-M groups comprised 24 DEGs (14 up and 10 down) and 59 DEGs (20 up and 39 down) expressed at 48 h and 96 h, respectively. A total of 35 DEGs (13 up and 22 down), 78 DEGs (51 up and 27 down) and 73 DEGs (46 up and 27 down) were identified at three time points after PTM infestation, respectively. A total of 15, 54, 32 up-regulated genes and 64, 38, 3 down-regulated genes were found respectively at 3 h, 48 h and 96 h within M-vs-P comparisons (Table 2, Supplemental Table S6). The total number of up-regulated genes encoding peroxidase (POD) was the largest group, whereas glutathione s-transferase (GST) had the opposite gene expression with the largest down-regulated group among all comparisons (Table 2).

Table 2. DEGs associated with reactive oxygen species (ROS) scavenging.

Category		CAT	APX	POD	Fd	TRX	GRX	GST	MDHAR	DHAR	GPX	PrxR	NRX	Total
CK3-vs-M3	up	0	0	0	0	0	0	0	0	0	0	0	0	0
	dw	0	0	0	0	0	0	0	0	0	0	0	0	0
CK48-vs-M48	up	0	1	3	3	1	0	5	0	1	0	0	0	14
	dw	0	1	2	0	2	3	2	0	0	0	0	0	10
CK96-vs-M96	up	1	0	6	1	5	6	0	0	0	1	0	0	20
	dw	1	3	5	3	10	4	10	0	0	1	1	1	39
M/CK Total	up	1	1	9	4	6	6	5	0	1	1	0	0	34
	dw	1	4	7	3	12	7	12	0	0	1	1	1	49
CK3-vs-P3	up	0	0	3	1	4	0	2	0	2	0	1	0	13
	dw	0	3	1	2	2	8	6	0	0	0	0	0	22

Table 2. *Cont.*

Category		CAT	APX	POD	Fd	TRX	GRX	GST	MDHAR	DHAR	GPX	PrxR	NRX	Total
CK48-vs-P48	up	0	1	22	4	2	15	0	1	1	1	4	0	51
	dw	1	2	1	3	4	5	10	0	0	0	0	1	27
CK96-vs-P96	up	1	0	22	1	5	14	1	0	0	1	1	0	46
	dw	1	1	5	2	8	2	7	0	0	0	1	0	27
P/CK Total	up	1	1	47	6	11	29	3	1	3	2	6	0	110
	dw	2	6	7	7	14	15	23	0	0	0	1	1	76
M3-vs-P3	up	0	0	3	3	5	0	2	0	1	0	1	0	15
	dw	1	2	9	3	10	14	24	0	0	0	0	1	64
M48-vs-P48	up	1	3	24	3	3	11	2	1	1	0	5	0	54
	dw	1	2	2	3	8	4	17	0	0	0	0	1	38
M96-vs-P96	up	0	0	15	3	2	9	2	0	1	0	0	0	32
	dw	0	0	0	0	0	3	0	0	0	0	0	0	3
P/M Total	up	1	3	42	9	10	20	6	1	3	0	6	0	101
	dw	2	4	11	6	18	21	41	0	0	0	0	2	105

Up—up-regulated genes; dw—down-regulated genes; CAT—catalase isozyme; APX—ascorbate peroxidase; POD—peroxidase; Fd—ferredoxin; TRX—thioredoxin; GRX—glutaredoxin; GST—glutathione s-transferase; MDHAR—monodehydroascorbate reductase; DHAR—dehydroascorbate reductase; GPX—glutathione peroxidase; PrxR—peroxiredoxin; NRX—nucleoredoxin.

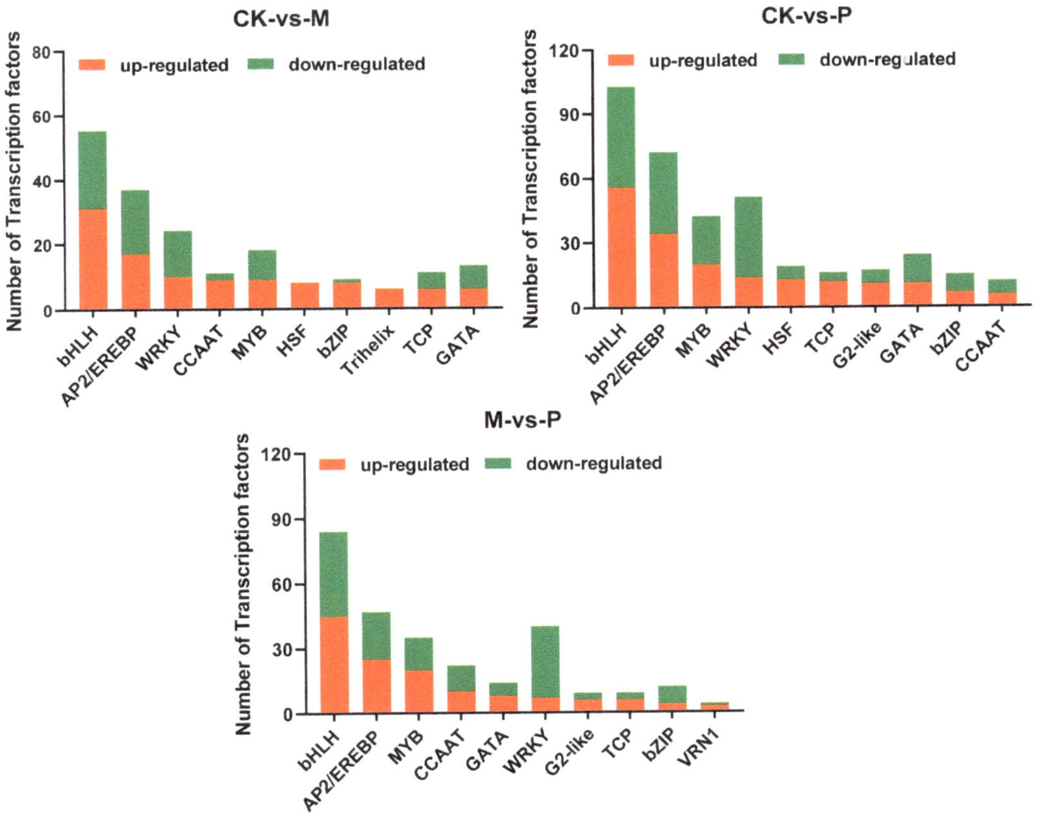

Figure 9. The total number of the top 10 transcription factors (TF) families with the largest number of up-regulated genes among the CK-vs-M, CK-vs-P and M-vs-P comparisons.

Among the top 10 TFs families, 9 TFs families, including bHLH, AP2/EREBP, WRKY, CCAAT, MYB, HSF, bZIP, TCP, and GATA, were commonly identified in three comparison groups and may have overlapping effects in mediating the defense responses to PTM infestation and mechanical damage (Figure 9). However, the difference is that the Trihelix family and G2-like family only appeared, respectively, in the CK-vs-M and CK-vs-P comparisons, which suggested the Trihelix family was mainly involved in defense responses to mechanical damage, while the G2-like family mainly mediated defense responses against PTM infestation (Figure 9). The dominant defensive role of the G2-like family against P treatment has also been demonstrated in the P/M comparison (Figure 9).

The up- and down-regulated transcription factor-related genes in CK-vs-M, CK-vs-P and M-vs-P comparisons were analyzed by KEGG. The results showed that up-regulated genes in CK-vs-M and CK-vs-P comparisons were enriched in circadian rhythm, MAPK signaling pathway and plant–pathogen interaction. The down-regulated genes of CK-vs-M are mainly related to protein processing in the endoplasmic reticulum, while the down-regulated genes of CK-vs-P are mainly associated with plant hormone signal transduction (Supplemental Table S8).

2.5. Expression Analysis of DEGs Involved in Plant Secondary Metabolites (PSMs)

Plant secondary metabolites have essential roles in the regulation defense signalings and function in the plant defense system against herbivore attacks and harsh environments [48]. The detailed information on PSMs-related DEGs are shown in Supplemental Table S9. The numbers of up-regulated genes related flavonoid biosynthesis were much smaller than that of down-regulated genes in a majority of comparisons, which indicated that the biosynthesis of flavonoids may have been inhibited in the process of plant resistance to two treatments. Compared to flavonoids, the biosynthesis of terpenoids in potatoes was gradually induced by PTM feeding from 3 h to 96 h (Tables 3 and S9). In CK-vs-M comparisons, there was no one up-regulated gene related alkaloids and steroids biosynthesis appeared for first two time points, which showed alkaloids and steroids were not involved in response against mechanical damage at 48 h post mechanical damage. Interestingly, 17 (6 up and 11 down), 24 (15 up and 9 down) and 25 (12 up and 13 down) DEGs related to alkaloids as well as 8 (5 up and3 down), 8 (4 up and 4 down) and 14 (9 up and 5 down) DEGs associated with steroids biosynthesis were detected at three time points respectively in CK-vs-P comparisons, which suggested that alkaloids and steroids have been involved in defense reactions of potato to PTM at 3 h and both remained until 96 h post-infestation. Compared with other PSMs, only several up-regulated genes were identified in each comparison, which showed that the potato may induce low concentrations of quinones in response to PTM feeding and mechanical damage (Table S9).

2.6. Expression Analysis of DEGs Involved in Plant–Pathogen Interactions and Defense Response

The incursion of pests and pathogens can elicit the up-regulation of related plant defensive gene families, including plant–pathogen interaction genes, lectin, proteinase inhibitors, chitinases and MAPK cascades signaling genes [49]. Detailed information about the various defense genes are presented in Tables S10 and S11.

PIs and CHIs both belong to pathogenesis-related (PR) proteins [50]. Proteinase inhibitors (PIs) families are a large and complex group of plant proteins and have the ability to form complexes with a widely accepted five classes of proteolytic enzymes, including serine, cysteine, threonine, aspartic and metallo-proteases, which suppress the normal assimilation of food proteins and bring the adverse effects for insects [51]. In our study, a total of five serine-, four cysteine-, three aspartic-type protease inhibitor-related genes and two protease inhibitor-related proteins were identified (Table S11). A majority of genes (50 genes) encoding PIs were up-regulated. The largest number of up-regulated genes both appeared at 48 h post two treatments relative to control treatments (Table 4).

Table 3. DEGs associated with plant secondary metabolites biosynthesis.

Category		Terpenoids	Flavonoids	Alkaloids	Steroids	Quinone	Total
CK3-vs-M3	up	2	0	0	0	0	2
	dw	0	0	0	0	0	0
CK48-vs-M48	up	5	5	0	0	1	11
	dw	10	8	2	0	2	22
CK96-vs-M96	up	12	2	7	7	2	30
	dw	28	23	16	8	10	85
CK3-vs-P3	up	9	2	6	5	2	24
	dw	18	13	11	3	5	50
CK48-vs-P48	up	12	6	15	4	3	40
	dw	24	16	9	4	6	59
CK96-vs-P96	up	24	6	12	9	4	55
	dw	23	21	13	5	10	72
M3-vs-P3	up	15	7	6	9	2	39
	dw	27	24	18	4	6	79
M48-vs-P48	up	16	8	20	7	5	56
	dw	24	11	8	3	7	53
M96-vs-P96	up	12	12	8	6	4	42
	dw	0	0	0	0	0	0

Up—up-regulated genes; dw—down-regulated genes.

Table 4. DEGs involved in proteinase inhibitors (PIs), chitinases and MAPK cascades.

Category		M3/CK3	M48/CK48	M96/CK96	P3/CK3	P48/CK48	P96/CK96	P3/M3	P48/M48	P96/M96	Total
PIs	up	0	9	1	5	8	7	10	4	6	50
	dw	0	1	0	1	0	0	1	3	0	6
Chitinases	up	0	1	3	1	2	6	2	7	4	26
	dw	0	2	5	7	3	1	11	4	0	33
MAPK	up	0	0	6	1	1	6	2	2	1	19
	dw	0	0	6	0	4	6	2	4	1	23

M3/CK3—abbreviation for CK3-vs-M3. The abbreviated forms of other comparisons as described for the first comparison.

Plant chitinase (CHIs) can attack chitin as the structural molecule in skeletons of insects as well as cell walls of fungi and hence chitinases play a potential role against pathogens and insects [52]. In CK-vs-M comparisons, up/down-regulated genes related to CHIs both appeared at 48 h and the number both peaked at 96 h. The number of up-regulated genes (one gene) were induced at 3 h by PTM infestation and the number (six genes) peaked at 96 h, while the down-regulated genes were the opposite case, maximum number of genes (seven genes) detected at 3 h and minimum number of genes (one gene) detected at 96 h (Table 4).

MAPK cascades serve as an important element for transcriptional activation of herbivore defense-related genes and the accumulation of defensive metabolites [4,53]. The MAPK-related up/down-regulated genes post two treatments showed similar expression patterns, with the largest number of genes appearing at 96 h (Table 4).

Previous studies showed that the gene expression level of attack from insect herbivory and infection by pathogens can have a partial or considerable overlap [54,55]. LecRLKs (possessing three subclasses: L-, G-, and C-type LecRLKs), as a subfamily of receptor-like protein kinases, consist of three domains: extracellular lectin domain, intermediate transmembrane domain and intracellular kinase domain, which play crucial roles in development, stress perception and pathogen detection [56,57]. The up-regulated DEGs involved in plant–pathogen interaction DEGs sequentially accounted for 51.8%, 50%, 17.3%, 22.7%, 47.3%, 22.3%, 33.7% and 87.5% of the total number of DEGs of each comparison

(CK3-vs-M3 comparison was excluded). GO analysis of each comparison group showed that the up-regulated gene function at 48 h post-mechanical damage were mainly associated with response to stimulus (BP), cellular protein metabolic process (BP) and protein binding (MF). The up-regulated genes of PTM infestation for 48 h were mainly enriched in single-organism process (BP) and catalytic activity (MF). However, compared with the control, the enriched GO terms of up-regulated genes related to plant–pathogen interaction were similar at 96 h post two treatments (Supplemental Table S12). The up-regulated DEGs screened from lecRLKs-related genes, respectively, accounted for 21.4%, 48.4%, 0%, 17.0%, 34.9%, 26.7%, 29.3% and 25.0% of total number of DEGs of each comparison (except for CK3-vs-M3 comparison) (Figure 10). This indicates that the number of up-regulated DEGs increased over time after two treatments. The GO analysis of up-regulated lecRLKs genes comprising each comparison group was mainly enriched in transferase activity (MF), kinase activity (MF), catalytic activity (MF) and biological processes (BP) (Supplemental Table S13).

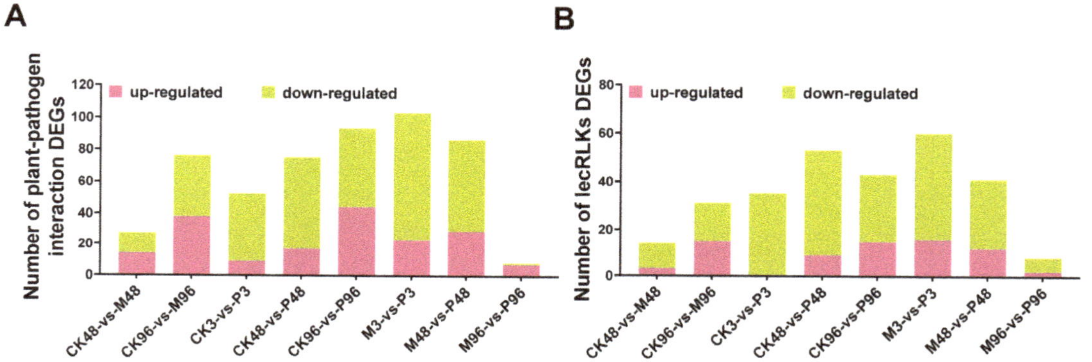

Figure 10. Genes differentially expressed in all comparisons. (**A**) The number of up- and down-regulated genes associated with plant–pathogen interactions. (**B**) The number of up- and down-regulated genes related to lectin receptor-like protein kinases (LecRLKs).

3. Discussion

Currently, the research about the defense mechanisms of the potato in response to *Phthorimaea operculella* infestation has remained limited. A previous study has revealed how potato plants integrate accumulatively a multitude of herbivory components from potato tuber moth and show the early defense responses of potato leaf to these reorganized stimuli [56,57]. In our study, the global transcriptomic response of potato plants to PTM infestation and mechanical damage was monitored by RNA-Seq in CK, M, P plants harvested at 3 h, 48 h and 96 h. Overall, the transcriptomes were significantly remodeled under the PTM attack and mechanical damage during the time course.

Except for the GO terms in common between mechanical damage and PTM infestation across three time points, two treatments at 3 h mainly activated the secondary metabolism of the potato, such as the S-glycoside and indoleacetic acid metabolic process (CK3-vs-M3) as well as the flavonoid metabolic process (CK3-vs-P3) (Table S2). At 48 h, GO terms included negative regulation of cellular metabolic process (CK48-vs-M48) and regulation of multicellular organismal development (CK48-vs-P48), which indicated two treatments mainly suppressed the growth and development of potatoes. At 96 h, there were significant enrichment in carbohydrate and polysaccharide metabolic process in M/CK comparison as well as organic acid transmembrane transport in P/CK comparison (Table S2), indicating the two treatments resulted in the redistribution of energy in potato plants. Physical damage can lead to the decline of the photosynthesis ability of plants and a decrease in ATP production by photosynthesis phosphorylation, resulting in energy deficiency [18].

Compared with mechanically wounding, insect feeding caused more severe damage to potato plants, which prompted the potato to obtain energy for growth and defense through other ways, such as organic acid metabolism.

The stimuli from insect feeding can be roughly divided into several types: mechanical damage, oral secretion stimuli OS, walking, frass depositions, and ovipositional fluids, etc. [19]. In M-vs-P, we identified different numbers of DEGs induced only by mechanical damage and PTM infestation, as well as some DEGs that may come from other potato tuber moth herbivorous stimuli by strategic comparative transcriptome analysis (Figure 4B–D). These genes may be key components and factors that determine the different gene expression patterns of downstream insect defense-related signaling pathways.

There were 11 up-regulated DEGs associated with ROS generation found at 3 h after PTM infestation and the number of genes was continuously increasing until 96 h (Table 1).

In the study of Mao et al. [20], ROS signaling was activated by PTM-derived herbivory-related cues as early as 1 h after treatment. In general, the process of ROS generation occurs within seconds to minutes of stress initiation [58,59], whereas it is possible that the subsequent results might be different in response to the age of PTMs or depending on treatment time. No genes were detected at 3 h, while the number of up-regulated genes induced by mechanical damage is slightly less than that induced by PTM infestation at 48 h and 96 h. This suggests that PTM feeding induced a faster and stronger signal than mechanical damage. Over-accumulation of ROS is toxic to plants, for which plants activate stringent ROS scavenging mechanisms, including antioxidant enzymes and non-enzymatic metabolites, to remove excess ROS and maintain normal levels of ROS [17,58–62]. Therefore, it is reasonable to see that there are more up-regulated genes related to ROS scavenging at 48 and 96 h after mechanical injury and PTM feeding, even though some were far more than down-regulated genes (Table 2). It has been observed that not only can ROS activate MAPK signaling, but also MAPK cascades can regulate ROS-related genes [17,60–62]. In our study, one and six MAPK-related up-regulated genes were observed at 3 h post PTM infestation and 96 h after mechanical damage, respectively, which may imply that MAPK cascades were activated at 3 h post PTM feeding, while MAPK signals were induced by mechanical damage after 48 h.

The activation of ROS signalings and MAPK cascades are usually accompanied by the rapid initiation of phytohormone networks and the complex interplay between ROS, MAPKs and hormones brings them to a level of mutual coordination [63–65]. In terms of JA biosynthesis, most genes were up-regulated under mechanical damage and PTM herbivory, which was largely consistent with the reported high accumulation of JA and JA–Ile under two treatments post one hour in Mao et al. study [20]. This further indicated that JA plays an essential role in mediating plant defenses against lepidopteran-chewing herbivores [66–68]. In our study, 13-LOX genes were more strongly induced than 9-LOX genes responsible for the production of 10-OPDA (Figure 5B). However, 9-lipoxygenase pathway is preferentially stimulated in cultured potato cells in response to treatment with P. infestans elicitor [69]. Feeding by S. frugiperda and beet armyworm Spodoptera exigua larvae on maize also induces the expression of 9-LOX genes to a greater extent than 13-LOX genes [70,71]. Species of herbivores/plants and different types of external stimuli could account for this disparity. Compared with MeJA, JA is inclined to be metabolized into JA–Ile as a major bioactive signal in the dynamic regulation of the JA signaling system [67,72], which is consistent with our results (Figure 5B). Moreover, we also identified several genes belonging to the GH3 family with the closest sequence homology to JAR1, but their role in JA metabolism is not known (Figure 5B). In Mao et al. study, accumulation of ABA was not induced significantly at 1 h by PTM actual herbivory and mechanical damage; even its induction was repressed by microbe from orally secreted bacteria (OSB) [20]. Most ABA biosynthesis genes were all down-regulated in CK-vs-M, CK-vs-P and M-vs-P comparisons in our study (Figure 7). The two research results emphasize the consistent performance of ABA-related genes in the potato's defense against PTM feeding. Interestingly, unlike JA and ABA, the expression patterns of SA biosynthesis-related genes in CK96-vs-M96 and

CK96-vs-P96 were different from that in M96-vs-P96 (Figure 6B). Accumulation of SA was also not induced significantly by all types of reorganized in a cumulative way potato tuber moth herbivory stimuli components in the Mao et al. study [20]. We previously mentioned that the M-vs-P groups possessed different proportions of genes from different potato tuber moth herbivorous stimuli (Figure 4B–D). Combining the above results, we speculated that mixed subsets of the genes from different potato tuber moth herbivory stimuli have different expression patterns in the synthesis of SA, which may result in the homeostasis of SA level. Ethylene biosynthesis-related genes were highly induced locally in M/CK, P/CK and P/M comparisons (Figure 8), which may imply ET pathway was involved in the defense against two treatments. In addition, we also detected the expression of all genes in the JA/SA/ABA/ETH signal transduction, which functions downstream from phytohormone production, and potato plants may have elevated constitutive expression of SA and ABA signaling cascade (Table S4).

A crosstalk between phytohormones depends on the plant–herbivore system has been studied. Many studies have reported that the combination of ET and JA can synergistically induce plant defense genes against biotic stress, including herbivores and pathogens [73–75]. In the direct defense of Nicotiana attenuata plants against chewing herbivores, jasmonic acid is a central mediator of defense gene expression, whose effect is modulated by ethylene [76]. Therefore, combining the research result of Mao et al. [20], we speculated that jasmonate and ethylene act in synergistic manners in the defense reaction of the potato to PTM feeding and mechanical damage, while salicylic acid and abscisic acid both play less significant roles in the defense response. ERF1 and EIN3 are downstream components of the ethylene signal transduction. It has been demonstrated that ERF1 is a key integrator of JA and ET signals, and its induction requires both signaling pathways simultaneously to be activated [77]. Zhu et al. speculated that EIN3/EIL1 is the direct molecular link for jasmonate-ethylene synergistic interactions based on the previously accumulated data [78]. In ET signal transduction, most genes, including the encoding ERF1 and EIN3/EIL were up-regulated; even the genes EIL3 (PGSC0003DMG400016747) and ERF1B (PGSC0003DMG400010285) had FC values exceeding 256 (Table S4). Moreover, ethylene plays an important accessory role in JA-mediated plant defenses against herbivores, and they can synergistically regulate the synthesis of downstream alkaloids etc., secondary metabolites [76,79], and the expression of protease inhibitors (PIs) genes [80]. From the number of up/down-regulated genes in the biosynthesis of various secondary metabolites and protease inhibitors, it can be concluded that the biosynthesis of terpenoids, alkaloids, steroids, and protease inhibitors in the potato is significantly induced by potato tuber moth feeding (Tables 3 and 4). Flavonoids are derived from the phenylpropanoid pathway and belong to the phenolics. However, their biosynthesis was repressed by both treatments, which may result from the down-regulation of the genes encoding phenylalanine ammonia lyase (PAL), a key enzyme in this pathway (Figure 6). Serine protease inhibitors play a more prominent role in lepidopteran insects (Lepidoptera) [8], which is in line with our results (Table S11).

Transcription factor families have important regulatory roles in stress tolerance and mediating the expression of downstream defense genes in plants. Many excellent reviews and literature have reported that several transcription factor families, such as NAC, WRKY, AP2/ERF, bHLH, MYB, TCP, and bZIP are associated with biotic stress response in the potato [81]. The G2-like (GLKs) family plays an active role in resisting pathogen invasion, regulating leaf senescence and chloroplast development, and responding to abiotic stresses [82]. Only three up-regulated genes associated with the bZIP family appeared at 48 h after mechanical damage, whereas only five down-regulated genes were detected at 48 h after PTM infestation. Interestingly, the TCP family showed opposite gene expression patterns at 48 h after both treatments (Table S7), which indicated that the TFs families may fine-tune the response of plants to different external stimuli.

In solanaceae crops (e.g., potato and tobacco), the LecRK/LecRLK or NbLRK1 genes generate resistance by sensing elicitors produced by insects and pathogens [57,83]. The plant–pathogen interaction can affect the induction rapidity and effectiveness of chitinase [83]. In our study, the up-regulated genes related to plant–pathogen interaction, lectin receptor-like protein kinases (LecRLKs) and chitinases showed a consistent change in expression over time (Figure 10, Table 4). It has been shown that infestation by chewing Pieris rapae larvae induces resistance in Arabidopsis not only to *P. rapae* itself, but also to several microbial pathogens [84], indicating an overlap in the signaling network of plant defense against insects and pathogens, which is consistent with our results. This overlap suggests that the regulation of adaptive responses in plants is a delicate balance between protection against microbial and insect invaders [55]. De Vos et al. found that consistent changes in Arabidopsis transcriptional profiles caused by pathogens and insects with different attack modes not only showed considerable overlap but also varied in number with different combinations of attackers [54]. Additionally, Nilaparvata lugens or Chilo suppressalis infestation caused extensive up-regulation of genes related to phytopathogenic interactions in rice at 3 h and 6 h with a much higher proportion of up-regulated genes compared with our results [18]. Therefore, the magnitude of genetic overlap triggered by pathogens and insects reflects the intensity of the plant response, which is dependent on treatment time and plant-attacker combinations [54]. It would be interesting to further elucidate the role of uncharacterized genes such as protease inhibitors, chitinases, and lectins in plant–insect interactions in the future through gene overexpression, silencing, mutant analysis, etc.

4. Materials and Methods

4.1. Insect Colony

Potato tuber moth (*Phthorimaea operculella*, PTM) populations were collected from potato fields in Kunming, Yunnan Province, China and further raised for several generations on potato variety, Hezuo 88 (HZ 88). The obtained PTM eggs were placed in Petri dishes lined with moist filter paper and incubated in a growth chamber (temperature of 28 $\pm$ 2 °C, relative humidity 80%, photoperiod 12L:12D). Newly hatched PTM larvae were used for the experiments.

4.2. Plant Growth, Treatment and Sample Collection

Hezuo 88 healthy seed potatoes (only one bud eye per tuber was kept) were grown in plastic pots (diameter $\times$ height = 20 cm $\times$ 30 cm) and placed in a greenhouse (temperature 27 $\pm$ 2 °C, RH 75%, photoperiod 14L:10D) for cultivation. To avoid other insect attacks and interactions between plants, each plastic pot was placed 2 m apart and covered with 100-mesh nylon nets. The potato plants were regularly watered and fertilized according to the same criteria and were left to grow to five fully expanded leaves for the experiment. PTM larvae were starved for 6 h before being used to infest potatoes. One freshly hatched PTM larva at the age of 2 days was inoculated manually on five fully expanded leaves of potato, respectively, which was the PTM infestation treatment (P). The mechanical damage treatment (M) was achieved by randomly pricking the potato leaves obliquely at an angle of about 45 degrees with an alcohol-sterilized No. 3 insect needle. Potato plants without both treatments were used as a control (CK). At 3 h, 48 h, and 96 h post CK, M, and P treatments, potato leaves with uniform size and leaf age were harvested in aluminum foil and flash-frozen in liquid nitrogen for RNA extraction. For each treatment, three replicate samples at each time point were obtained from three uniformly growing and independent potato plants.

4.3. RNA Extraction, Illumina Library Construction and Sequencing

Total RNA was extracted from every sample using Trizol reagent kit (Invitrogen, Carlsbad, CA, USA) according to the manufacturer's protocol and treated with RNase-free DNase I to remove genomic DNA contamination. RNA quality was assessed on an Agilent 2100 Bioanalyzer (Agilent Technologies, Palo Alto, CA, USA) and checked using RNase free agarose gel electrophoresis. After total RNA was extracted, eukaryotic mRNA was enriched by Oligo(dT) beads. Then, the enriched mRNA was fragmented into short fragments using a fragmentation buffer and reverse transcripted into cDNA with random primers. Second-strand cDNA was synthesized by DNA polymerase I, RNase H, dNTP and buffer. Then the cDNA fragments were purified with a QiaQuick PCR extraction kit (Qiagen, Venlo, The Netherlands), end-repaired, poly(A) added, and ligated to Illumina sequencing adapters. The cDNA library fragments were purified with the AMPure XP system to obtain cDNA fragments with a preferred around 200 bp, PCR amplified, and sequenced using Illumina HiSeq2500 by Gene Denovo Biotechnology Co. (Guangzhou, China).

4.4. Processing and Analysis of RNA-Seq Data

To obtain high-quality clean reads, raw reads were further filtered by fastp V0.18.0 [85] (https://github.com/OpenGene/fastp, accessed on 9 April 2023) to remove reads containing adapters, poly(N) and low-quality reads containing more than 50% of low-quality (Q-value $\leq$ 20) bases. An index of the potato reference genome was built, and paired-end clean reads were mapped to the reference genome SolTub_3.0 (https://www.ncbi.nlm.nih.gov/assembly/GCF_000226075.1, accessed on 10 April 2023) using HISAT2.2.4 [86] with "-rna-strandness RF" and other parameters set as a default. The mapped reads of each sample were assembled by using StringTie v1.3.1 [87,88] in a reference-based approach. For each transcription region, a FPKM (fragment per kilobase of transcript per million mapped reads) value was calculated to quantify its expression abundance and variations using StringTie software.

RNA differential expression analysis was performed by DESeq2 [89] software between two different groups (and by edgeR [90] between two samples). The Benjamini–Hochberg algorithm was used to adjust the p-value and hence control the false discovery rate (FDR). The genes with the parameter of false discovery rate (FDR) below 0.05 and absolute fold change $\geq$ 2 were considered differentially expressed genes (DEGs). Related genes between different treatment groups were identified based on gene annotation and differential expression analysis.

4.5. GO Functional and KEGG Pathway Enrichment Analysis

GO enrichment analysis provides all GO terms that are significantly enriched in DEGs compared to the genome background and filters the DEGs that correspond to biological functions. Firstly, all DEGs were mapped to GO terms in the Gene Ontology database (http://www.geneontology.org/, accessed on 20 April 2023), gene numbers were calculated for every term, and significantly enriched GO terms in DEGs compared to the genome background were defined by hypergeometric test. This analysis was able to recognize the main biological functions that DEGs exercise. Genes usually interact with each other to play roles in certain biological functions. Pathway-based analysis helps to further understand the genes' biological functions. KEGG is the major public pathway-related database [90]. Pathway enrichment analysis identified significantly enriched metabolic pathways or signal transduction pathways in DEGs compared with the whole genome background.

The calculated p-values were processed through FDR Correction, taking FDR $\leq$ 0.05 as a threshold. GO terms and pathways meeting this condition were defined as significantly enriched GO terms and pathways in DEGs, respectively.

5. Conclusions

In conclusion, the defense mechanism of the potato is activated after the initiation of mechanical damage and PTM infestation, and response to two treatments built up over time.

GO functional enrichment showed PTM feeding and mechanical damage activated primary and secondary metabolism over time and suppressed the growth and development of the potato. Phytohormone analysis showed that genes involved in JA and ET signaling pathways were strongly induced, and both may synergistically induce defense responses against herbivores, with JA having a more dominant role than ET. At 48 h and 96 h, more up-regulated genes linked to ROS scavenging were observed, as were the genes encoding secondary metabolites. This suggests that the antioxidant system plays an indispensable role in plant adaptation to both stimuli, while the production of secondary metabolites such as terpenoids, alkaloids, and steroids would presumably benefit only PTM-challenged plants from the proportion of up- and down-regulated genes. Several TF families, such as bHLH, AP2/EREBP, WRKY and MYB, were differentially regulated, while the Trihelix family and the G2-like family appeared to be particularly active in mediating the response to mechanical damage and PTM infestation, respectively. Both treatments also induced gene expression of many pathogenesis-related (PR) proteins (e.g., protease inhibitors and chitinases) and lectins.

Our exploration of mechanical damage and actual PTM infestation at multiple time points makes a strong addition to the gene expression patterns of potato–PTM interactions over the time span.

Supplementary Materials: The following supporting information can be downloaded at: https://www.mdpi.com/article/10.3390/plants12173092/s1, Table S1: Sunmary of RNA-Seq data; Table S2: The top 20 GO enrichment terms in each comparison group; Table S3: The top 20 KEGG enrichment pathways in each comparison group; Table S4: The KEGG enrichment pathways of up- and down-regulated genes in each comparison group; Table S5: DEGs involved in the phytohormones pathways; Table S6: DEGs involved in the reactive oxygen species; Table S7: Number of DEGs involved in transcription factors; Table S8: The KEGG enrichment pathways of up- and down-regulated TFs genes in each comparison group; Table S9: DEGs involved in secondary metabolites; Table S10: DEGs involved in plant–pathogen interaction; Table S11: DEGs involved in PI, chitinase, lectin and MAPK; Table S12: The GO enrichment terms of up- and down-regulated genes associated with plant-pathogen interactions in each comparison group; Table S13: The GO enrichment terms of up- and down-regulated genes associated with LecRLKs in each comparison group; Figure S1: Differentially expressed genes (Padj < 0.05 and an absolute value of Log_2FoldChange (relative to the control) ≥ 2 was used as the standard for screening DEGs).

Author Contributions: Conceptualization, C.Z.; data curation, C.Z.; funding acquisition, G.X. and B.C.; resources, M.Y., S.Z., X.Y., Y.L. and Y.Y.; software, C.Z.; writing—original draft, C.Z.; writing—review and editing, C.Z. All authors have read and agreed to the published version of the manuscript.

Funding: This research was supported by the National Natural Science Foundation Programs: No. 32360711 and No. 31760519, the Scientific Research Fund Project of Yunnan Provincial Education Department, China (Project No. 2023Y1026), the project from the Science and Technology Innovation Team of Yunnan Provincial Department of Education [2022ZKY223].

Data Availability Statement: The transcriptomics dataset is available under Bioproject: PRJNA978946.

Acknowledgments: We thank Wenqian Wang for supplying potato tuber moths. We also thank Guang Wang and Yanyan He for their help with data analysis.

Conflicts of Interest: The authors declare no conflict of interest.

References

1. Howe, G.A.; Jander, G. Plant immunity to insect herbivores. *Annu. Rev. Plant Biol.* **2008**, *59*, 41–66.
2. Pingault, L.; Basu, S.; Vellichirammal, N.N.; Williams, W.P.; Sarath, G.; Louis, J. Co-Transcriptomic Analysis of the Maize-Western Corn Rootworm Interaction. *Plants* **2022**, *11*, 2335.
3. Gibson, R.W. Glandular hairs providing resistance to aphids in certain wild potato species. *Ann. Appl. Biol.* **1971**, *68*, 113–119. [CrossRef]
4. Hettenhausen, C.; Schuman, M.C.; Wu, J. MAPK signaling: A key element in plant defense response to insects. *Insect Sci.* **2015**, *22*, 157–164. [PubMed]
5. Paiva, N.L. An Introduction to the Biosynthesis of Chemicals Used in Plant-Microbe Communication. *J. Plant Growth Regul.* **2000**, *19*, 131–143. [PubMed]
6. Gatehouse, J.A. Plant resistance towards insect herbivores: A dynamic interaction. *New Phytol.* **2002**, *156*, 145–169.
7. Zhan, X.; Chen, Z.; Chen, R.; Shen, C. Environmental and Genetic Factors Involved in Plant Protection-Associated Secondary Metabolite Biosynthesis Pathways. *Front. Plant Sci.* **2022**, *13*, 877304.

8. Mosolov, V.V.; Valueva, T.A. Proteinase inhibitors and their function in plants: A review. *Appl. Biochem. Microbiol.* **2005**, *41*, 227–246. [CrossRef]
9. Vandenborre, G.; Smagghe, G.; Van Damme, E.J. Plant lectins as defense proteins against phytophagous insects. *Phytochemistry* **2011**, *72*, 1538–1550.
10. Leitner, M.; Boland, W.; Mithöfer, A. Direct and indirect defences induced by piercing-sucking and chewing herbivores in Medicago truncatula. *New Phytol.* **2005**, *167*, 597–606. [PubMed]
11. Chen, C.Y.; Mao, Y.B. Research advances in plant-insect molecular interaction. *F1000Research* **2020**, *9*, F1000 Faculty Rev-198.
12. Erb, M.; Reymond, P. Molecular Interactions between Plants and Insect Herbivores. *Annu. Rev. Plant Biol.* **2019**, *70*, 527–557. [CrossRef]
13. Maffei, M.E.; Mithöfer, A.; Boland, W. Before gene expression: Early events in plant-insect interaction. *Trends Plant Sci.* **2007**, *12*, 310–316. [CrossRef] [PubMed]
14. Vincent, T.R.; Avramova, M.; Canham, J.; Higgins, P.; Bilkey, N.; Mugford, S.T.; Pitino, M.; Toyota, M.; Gilroy, S.; Miller, A.J.; et al. Interplay of Plasma Membrane and Vacuolar Ion Channels, Together with BAK1, Elicits Rapid Cytosolic Calcium Elevations in Arabidopsis during Aphid Feeding. *Plant Cell* **2017**, *29*, 1460–1479. [CrossRef] [PubMed]
15. Maffei, M.E.; Mithöfer, A.; Arimura, G.; Uchtenhagen, H.; Bossi, S.; Bertea, C.M.; Starvaggi Cucuzza, L.; Novero, M.; Volpe, V.; Quadro, S.; et al. Effects of feeding Spodoptera littoralis on lima bean leaves. III. Membrane depolarization and involvement of hydrogen peroxide. *Plant Physiol.* **2006**, *140*, 1022–1035. [CrossRef] [PubMed]
16. Vadassery, J.; Reichelt, M.; Hause, B.; Gershenzon, J.; Boland, W.; Mithöfer, A. CML42-mediated calcium signaling coordinates responses to Spodoptera herbivory and abiotic stresses in Arabidopsis. *Plant Physiol.* **2012**, *159*, 1159–1175. [PubMed]
17. Kandoth, P.K.; Ranf, S.; Pancholi, S.S.; Jayanty, S.; Walla, M.D.; Miller, W.; Howe, G.A.; Lincoln, D.E.; Stratmann, J.W. Tomato MAPKs LeMPK1, LeMPK2, and LeMPK3 function in the systemin-mediated defense response against herbivorous insects. *Proc. Natl. Acad. Sci. USA* **2007**, *104*, 12205–12210. [CrossRef]
18. Li, H.; Zhou, Z.; Hua, H.; Ma, W. Comparative transcriptome analysis of defense response of rice to *Nilaparvata lugens and Chilo suppressalis* infestation. *Int. J. Biol. Macromol.* **2020**, *163*, 2270–2285. [CrossRef]
19. Wang, D.; Li, C.-F.; Ma, C.-L.; Chen, L. Novel insights into the molecular mechanisms underlying the resistance of Camellia sinensis to *Ectropis oblique* provided by strategic transcriptomic comparisons. *Sci. Hortic.* **2015**, *192*, 429–440. [CrossRef]
20. Mao, Z.; Ge, Y.; Zhang, Y.; Zhong, J.; Munawar, A.; Zhu, Z.; Zhou, W. Disentangling the Potato Tuber Moth-Induced Early-Defense Response by Simulated Herbivory in Potato Plants. *Front. Plant Sci.* **2022**, *13*, 902342. [CrossRef]
21. Li, R.; Zhang, J.; Li, J.; Zhou, G.; Wang, Q.; Bian, W.; Erb, M.; Lou, Y. Prioritizing plant defence over growth through WRKY regulation facilitates infestation by non-target herbivores. *eLife* **2015**, *4*, e04805. [CrossRef] [PubMed]
22. Meraj, T.A.; Fu, J.; Raza, M.A.; Zhu, C.; Shen, Q.; Xu, D.; Wang, Q. Transcriptional Factors Regulate Plant Stress Responses through Mediating Secondary Metabolism. *Genes* **2020**, *11*, 346. [CrossRef] [PubMed]
23. Reymond, P.; Weber, H.; Damond, M.; Farmer, E.E. Differential gene expression in response to mechanical wounding and insect feeding in Arabidopsis. *Plant Cell* **2000**, *12*, 707–720. [CrossRef] [PubMed]
24. Korth, K.L.; Dixon, R.A. Evidence for Chewing Insect-Specific Molecular Events Distinct from a General Wound Response in Leaves. *Plant Physiol.* **1997**, *115*, 1299–1305. [CrossRef] [PubMed]
25. Li, Y.; Wang, Q.; Xu, X.; Guo, H. UHPLC-MS/MS Analysis of the Accumulation and Excretion of Steroidal Glycoalkaloids Consumed by Potato Tuber Moth (*Phthorimaea operculella*) Larvae under Different Feeding Treatments. *Insects* **2022**, *14*, 26. [CrossRef]
26. Zhang, M.; Yan, J.; Ali, A.; Gao, Y. Potato plant variety affects the performance and oviposition preference of *Phthorimaea operculella* Zeller (*Lepidoptera: Gelechiidae*). *Pest Manag. Sci.* **2022**, *78*, 3912–3919. [CrossRef] [PubMed]
27. Gallego, J.R.; Caicedo, O.; Gamez, M.; Hernandez, J.; Cabello, T. Selection of Predatory Mites for the Biological Control of Potato Tuber Moth in Stored Potatoes. *Insects* **2020**, *11*, 196. [CrossRef] [PubMed]
28. Richardson, M.E.; Rose, D.J.W. Chemical control of potato tuber moth, *phthorimaea operculella* (zell.), in rhodesia. *Bull. Entomol. Res.* **1967**, *57*, 271. [CrossRef]
29. del Pulgar, J.S.; Lucarini, M.; Aguzzi, A.; Gabrielli, P.; Parisi, B.; Pacifico, D.; Mandolino, G.; Lombardi-Boccia, G. Glycoalkaloid Content in Italian Potato Breeding Clones Improved for Resistance against Potato Tuber Moth (*Phthorimaea operculella* Zeller). *Potato Res.* **2021**, *64*, 229–240. [CrossRef]
30. Meiyalaghan, S.; Barrell, P.J.; Jacobs, J.M.; Conner, A.J. Regeneration of multiple shoots from transgenic potato events facilitates the recovery of phenotypically normal lines: Assessing a cry9Aa2 gene conferring insect resistance. *BMC Biotechnol.* **2011**, *11*, 93. [CrossRef]
31. Ito, Y.; Sasaki, K.; Ogihara, T.; Kitaoka, N.; Takahahi, K.; Matsuura, H. Verification of the versatility of the in vitro enzymatic reaction giving (+)-cis-12-Oxo-phytodienoic acid. *Bioorganic Med. Chem. Lett.* **2021**, *49*, 128284. [CrossRef]
32. Royo, J.; Vancanneyt, G.; Pérez, A.G.; Sanz, C.; Störmann, K.; Rosahl, S.; Sánchez-Serrano, J.J. Characterization of three potato lipoxygenases with distinct enzymatic activities and different organ-specific and wound-regulated expression patterns. *J. Biol. Chem.* **1996**, *271*, 21012–21019. [CrossRef]
33. Wasternack, C. Jasmonates: An update on biosynthesis, signal transduction and action in plant stress response, growth and development. *Ann. Bot.* **2007**, *100*, 681–697. [CrossRef] [PubMed]
34. Ding, P.; Ding, Y. Stories of Salicylic Acid: A Plant Defense Hormone. *Trends Plant Sci.* **2020**, *25*, 549–565. [CrossRef] [PubMed]

35. Zhou, J.M.; Zhang, Y. Plant Immunity: Danger Perception and Signaling. *Cell* **2020**, *181*, 978–989. [CrossRef] [PubMed]
36. Vishwakarma, K.; Upadhyay, N.; Kumar, N.; Yadav, G.; Singh, J.; Mishra, R.K.; Kumar, V.; Verma, R.; Upadhyay, R.G.; Pandey, M.; et al. Abscisic Acid Signaling and Abiotic Stress Tolerance in Plants: A Review on Current Knowledge and Future Prospects. *Front. Plant Sci.* **2017**, *8*, 161. [CrossRef]
37. Park, H.Y.; Seok, H.Y.; Park, B.K.; Kim, S.H.; Goh, C.H.; Lee, B.H.; Lee, C.H.; Moon, Y.H. Overexpression of Arabidopsis ZEP enhances tolerance to osmotic stress. *Biochem. Biophys. Res. Commun.* **2008**, *375*, 80–85. [CrossRef]
38. Al-Babili, S.; Hugueney, P.; Schledz, M.; Welsch, R.; Frohnmeyer, H.; Laule, O.; Beyer, P. Identification of a novel gene coding for neoxanthin synthase from Solanum tuberosum. *FEBS Lett.* **2000**, *485*, 168–172. [CrossRef]
39. Tan, B.C.; Schwartz, S.H.; Zeevaart, J.A.; McCarty, D.R. Genetic control of abscisic acid biosynthesis in maize. *Proc. Natl. Acad. Sci. USA* **1997**, *94*, 12235–12240. [CrossRef]
40. Seo, M.; Peeters, A.J.; Koiwai, H.; Oritani, T.; Marion-Poll, A.; Zeevaart, J.A.; Koornneef, M.; Kamiya, Y.; Koshiba, T. The Arabidopsis aldehyde oxidase 3 (AAO3) gene product catalyzes the final step in abscisic acid biosynthesis in leaves. *Proc. Natl. Acad. Sci. USA* **2000**, *97*, 12908–12913. [CrossRef]
41. Cheng, W.H.; Endo, A.; Zhou, L.; Penney, J.; Chen, H.C.; Arroyo, A.; Leon, P.; Nambara, E.; Asami, T.; Seo, M.; et al. A unique short-chain dehydrogenase/reductase in Arabidopsis glucose signaling and abscisic acid biosynthesis and functions. *Plant Cell* **2002**, *14*, 2723–2743. [CrossRef]
42. Rock, C.D.; Heath, T.G.; Gage, D.A.; Zeevaart, J.A. Abscisic alcohol is an intermediate in abscisic Acid biosynthesis in a shunt pathway from abscisic aldehyde. *Plant Physiol.* **1991**, *97*, 670–676. [CrossRef]
43. Rodrigues, M.A.; Bianchetti, R.E.; Freschi, L. Shedding light on ethylene metabolism in higher plants. *Front. Plant Sci.* **2014**, *5*, 665. [CrossRef] [PubMed]
44. Adie, B.; Chico, J.M.; Rubio-Somoza, I.; Solano, R. Modulation of Plant Defenses by Ethylene. *J. Plant Growth Regul.* **2007**, *26*, 160–177. [CrossRef]
45. Czarny, J.C.; Grichko, V.P.; Glick, B.R. Genetic modulation of ethylene biosynthesis and signaling in plants. *Biotechnol. Adv.* **2006**, *24*, 410–419. [CrossRef] [PubMed]
46. Singh, R.; Singh, S.; Parihar, P.; Mishra, R.K.; Tripathi, D.K.; Singh, V.P.; Chauhan, D.K.; Prasad, S.M. Reactive Oxygen Species (ROS): Beneficial Companions of Plants' Developmental Processes. *Front. Plant Sci.* **2016**, *7*, 1299. [CrossRef]
47. Hrmova, M.; Hussain, S.S. Plant Transcription Factors Involved in Drought and Associated Stresses. *Int. J. Mol. Sci.* **2021**, *22*, 5662. [CrossRef] [PubMed]
48. Wink, M. Plant Secondary Metabolites Modulate Insect Behavior-Steps Toward Addiction? *Front. Physiol.* **2018**, *9*, 364. [CrossRef]
49. Donze-Reiner, T.; Palmer, N.A.; Scully, E.D.; Prochaska, T.J.; Koch, K.G.; Heng-Moss, T.; Bradshaw, J.D.; Twigg, P.; Amundsen, K.; Sattler, S.E.; et al. Transcriptional analysis of defense mechanisms in upland tetraploid switchgrass to greenbugs. *BMC Plant Biol.* **2017**, *17*, 46. [CrossRef] [PubMed]
50. Kang, J.N.; Lee, W.H.; Won, S.Y.; Chang, S.; Hong, J.P.; Oh, T.J.; Lee, S.M.; Kang, S.H. Systemic Expression of Genes Involved in the Plant Defense Response Induced by Wounding in *Senna tora*. *Int. J. Mol. Sci.* **2021**, *22*, 10073. [CrossRef]
51. Lawrence, P.K.; Koundal, K.R. Plant protease inhibitors in control of phytophagous insects. *Electron. J. Biotechnol.* **2002**, *5*, 5–6. [CrossRef]
52. Vaghela, B.; Vashi, R.; Rajput, K.; Joshi, R. Plant chitinases and their role in plant defense: A comprehensive review. *Enzym. Microb. Technol.* **2022**, *159*, 110055. [CrossRef]
53. Zhang, S.; Klessig, D.F. MAPK cascades in plant defense signaling. *Trends Plant Sci.* **2001**, *6*, 520–527. [CrossRef] [PubMed]
54. De Vos, M.; Van Oosten, V.R.; Van Poecke, R.M.; Van Pelt, J.A.; Pozo, M.J.; Mueller, M.J.; Buchala, A.J.; Métraux, J.P.; Van Loon, L.C.; Dicke, M.; et al. Signal signature and transcriptome changes of Arabidopsis during pathogen and insect attack. *Mol. Plant-Microbe Interact. MPMI* **2005**, *18*, 923–937. [CrossRef] [PubMed]
55. De Vos, M.; Van Oosten, V.R.; Jander, G.; Dicke, M.; Pieterse, C.M. Plants under attack: Multiple interactions with insects and microbes. *Plant Signal. Behav.* **2007**, *2*, 527–529. [CrossRef] [PubMed]
56. Bellande, K.; Bono, J.J.; Savelli, B.; Jamet, E.; Canut, H. Plant Lectins and Lectin Receptor-Like Kinases: How Do They Sense the Outside? *Int. J. Mol. Sci.* **2017**, *18*, 1164. [CrossRef] [PubMed]
57. Vaid, N.; Macovei, A.; Tuteja, N. Knights in action: Lectin receptor-like kinases in plant development and stress responses. *Mol. Plant* **2013**, *6*, 1405–1418. [CrossRef]
58. Mittler, R.; Zandalinas, S.I.; Fichman, Y.; Van Breusegem, F. Reactive oxygen species signalling in plant stress responses. *Nat. Rev. Mol. Cell Biol.* **2022**, *23*, 663–679. [CrossRef] [PubMed]
59. Tyagi, S.; Shah, A.; Karthik, K.; Rathinam, M.; Rai, V.; Chaudhary, N.; Sreevathsa, R. Reactive oxygen species in plants: An invincible fulcrum for biotic stress mitigation. *Appl. Microbiol. Biotechnol.* **2022**, *106*, 5945–5955. [CrossRef]
60. Liu, Y.; He, C. A review of redox signaling and the control of MAP kinase pathway in plants. *Redox Biol.* **2017**, *11*, 192–204. [CrossRef] [PubMed]
61. Kovtun, Y.; Chiu, W.L.; Tena, G.; Sheen, J. Functional analysis of oxidative stress-activated mitogen-activated protein kinase cascade in plants. *Proc. Natl. Acad. Sci. USA* **2000**, *97*, 2940–2945. [CrossRef] [PubMed]

62. Pitzschke, A.; Djamei, A.; Bitton, F.; Hirt, H. A major role of the MEKK1-MKK1/2-MPK4 pathway in ROS signalling. *Mol. Plant* **2009**, *2*, 120–137. [CrossRef] [PubMed]
63. Mansoor, S.; Ali Wani, O.; Lone, J.K.; Manhas, S.; Kour, N.; Alam, P.; Ahmad, A.; Ahmad, P. Reactive Oxygen Species in Plants: From Source to Sink. *Antioxidants* **2022**, *11*, 225. [CrossRef] [PubMed]
64. Sewelam, N.; Kazan, K.; Schenk, P.M. Global Plant Stress Signaling: Reactive Oxygen Species at the Cross-Road. *Front. Plant Sci.* **2016**, *7*, 187. [CrossRef] [PubMed]
65. Raja, V.; Majeed, U.; Kang, H.; Andrabi, K.I.; John, R. Abiotic stress: Interplay between ROS, hormones and MAPKs. *Environ. Exp. Bot.* **2017**, *137*, 142–157. [CrossRef]
66. Lu, J.; Robert, C.A.; Riemann, M.; Cosme, M.; Mène-Saffrané, L.; Massana, J.; Stout, M.J.; Lou, Y.; Gershenzon, J.; Erb, M. Induced jasmonate signaling leads to contrasting effects on root damage and herbivore performance. *Plant Physiol.* **2015**, *167*, 1100–1116. [CrossRef] [PubMed]
67. McConn, M.; Creelman, R.A.; Bell, E.; Mullet, J.E.; Browse, J. Jasmonate is essential for insect defense in Arabidopsis. *Proc. Natl. Acad. Sci. USA* **1997**, *94*, 5473–5477. [CrossRef] [PubMed]
68. Glauser, G.; Dubugnon, L.; Mousavi, S.A.; Rudaz, S.; Wolfender, J.L.; Farmer, E.E. Velocity estimates for signal propagation leading to systemic jasmonic acid accumulation in wounded Arabidopsis. *J. Biol. Chem.* **2009**, *284*, 34506–34513. [CrossRef]
69. Göbel, C.; Feussner, I.; Schmidt, A.; Scheel, D.; Sanchez-Serrano, J.; Hamberg, M.; Rosahl, S. Oxylipin profiling reveals the preferential stimulation of the 9-lipoxygenase pathway in elicitor-treated potato cells. *J. Biol. Chem.* **2001**, *276*, 6267–6273. [CrossRef]
70. Woldemariam, M.G.; Ahern, K.; Jander, G.; Tzin, V. A role for 9-lipoxygenases in maize defense against insect herbivory. *Plant Signal. Behav.* **2018**, *13*, e1422462. [CrossRef] [PubMed]
71. Ye, W.; Bustos-Segura, C.; Degen, T.; Erb, M.; Turlings TC, J. Belowground and aboveground herbivory differentially affect the transcriptome in roots and shoots of maize. *Plant Direct* **2022**, *6*, e426. [CrossRef] [PubMed]
72. Koo Abraham, J. Metabolism of the plant hormone jasmonate: A sentinel for tissue damage and master regulator of stress response. *Phytochem. Rev.* **2017**, *17*, 51–80.
73. Diezel, C.; von Dahl, C.C.; Gaquerel, E.; Baldwin, I.T. Different lepidopteran elicitors account for cross-talk in herbivory-induced phytohormone signaling. *Plant Physiol.* **2009**, *150*, 1576–1586. [CrossRef] [PubMed]
74. Xu, Y.; Chang, P.; Liu, D.; Narasimhan, M.L.; Raghothama, K.G.; Hasegawa, P.M.; Bressan, R.A. Plant Defense Genes Are Synergistically Induced by Ethylene and Methyl Jasmonate. *Plant Cell* **1994**, *6*, 1077–1085. [CrossRef] [PubMed]
75. Penninckx, I.A.; Thomma, B.P.; Buchala, A.; Métraux, J.P.; Broekaert, W.F. Concomitant activation of jasmonate and ethylene response pathways is required for induction of a plant defensin gene in Arabidopsis. *Plant Cell* **1998**, *10*, 2103–2113. [CrossRef] [PubMed]
76. Onkokesung, N.; Baldwin, I.T.; Gális, I. The role of jasmonic acid and ethylene crosstalk in direct defense of *Nicotiana attenuata* plants against chewing herbivores. *Plant Signal. Behav.* **2010**, *5*, 1305–1307. [CrossRef] [PubMed]
77. Lorenzo, O.; Piqueras, R.; Sánchez-Serrano, J.J.; Solano, R. ETHYLENE RESPONSE FACTOR1 integrates signals from ethylene and jasmonate pathways in plant defense. *Plant Cell* **2003**, *15*, 165–178. [CrossRef]
78. Zhu, Z.; Lee, B. Friends or foes: New insights in jasmonate and ethylene co-actions. *Plant Cell Physiol.* **2015**, *56*, 414–420. [CrossRef]
79. Zhang, X.N.; Liu, J.; Liu, Y.; Wang, Y.; Abozeid, A.; Yu, Z.G.; Tang, Z.H. Metabolomics Analysis Reveals that Ethylene and Methyl Jasmonate Regulate Different Branch Pathways to Promote the Accumulation of Terpenoid Indole Alkaloids in *Catharanthus roseus*. *J. Nat. Prod.* **2018**, *81*, 335–342. [CrossRef]
80. O'Donnell, P.J.; Calvert, C.; Atzorn, R.; Wasternack, C.; Leyser HM, O.; Bowles, D.J. Ethylene as a Signal Mediating the Wound Response of Tomato Plants. *Science* **1996**, *274*, 1914–1917. [CrossRef]
81. Ng, D.W.; Abeysinghe, J.K.; Kamali, M. Regulating the Regulators: The Control of Transcription Factors in Plant Defense Signaling. *Int. J. Mol. Sci.* **2018**, *19*, 3737. [CrossRef]
82. Chen, M.; Ji, M.; Wen, B.; Liu, L.; Li, S.; Chen, X.; Gao, D.; Li, L. GOLDEN 2-LIKE Transcription Factors of Plants. *Front. Plant Sci.* **2016**, *7*, 1509. [CrossRef]
83. Bonaventure, G. The *Nicotiana attenuata* LECTIN RECEPTOR KINASE 1 is involved in the perception of insect feeding. *Plant Signal. Behav.* **2011**, *6*, 2060–2063. [CrossRef]
84. De Vos, M.; Van Zaanen, W.; Koornneef, A.; Korzelius, J.P.; Dicke, M.; Van Loon, L.C.; Pieterse, C.M. Herbivore-induced resistance against microbial pathogens in Arabidopsis. *Plant Physiol.* **2006**, *142*, 352–363. [CrossRef]
85. Chen, S.; Zhou, Y.; Chen, Y.; Gu, J. fastp: An ultra-fast all-in-one FASTQ preprocessor. *Bioinformatics* **2018**, *34*, i884–i890. [CrossRef] [PubMed]
86. Kim, D.; Langmead, B.; Salzberg, S.L. HISAT: A fast spliced aligner with low memory requirements. *Nat. Methods* **2015**, *12*, 357–360. [CrossRef] [PubMed]
87. Pertea, M.; Pertea, G.M.; Antonescu, C.M.; Chang, T.C.; Mendell, J.T.; Salzberg, S.L. StringTie enables improved reconstruction of a transcriptome from RNA-seq reads. *Nat. Biotechnol.* **2015**, *33*, 290–295. [CrossRef]
88. Pertea, M.; Kim, D.; Pertea, G.M.; Leek, J.T.; Salzberg, S.L. Transcript-level expression analysis of RNA-seq experiments with HISAT, StringTie and Ballgown. *Nat. Protoc.* **2016**, *11*, 1650–1667. [CrossRef] [PubMed]

89. Love, M.I.; Huber, W.; Anders, S. Moderated estimation of fold change and dispersion for RNA-seq data with DESeq2. *Genome Biol.* **2014**, *15*, 550. [CrossRef]
90. Robinson, M.D.; McCarthy, D.J.; Smyth, G.K. edgeR: A Bioconductor package for differential expression analysis of digital gene expression data. *Bioinformatics* **2010**, *26*, 139–140. [CrossRef]

plants
MDPI

Article

Molecular Characterization Analysis and Adaptive Responses of *Spodoptera frugiperda* (Lepidoptera: Noctuidae) to Nutritional and Enzymatic Variabilities in Various Maize Cultivars

Qiangyan Zhang [1], Yanlei Zhang [1,2], Kexin Zhang [1], Huiping Liu [1], Yuping Gou [1], Chunchun Li [1,*], Inzamam Ul Haq [3], Peter Quandahor [4] and Changzhong Liu [1,*]

1 Biocontrol Engineering Laboratory of Crop Diseases and Pests of Gansu Province, College of Plant Protection, Gansu Agricultural University, Lanzhou 730070, China; zhangqiangyan2@163.com (Q.Z.); 17752269426@163.com (Y.Z.); zhangkx199404@163.com (K.Z.); liuhp110823@163.com (H.L.); gouyp@gsau.edu.cn (Y.G.)
2 Dingxi Plant Protection and Quarantine Station, Dingxi 743099, China
3 State Key Laboratory of Ecological Pest Control for Fujian and Taiwan Crops, Fujian Agriculture and Forestry University, Fuzhou 350002, China; 000b370305@fafu.edu.cn
4 CSIR—Savanna Agricultural Research Institute, Tamale P.O. Box 52, Ghana; quandooh@yahoo.com
* Correspondence: lcc@gsau.edu.cn (C.L.); liuchzh@gsau.edu.cn (C.L.)

Abstract: The fall armyworm, *Spodoptera frugiperda* Smith (Lepidoptera: Noctuidae), a common agricultural pest known for its extensive migration and wide host ranges, causes considerable harm to maize (*Zea mays* L.). In this study, we utilized two molecular marker genes, *COI* and *Tpi*, to compare the genetic characteristics of the collected original samples. Additionally, through an interactive study between *S. frugiperda* larvae and six maize varieties aiming to understand the insect's adaptability and resistance mechanisms, our analysis revealed that both the *COI* and *Tpi* genes identified *S. frugiperda* as the corn strain. Further examination of the larvae showed significant differences in nutritional indices, digestive, and detoxification enzyme activities. Special maize varieties were found to offer higher efficiency in nutrient conversion and assimilation compared with common varieties. This study revealed adaptations in *S. frugiperda*'s digestive and detoxification processes in response to the different maize varieties. For instance, larvae reared on common maize exhibited elevated amylase and lipase activities. Interestingly, detoxification enzyme activities exhibited different patterns of variation in different maize varieties. The Pearson correlation analysis between nutritional indices, enzyme activities, and the nutritional content and secondary metabolites of maize leaves provided deeper insights into the pest's adaptability. The results highlighted significant relationships between specific nutritional components in maize and the physiological responses of *S. frugiperda*. Overall, our findings contribute substantially to the understanding of *S. frugiperda*'s host plant adaptability, offering critical insights for the development of sustainable pest management strategies.

Keywords: *Spodoptera frugiperda*; genotype; special maize; common maize; nutritional indexes; enzyme activities; host adaptation

1. Introduction

The fall armyworm, *Spodoptera frugiperda* Smith (Lepidoptera: Noctuidae), has emerged as a formidable global agricultural menace, originating from the tropical and subtropical zones of the Americas [1,2]. Its remarkable adaptability to various ecological conditions and its ability to cover extensive distances through wind currents have escalated its status to a critical threat to agriculture worldwide. Maize (*Zea mays* L.) is the principal crop worldwide and ranks first among grain crops in terms of production, and pests are the main factors affecting its quality and yield [3]. Special maize generally includes high-lysine maize, glutinous maize, sweet maize, popping maize, high-oil maize, and others. Maize varieties outside of special maize are referred to as common maize [4]. As a polyphagous pest, *S. frugiperda* exhibits a notable preference for maize, with its larval stages causing considerable damage due to its voracious feeding habits across a multitude of host plants [5,6]. The impact on maize is profound, with potential global yield losses estimated between 15% to 73% [7]. In China, the pest was first reported in the southeastern region of Yunnan Province in January 2019. It has since expanded rapidly across the majority of agricultural provinces, inflicting substantial damage to crops, particularly to maize [8]. By the end of the same year, *S. frugiperda* had invaded 26 provinces, afflicting over 1.08 million hectares of crops, thereby presenting a considerable threat to the nation's maize production [9].

Two biotypes of *S. frugiperda*, namely the corn and rice strains, have been identified, exhibiting significant differences in host selection, feeding tendencies, and physiological and behavioral characteristics [10]. The rice type primarily feeds on crops such as rice (*Oryza sativa* L.), alfalfa (*Medicago sativa* L.), and forage grass, while the corn type primarily feeds on maize, sorghum (*Sorghum bicolor* (L.) Moench), and cotton (*Gossypium hirsutum* L.) [11]. Traditional morphological identification poses challenges in distinguishing *S. frugiperda* biotypes, while molecular identification offers a precise and efficient means of achieving accurate classification [12]. Rapidly accurate identification and classification of invasive pests are helpful to advance scientific layout and carry out targeted control measures. Currently, the primary method for identifying *S. frugiperda* biotypes involves the use of molecular marker techniques [13]. The mitochondrial Cytochrome oxidase subunit I gene (*COI*) and Z-chromosome-linked Triose phosphate isomerase gene (*Tpi*) are presently the most commonly used strain markers [14]. Zhang et al. [15] identified 83 samples collected from Yunnan Province, confirming that the invading *S. frugiperda* in China is consistently of the corn type. Upon organically combining these two identification methods and mutually validating them, the identification results can be made more reliable.

The dynamic interaction between plants and herbivorous insects has garnered increasing attention in ecological research because of its complexity and significance [16]. Plants are not passive in the face of herbivory; they can promptly trigger defense responses upon attack, altering the feeding behaviors and oviposition strategies of their assailants. This defensive cascade is balanced by the evolutionary adaptations of phytophagous insects, which employ a repertoire of chemical effectors to neutralize plant chemical defenses through mechanisms such as selective storage, detoxification, and desensitization [17]. The relationship between herbivorous insects and their host plants is shaped by myriad factors, including host species, nutritional conditions, and the insect's inherent detoxification ability [18,19]. Host plants serve as a source of both nutrition, which varies in quality and quantity [20], and secondary metabolites, which function as a chemical shield against herbivory [21]. Nutritional indices, therefore, become pivotal in evaluating the fitness of herbivores, particularly in species such as *S. frugiperda* [22]. Previous studies have reported that larval consumption and nutrient utilization of host plants can significantly affect the growth, development, and reproduction of *S. frugiperda*, ultimately influencing its host plant adaptability [23,24]. Previous research by Scriber and Slansky Jr [25] highlighted the profound implications of food quality on the physiological processes of insects post-ingestion. Nonetheless, *S. frugiperda* has developed a variety of strategies to cope with the nutritive and defensive properties of host plants, notably through modulating the activities of digestive and detoxification enzymes. These enzymatic adjustments allow the insect to fine-tune its digestive efficiency, broadening its range of potential host plants [26]. Critical enzymes, such as carboxylesterase (CarE), glutathione-S-transferase (GST), and cytochrome P450 (CYP450), are instrumental in the detoxification of plant secondary metabolites, underscoring their importance in the metabolic detoxification processes of insects [27,28].

The diversity in nutritional content across various crop species is a critical determinant in the growth, development, and overall survival of insect populations [29]. This is particularly evident in insect larvae, where the quality of the host plant has been shown to significantly differ among plant species, thereby impacting larval health and development [30]. Extensive research into the dietary preferences during the larval stage of *S. frugiperda* has highlighted the significant role of these diets in shaping the pest's development and reproduction, with maize emerging as a notably suitable host [31–34]. The influence of different maize varieties and their respective plant tissues on the development and nutritional indices of *S. frugiperda* has been substantiated by numerous studies [35–37]. In our previous research, we examined the significant effects of the same maize varieties as in this study on various life history parameters of *S. frugiperda*, such as oviposition preference, developmental duration, pupal weight, survival rate, and fecundity [38].

Here, we have identified the collected *S. frugiperda* as the corn strain, and significant variations were observed in their nutritional indices, as well as the digestive and detoxification enzyme activities of *S. frugiperda* on six maize varieties. Furthermore, our research has emphasized the correlation of these parameters with specific nutritional components in maize. These findings elucidate the proficiency of pests in nutrient absorption and their ability to regulate digestion and detoxification processes. The research results are expected to provide valuable insights for effective agricultural practices and pest management, guiding toward more sustainable agricultural approaches.

2. Results

2.1. Molecular Characterization Analysis of Host Strain of S. frugiperda

Five *COI* (accession no. PP301981-PP301985, GS-1~GS-5) and *Tpi* (accession no. PP331838-PP331842, GS-1~GS-5) gene sequences were obtained from collected *S. frugiperda* in Gansu, and there was no intraspecific sequence variation. Therefore, the first sequence (GS-1) will be chosen for analysis (Figure 1). The *COI* (Figure 1A) and *Tpi* (Figure 1B) gene sequences from GS-1 were closest to the corn strain and 0 bp variation. However, they differed from the rice strain by 17 bp and 10 bp, respectively. Thus, our analysis revealed that both the *COI* and *Tpi* genes identified the collected *S. frugiperda* as the corn strain.

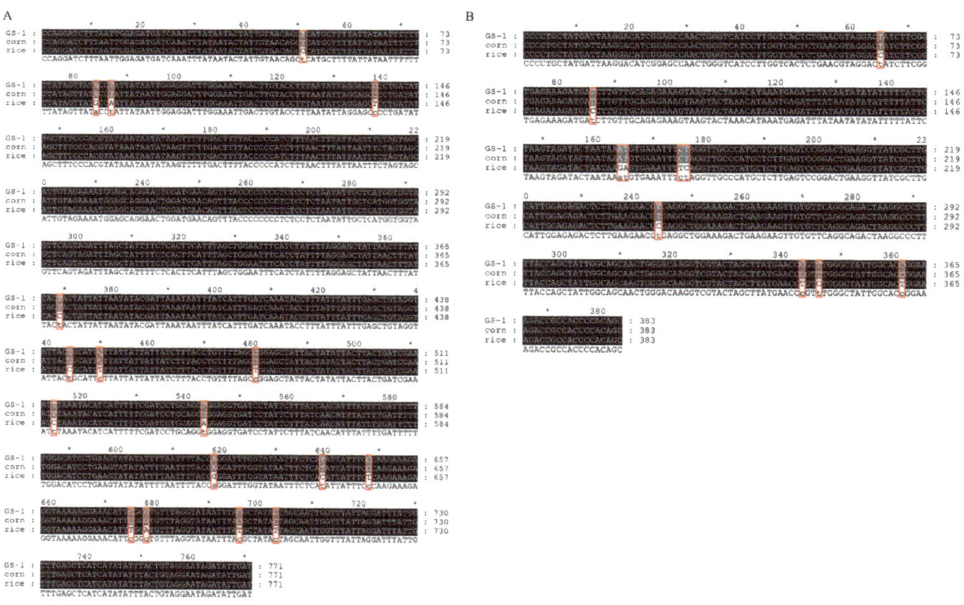

Figure 1. Multiple alignments of *COI* and *Tpi* gene fragments in *S. frugiperda* individuals. (**A**) *COI* gene fragments; (**B**) *Tpi* gene fragments. GS-1 represents the first sequence of collected *S. frugiperda* in Gansu. Corn and rice represent the corn and rice strain. In the last line, capital letters represent identical sequences, while lowercase letters represent different sequences. The asterisk (*) indicates a difference of 10 base pairs compared to the preceding sequence, and the red box represents variant bases.

2.2. Effects of Various Maize Varieties on Nutritional Indices of 3rd Instar Larvae of S. frugiperda

Relative Growth Rate (RGR): Our study revealed a significant influence of maize variety on the RGR of *S. frugiperda* larvae (Figure 2A). Larvae feeding on the special maize varieties demonstrated notably higher RGRs compared with those on the common maize varieties ($p < 0.05$), indicating faster growth on the former. Among them, Baitiannuo, Ziyunuo, and Zaocuiwang exhibited the most pronounced RGRs, markedly exceeding those on Wuke 113, Wuke No. 4, and Zhengdan 958 ($p < 0.05$). However, the RGRs on Baitiannuo, Ziyunuo, and Zaocuiwang were statistically indistinguishable from each other ($p > 0.05$). Wuke 113 and Wuke No. 4 had a significantly higher difference in RGR than Zhengdan 958 ($p < 0.05$), but there was no significant difference between Wuke 113 and Wuke No. 4 ($p > 0.05$). The lowest RGR was recorded on Zhengdan 958 (0.395 g/g/d) (Figure 2A).

Relative Consumption Rate (RCR): *S. frugiperda* displayed the highest RCR on Wuke 113 (7.392 g/g/d), significantly higher than on the other varieties ($p < 0.05$) (Figure 2B). The lowest RCR was observed on Ziyunuo (3.740 g/g/d). However, no statistically significant differences ($p > 0.05$) were found among Wuke No. 4, Baitiannuo, Zaocuiwang, Zhengdan 958, and Ziyunuo (Figure 2B).

Efficiency of Conversion of Digested Food (ECD): ECD values varied significantly among the maize varieties (Figure 2C). Ziyunuo exhibited the highest ECD (28.620%), displaying a significantly greater increase compared with the others ($p < 0.05$). Zaocuiwang ranked second (21.932%), with significantly higher ECD compared with Zhengdan 958, Wuke No. 4, and Wuke 113 ($p < 0.05$). Baitiannuo also showed a significantly higher ECD than Wuke 113. No significant differences ($p > 0.05$) were observed among Baitiannuo, Wuke No. 4, and Zhengdan 958. ECD values for common maize varieties were significantly lower than those for special maize varieties, particularly compared with Zaocuiwang and Ziyunuo ($p < 0.05$). The lowest ECD was on Wuke 113 (8.906%) (Figure 2C).

Efficiency of Conversion of Ingested Food (ECI): ECI was significantly higher in larvae fed on the special maize varieties compared with common ones ($p < 0.05$) (Figure 2D). Ziyunuo showed the highest ECI (19.395%), significantly surpassing Baitiannuo, Zhengdan 958, Wuke No. 4, and Wuke 113 ($p < 0.05$). Zaocuiwang and Baitiannuo also exhibited significantly higher ECIs compared with Zhengdan 958, Wuke No. 4, and Wuke 113 ($p < 0.05$), with no significant difference between Zaocuiwang and Baitiannuo. The ECIs for common maize varieties were considerably lower, with no significant differences ($p > 0.05$) among Zhengdan 958, Wuke No. 4, and Wuke 113. The lowest ECI was on Wuke 113 (6.670%) (Figure 2D).

Approximate Digestibility (AD): Baitiannuo exhibited the highest AD at 78.559%, whereas Ziyunuo showed the lowest AD at 67.652% (Figure 2E). The AD on Ziyunuo was significantly lower than those on other varieties ($p < 0.05$), but no significant differences ($p > 0.05$) were observed among the cultivars Baitiannuo, Zaocuiwang, Zhengdan 958, Wuke 113, and Wuke No. 4 (Figure 2E).

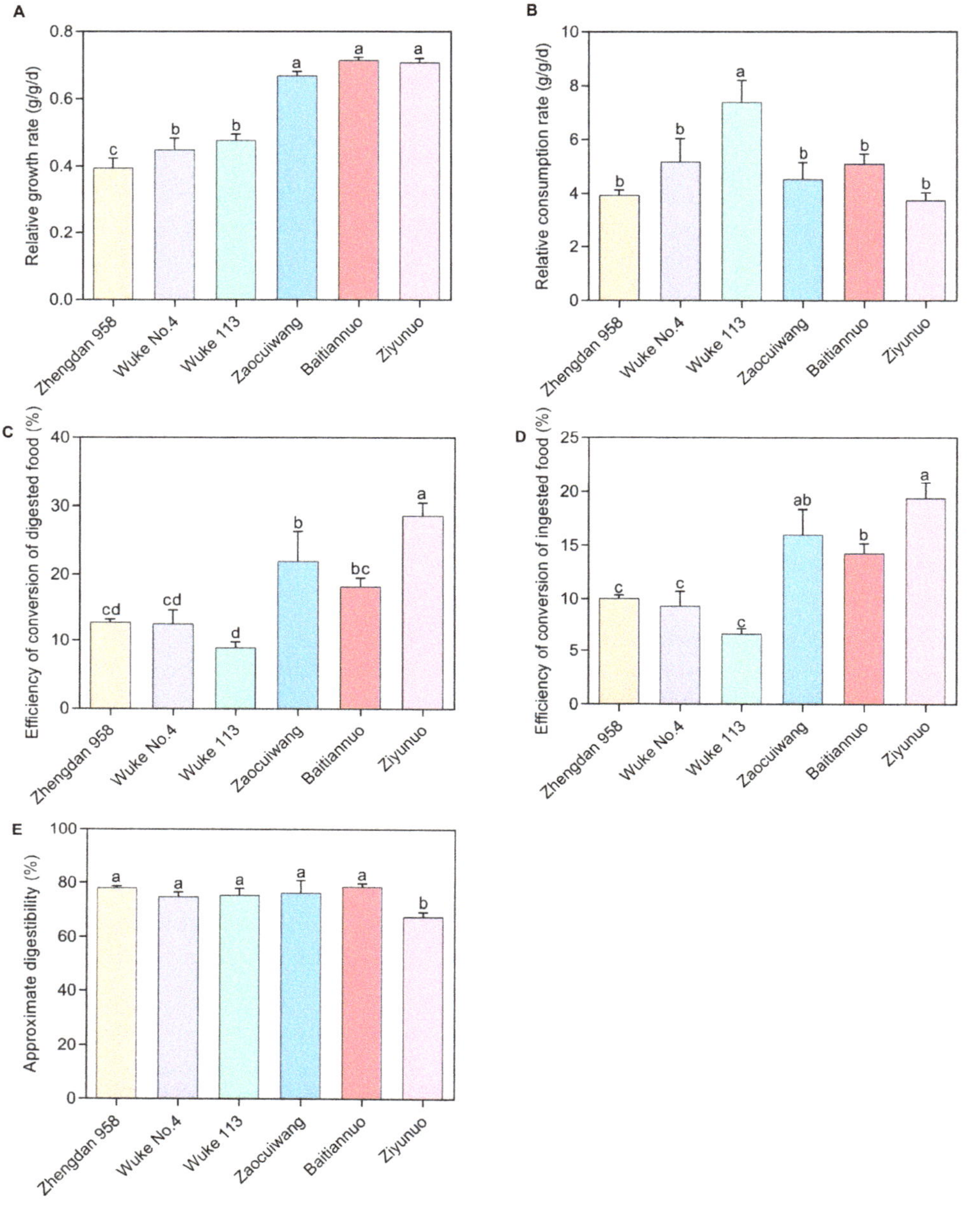

Figure 2. Effects of six maize varieties on nutritional indices of 3rd-instar larvae of *S. frugiperda*. (**A**) Relative growth rate; (**B**) Relative consumption rate; (**C**) Efficiency of conversion of digested food; (**D**) Efficiency of conversion of ingested food; (**E**) Approximate digestibility. Data presented are the mean ± standard error (SE) of five replicates, $n = 25$. Distinct lowercase letters within a histogram indicate significant differences among maize varieties (Duncan's test, $p < 0.05$). The X-axis represents various maize varieties. The common maize varieties include Zhengdan 958, Wuke No. 4, and Wuke 113, while the special maize varieties consist of Zaocuiwang, Baitiannuo, and Ziyunuo.

2.3. Effects of Six Maize Varieties on the Digestive Enzyme Activity of *S. frugiperda*

The digestive enzyme was significantly influenced by maize varieties (Figure 3). The α-amylase activity varied, with a decrease noted from 0.464 U/g in Zaocuiwang to 0.306 U/g in Wuke 113 (Figure 3A). However, this study found no significant differences in α-amylase activity among the larvae reared on the six maize varieties. The pattern of total-amylase activity closely matched that of β-amylase activity across the maize cultivars (Figure 3B,C). The common maize varieties presented a higher β-amylase and total-amylase activity compared with the special maize varieties. Notably, Wuke 113 displayed the highest β-amylase activity (29.370 U/g) and total-amylase (29.677 U/g), which were significantly greater than that in the other varieties ($p < 0.05$). There was a significant increase in β-amylase activity and total-amylase on Wuke No. 4 when compared with Zhengdan 958, Zaocuiwang, Baitiannuo, and Ziyunuo ($p < 0.05$). In contrast, Zaocuiwang exhibited the lowest activity, significantly lower than the others ($p < 0.05$) (Figure 3B,C). The larvae reared on Zhengdan 958 showed the most elevated lipase activity at 19.376 U/g, significantly surpassing those on the other maize varieties ($p < 0.05$) (Figure 3D). Larvae on Wuke 113, Zaocuiwang, Baitiannuo, and Ziyunuo exhibited slightly lower lipase activities, with no significant differences among them. Additionally, the lowest lipase activity was recorded in Wuke No. 4 at 11.695 U/g, significantly less than Zhengdan 958, Wuke 113, and Baitiannuo ($p < 0.05$) (Figure 3D).

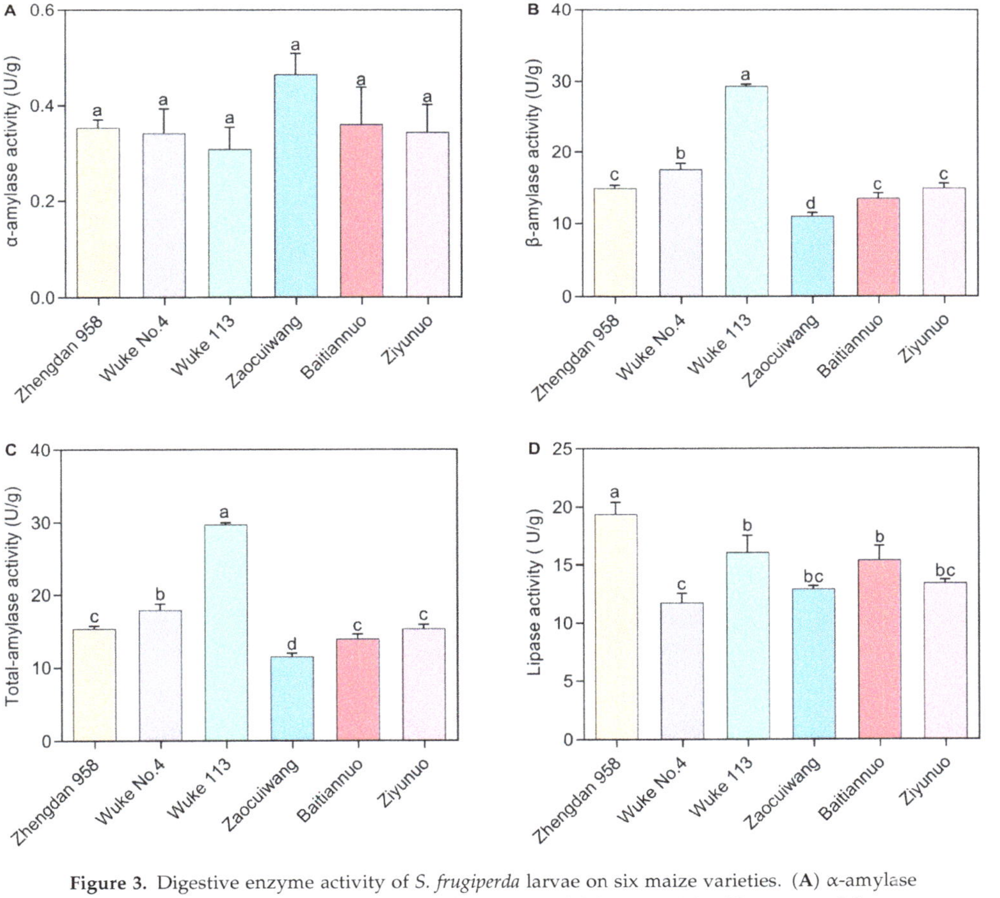

Figure 3. Digestive enzyme activity of *S. frugiperda* larvae on six maize varieties. (**A**) α-amylase activity; (**B**) β-amylase activity; (**C**) Total-amylase activity; (**D**) Lipase activity. The presented data are

expressed as the mean ± SE of three replicates, $n = 10$. Significant distinct values at the 0.05 level are marked by different lowercase letters. The X-axis represents various maize varieties. The common maize varieties include Zhengdan 958, Wuke No. 4, and Wuke 113, while the special maize varieties consist of Zaocuiwang, Baitiannuo, and Ziyunuo.

2.4. Effects of Six Maize Varieties on Detoxification Enzyme Activity of S. frugiperda

Glutathione-S-transferase (GST) Activity: Different maize varieties exerted a significant effect on the GST activity of *S. frugiperda* ($p < 0.05$) (Figure 4A). GST activity notably decreased from 0.205 U/g (Ziyunuo) to 0.112 U/g (Zhengdan 958). The GST activities in Ziyunuo, Wuke 113, and Baitiannuo were significantly higher than those in Zhengdan 958 and Zaocuiwang ($p < 0.05$). Significant variations were observed between Ziyunuo, Wuke 113, and Wuke No. 4 ($p < 0.05$). However, the differences in GST activity among Ziyunuo, Wuke 113, and Baitiannuo, as well as between Baitiannuo and Wuke No. 4, were not significant ($p > 0.05$). Similarly, there were no significant differences between Zhengdan 958, Zaocuiwang, and Wuke No. 4 ($p > 0.05$).

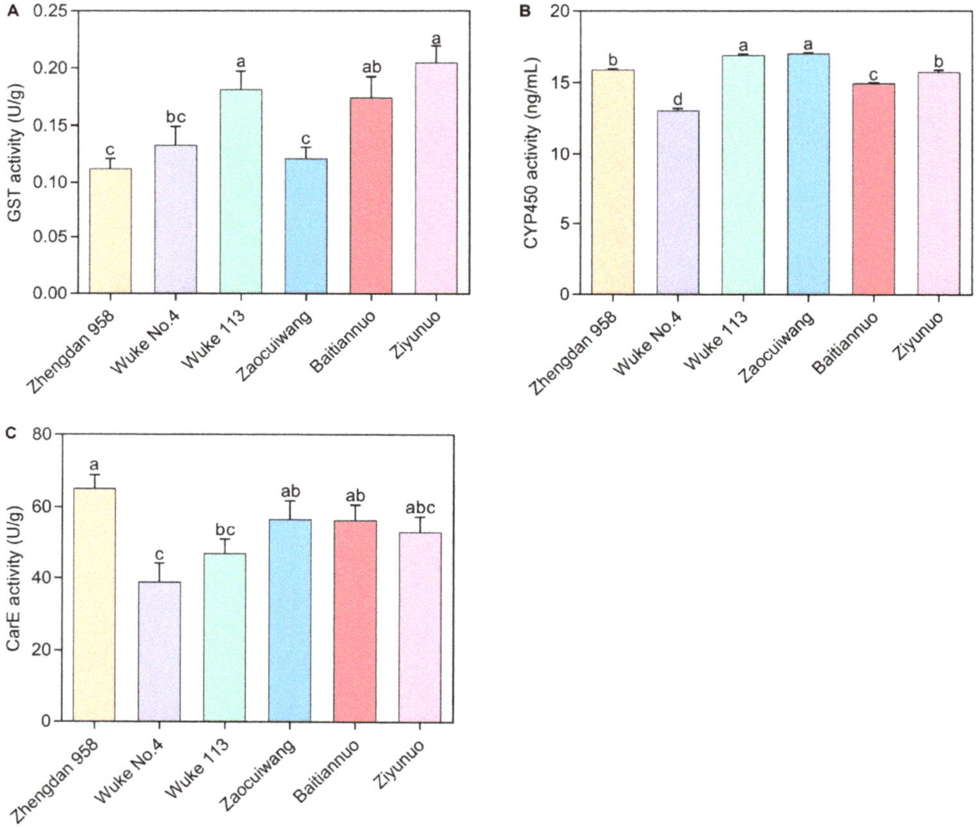

Figure 4. Effects of six maize varieties on detoxification enzyme activity of *S. frugiperda* larvae. (**A**) GST activity; (**B**) CYP450 activity; (**C**) CarE activity. The data represent the mean ± SE of three replicates, $n = 5$. Lowercase letters indicate a significant difference among the six maize varieties (Duncan's test, $p < 0.05$). The X-axis represents various maize varieties. The common maize varieties include Zhengdan 958, Wuke No. 4, and Wuke 113, while the special maize varieties consist of Zaocuiwang, Baitiannuo, and Ziyunuo.

Cytochrome P450 (CYP450) Activity: Figure 4B reveals the CYP450 activity was significantly affected by the maize cultivars. Wuke 113 (16.895 ng/mL) and Zaocuiwang (17.034 ng/mL) showed the highest activities, significantly surpassing those in Zhengdan 958, Wuke No. 4, Baitiannuo, and Ziyunuo ($p < 0.05$). Both Zhengdan 958 (15.885 ng/mL) and Ziyunuo (15.725 ng/mL) also showed significantly higher CYP450 activity than Wuke No. 4 and Baitiannuo ($p < 0.05$), with Wuke No. 4 showing the lowest activity. The common maize varieties exhibited significant differences in CYP450 activity, with the highest on Wuke 113, followed by Zhengdan 958, and the lowest on Wuke No. 4. For the special maize varieties, Zaocuiwang exhibited the highest activity, followed by Ziyunuo, with the lowest found in Baitiannuo ($p < 0.05$) (Figure 4B).

Carboxylesterase (CarE) Activity: CarE activity also varied significantly among the maize varieties (Figure 4C). Zhengdan 958 had the highest CarE activity (65.127 U/g), significantly exceeding Wuke No. 4 and Wuke 113 ($p < 0.05$). Zaocuiwang (56.606 U/g) and Baitiannuo (56.425 U/g) exhibited significantly greater CarE activities compared with Wuke No. 4 ($p < 0.05$). There was no significant difference in CarE activity among Zhengdan 958, Zaocuiwang, Baitiannuo, and Ziyunuo or among Wuke No. 4, Wuke 113, and Ziyunuo ($p > 0.05$). Wuke No. 4 had the lowest CarE activity of 38.938 U/g, significantly lower than Zhengdan 958, Zaocuiwang, and Baitiannuo ($p < 0.05$). Among the common maize varieties, the highest activity was on Zhengdan 958, significantly greater than Wuke No. 4 and Wuke 113, with the lowest on Wuke No. 4 ($p < 0.05$). Surprisingly, among the special maize varieties, CarE activity did not differ significantly among Zaocuiwang, Baitiannuo, and Ziyunuo ($p > 0.05$) (Figure 4C).

2.5. Correlation among Chemical Substances in Maize Leaves, the Nutritional Indices, Digestive, and Detoxifying Enzymes of S. frugiperda

Pearson's correlation analysis was employed to discern the relationships between various biochemical parameters in maize leaves, as measured in our previous study (Tables S1 and S2 of the Supplementary Materials) [39], and the corresponding physiological responses of *S. frugiperda* (Figure 5). This analysis yielded several noteworthy correlations:

RGR was extremely significantly correlated with fatty acid ($p \leq 0.01$, $r = 0.915$), ECD ($p \leq 0.01$, $r = 0.727$), and ECI ($p \leq 0.01$, $r = 0.763$). Reducing sugar was also significantly positively linked to RGR ($p \leq 0.05$, $r = 0.476$). Conversely, RGR had a significantly negative correlation with protein ($p \leq 0.05$, $r = -0.559$), chlorophyll a ($p \leq 0.05$, $r = -0.483$), chlorophyll b ($p \leq 0.05$, $r = -0.535$), β-amylase ($p \leq 0.05$, $r = -0.485$), and total-amylase ($p \leq 0.05$, $r = -0.483$), and an extremely significantly negative correlation with amino acids ($p \leq 0.01$, $r = -0.715$) and moisture ($p \leq 0.01$, $r = -0.67$), respectively. Similarly, RCR exhibited a significant positive correlation with AD ($p \leq 0.05$, $r = 0.512$), β-amylase ($p \leq 0.05$, $r = 0.515$), and total-amylase activities ($p \leq 0.05$, r = 0.519), and an extremely significant positive correlation with moisture ($p \leq 0.01$, r = 0.695). However, there were significant negative correlations with fatty acid ($p \leq 0.05$, $r = -0.518$) and flavone ($p \leq 0.05$, $p \leq 0.05$, $r = -0.487$), and a highly significant negative relationship with reducing sugar ($p \leq 0.01$, $r = -0.697$), ECI ($p \leq 0.01$, $r = -0.792$), and ECD ($p \leq 0.01$, $r = -0.819$). ECD showed a highly significant positive correlation with fatty acid ($p \leq 0.01$, $r = 0.798$), reducing sugar ($p \leq 0.01$, $r = 0.666$), and ECI ($p \leq 0.01$, $r = 0.986$), respectively. ECD showed extremely significant negative correlations with amino acids ($p \leq 0.01$, $r = -0.607$) and moisture ($p \leq 0.01$, $r = -0.833$) and significant negative correlations with protein ($p \leq 0.05$, $r = -0.577$), AD ($p \leq 0.05$, $r = -0.531$), β-amylase ($p \leq 0.05$, $r = -0.547$), and total-amylase ($p \leq 0.05$, $r = -0.55$). Extremely significantly positive correlations were present between ECI and fatty acid ($p \leq 0.01$, $r = 0.834$) and reducing sugar ($p \leq 0.01$, $r = 0.618$). Conversely, ECI exhibited highly significant negative correlations with amino acids ($p \leq 0.01$, $r = -0.6$), moisture ($p \leq 0.01$, $r = -0.814$), β-amylase ($p \leq 0.01$, $r = -0.625$), and total-amylase ($p \leq 0.01$, $r = -0.627$), and a significant negative correlation with protein ($p \leq 0.05$, $r = -0.553$). A significant negative correlation was found between AD and reducing sugar ($p \leq 0.05$, $r = -0.583$).

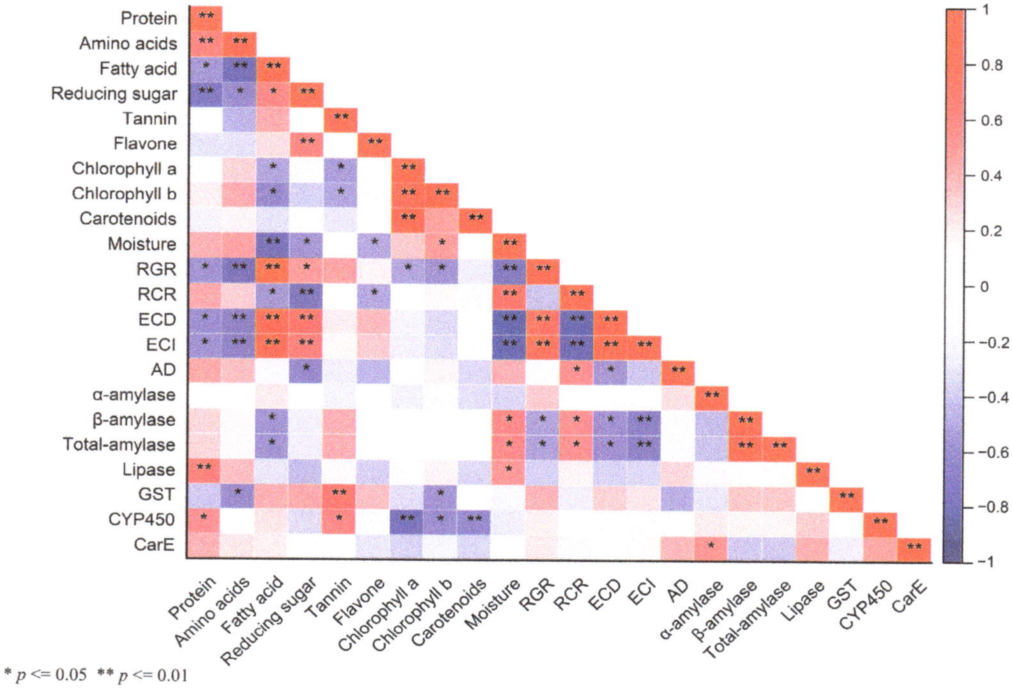

Figure 5. The correlation matrix of different indices is based on the Pearson correlation coefficient. Colors from blue to red represent a correlation from extremely negative to positive. Significance or extreme significance are indicated by $p \leq 0.05$ or $p \leq 0.01$, respectively.

α-amylase and CarE ($p \leq 0.05$, $r = 0.51$) showed a significant positive correlation. β-amylase and total-amylase activities showed significant positive correlations with moisture and a negative correlation with fatty acid content. Lipase activity was positively correlated with moisture content ($p \leq 0.05$, $r = 0.498$) and protein levels ($p \leq 0.01$, $r = 0.639$), the latter being highly significant.

GST activity had an extremely significant positive correlation with tannin content ($p \leq 0.01$, $r = 0.657$) but significant negative correlations with amino acids ($p \leq 0.05$, $r = -0.578$) and chlorophyll b ($p \leq 0.05$, $r = -0.53$). CYP450 activity had significant positive associations with protein ($p \leq 0.05$, $r = 0.539$) and tannin ($p \leq 0.05$, $r = 0.564$) but was significantly negatively correlated with chlorophyll b ($p \leq 0.05$, $r = -0.543$), and an extremely significant negative correlation with chlorophyll a ($p \leq 0.01$, $r = -0.744$) and carotenoids ($p \leq 0.01$, $r = -0.6$). These correlations suggest intricate interactions between the maize plant's chemical composition and the physiological and metabolic responses of *S. frugiperda*, highlighting the complexity of host–pest interactions.

3. Discussion

S. frugiperda, due to its wide host range and strong migratory and reproductive ability, has caused serious damage to many crops in China [40]. Genetic studies have identified the invasive *S. frugiperda* populations in China as the corn strain [41]. Guo et al. [23] determined the corn genotype of *S. frugiperda* after 13 generations of feeding on rice or corn. This is consistent with the findings of our study, where alignment analysis based on *COI* and *Tpi* gene fragment sequences revealed complete consistency with the corn type at all sites. Considering that China is the second-largest global maize producer, with cultivation spanning all provinces [42,43], the recurrent incursions of *S. frugiperda* into the country's maize belts significantly jeopardize its agricultural output and food security. Therefore, it is crucial to understand the physiological changes in *S. frugiperda* after feeding on maize. However, due to limitations in sampling locations and sample numbers, the existence of the rice type of *S. frugiperda* cannot be completely ruled out. Therefore, it is necessary to conduct larger-scale monitoring and investigations.

The nutritional indices play a pivotal role in reflecting the nutrient utilization of insects to host plants [25]. Polyphagous insects, which exhibit a diverse range of host plants, demonstrate significant differences in their adaptability and ability to utilize nutrients [44]. Our study elucidated notable distinctions in the nutritional indices of *S. frugiperda* reared on six maize varieties. These variations were further reflected in the fitness levels observed among the maize varieties. Particularly, the special maize varieties exhibited markedly higher values for ECD, ECI, and RGR compared with the common maize varieties. *S. frugiperda* larvae showed a more efficient conversion of digested food into biomass when feeding on the special maize varieties compared with common ones. Similarly, Hoo and Fraenkel [45] reported that larvae consume smaller quantities of specific food sources due to their increased efficiency in converting it, thereby achieving the desired level of growth without the need for large amounts of food. Interestingly, Wuke 113 displayed the highest RCR yet the lowest ECD and ECI, which is similar to the findings of Behmer [46] and Silva et al. [47]. This shows that the compensatory feeding behavior exhibited by insects when consuming hosts with lower nutritional value, such as prolonged feeding time or increased food intake. More importantly, Pearson correlation analysis revealed that nutrient substance contents and secondary metabolites of maize leaf can significantly affect nutritional indexes. For instance, higher fatty acid and reducing sugar, and lower moisture and protein can lead to higher RGR, ECD, and ECI, yet the lowest RCR of the special maize varieties. The observed phenomenon suggests that the special maize varieties probably contained sufficient nutrients, which helped *S. frugiperda* larvae to adapt and efficiently utilize its nutrients. This was demonstrated by a higher relative growth rate of *S. frugiperda* in comparison with the common maize varieties. Other studies have also demonstrated lower ECI and ECD in *Spodoptera littoralis* Boisd (Lepidoptera: Noctuidae) larvae reared on Mashhad cowpea (*Vigna sinensis* L.) cultivar [48]. Similarly, Kang [49] found that a high proportion of reducing sugars in plants was conducive to feeding larvae and adults, and promoted the growth and metabolism of insects. Moreover, low-protein diets influenced the developmental period, larval weight and mortality, pupation rate, percentage of adult emergence, and nutritional indices [50,51]. In our study, Ziyunuo showed the lowest AD and RCR but the highest ECD and ECI among the tested maize varieties. This discrepancy could be due to the high content of flavonoid metabolites in Ziyunuo, which possibly inhibited the approximate digestibility and consumption of *S. frugiperda* larvae [39]. Similar to *Hyphantria cunea* Drury (Lepidoptera: Arctiidae) larvae, *S. frugiperda* larvae might reduce the toxicity of phenolic secondary metabolites by reducing their food intake and digestibility [52]. Interestingly, the present study observed a positive correlation between ECI and ECD but a negative correlation with AD. This intriguing finding indicates that the elevated conversion and utilization rates might serve as a physiological compensation for the decreased digestibility, which is consistent with previous research [19,53]. These research studies indicated that the nutrients and secondary metabolites between maize varieties stimulate changes in nutritional indexes, as they play a direct role in the growth and development of larvae [39,54].

Nutrient utilization and conversion in plant-feeding insects hinge on digestive enzyme activities in the midgut, reflecting their capacity for digestion, absorption and their consequent impact on growth and development [55,56]. According to differences analysis, different maize varieties can significantly affect enzyme activities. As confirmed by previous studies [57], changes in digestive enzyme activity have been shown to affect the adaptability of *S. frugiperda* to different hosts. In our study, *S. frugiperda* reared on the common maize varieties exhibited higher activities of β-amylase, total-amylase, and lipase. For example, Wuke 113 showed the highest levels of β-amylase and total-amylase activities, whereas Zaocuiwang exhibited the lowest activities for both β-amylase and total-amylase. Pearson correlation analysis showed that β-amylase and total-amylase exhibited significantly negative and positive correlations with fatty acid and moisture, respectively. Similarity exists with previous studies that have shown the amylase activity of *Helicoverpa armigera* Hübner (Lepidoptera: Noctuidae) decreases when reared on plants with high carbohydrate content [58]. Furthermore, higher protein in common maize varieties may lead to higher lipase activity. These observed variations in enzyme levels could be attributed to differences in nutrient composition or secondary metabolite content in the host plants. Our research showed that nutrient utilization efficiency and digestive enzyme activity can interact with each other. Common maize varieties, for example, had higher digestive enzyme activities and RCR but lower ECD and ECI. These results suggest that *S. frugiperda* perhaps needs to enhance digestive enzyme activities in order to digest and absorb food nutrients more effectively, thereby maintaining normal development and aiding in coping with plant defense mechanisms, likely as an adaptive response. Furthermore, we observed that, under conditions of high food intake, *S. frugiperda* gut amylase activity reached its peak [59]. Similar findings were reported on *H. armigera* reared on seven bean cultivars [60]. Previous studies reported that the efficiency of converting digested food into larval biomass relies on the levels of digestive enzyme activities [61]. Therefore, insects possess a mechanism for accurately detecting and quantifying food contents, allowing them to regulate essential digestive enzyme levels [58]. Our previous research found that *S. frugiperda* showed lower fitness in common maize varieties [38]. In general, plants defend themselves against herbivores by producing various digestion inhibitors that limit nutrient uptake. In response to plants' defense mechanisms, insect herbivores may counter by augmenting digestive enzyme activities or upregulating inhibitor-insensitive enzyme levels in their midgut [62,63]. Moreover, Namin et al. [60] indicated that enzyme-inhibiting components in red kidney beans (*Phaseolus vulgaris* L.) Sayyad could reduce the food conversion rate in *H. armigera* larvae. Consequently, variations in enzyme levels might also arise from differences in insect responses to enzyme inhibitors within host plants.

Plants typically defend themselves against herbivores by producing toxic secondary metabolites, either constitutively or in response to attack. However, insects have evolved detoxifying enzymes to counteract these defenses [64]. Furthermore, herbivorous insects tend to increase the expression of detoxification enzyme isoforms when feeding on host plants [65–67]. In our study, all six tested maize varieties exerted a significant influence on GST, CYP450, and CarE activities in *S. frugiperda*. The activity of detoxification enzymes changes with different host plants, as evidenced by numerous studies. For example, Glutathione S-transferase Slgste1 has been identified as a crucial detoxification enzyme induced by phytochemicals in host plants, potentially playing a role in the host plant adaptation of *Spodoptera litura* Fabricius (Lepidoptera: Noctuidae) [68], which is consistent with the results of this study. Pearson correlation analysis showed that detoxification enzyme activities had a certain internal relationship between nutrient composition and secondary metabolite content. For example, protein, amino acids, tannin, chlorophyll a, chlorophyll b, and carotenoids can significantly influence the activities of detoxification enzymes. Additionally, Gossypol, a compound found in cotton leaves, has been shown to induce several P450 enzymes (CYP9A12, CYP9A14, CYP6B7, CYP6AE14, and CYP321A1) in *H. armigera* [67]. Similarly, in *Manduca sexta* Linnaeus (Lepidoptera: Sphingidae) larvae, increased P450 levels allow for improved detoxification of secondary metabolites [69], leading to enhanced adaptation to host plants [70]. As a result, we hypothesized that differences in GST, CYP450, and CarE activities could be explained by differences in nutrient content and secondary metabolites of different maize varieties and that higher tannin and protein content of maize varieties could lead to increased GST and CYP450 activities. This may also be an adaptation of insects to different host plants during long-term evolutionary development. Insects can actively regulate enzyme activity to maximize the digestion and utilization of the compounds they feed on. Our previous research has shown that the nutrient content and secondary metabolites of different maize varieties can affect the feeding selection, development, and reproduction of *S. frugiperda* [38,39]. In this study, we also found that there are significant differences in nutrient utilization efficiency and the digestive and detoxification enzyme activities of *S. frugiperda* feeding on different maize varieties, and there is a significant correlation between the nutrient content and secondary metabolites of maize. Thus, the nutrient content and secondary metabolites of different corn varieties may be key factors affecting the adaptability of the *S. frugiperda*.

4. Materials and Methods

4.1. Insect Rearing

S. frugiperda larvae were collected from maize crops in Jingyuan County, Gansu, China (36°34′16″ N, 104°40′36″ E) in September 2019. Subsequently, these larvae were exposed to an artificial diet, as described by Wang et al. [71], while the adults were fed with a 10% honey solution. All the insects were reared in a 40 × 40 × 40 cm^3 insect cage, set under controlled laboratory conditions with a temperature of 25 ± 1 °C, relative humidity of 75 ± 5%, and a photoperiod of 16 h of light and 8 h of darkness. After acclimatization over ten generations, neonate larvae (<6 h old) were isolated into individual 9.0 cm diameter disposable plastic Petri dishes nurtured on a diet of maize seedlings.

4.2. Host Plants and Growth Condition

Six maize varieties were selected for early oviposition selectivity based on the high, medium, and low egg-laying rates of *S. frugiperda* to common and special maize varieties [38]. Seeds of common maize varieties (Zhengdan 958, Wuke No. 4, and Wuke 113) and special maize varieties (Zaocuiwang, Baitiannuo, and Ziyunuo) were supplied by Jiuquan Jinhui Agricultural Development Co., Ltd. in Jiuquan, China.

The maize seeds were sterilized separately in 75% (*v*/*v*) ethanol and 1% (*w*/*v*) NaClO for 3 min and 15 min, respectively, then washed five times with sterile water. Subsequently, they were planted on 1/2 Murashige and Skoog (MS) medium containing 1% (*w*/*v*) sucrose and 0.7% (*w*/*v*) agar, with a pH of 5.8 and cultured at 28 °C under a 16-h light/8-h dark regime for the specified number of days until germination. Then, the seeds were sown in pots (15 cm in height by 20 cm in diameter) containing a potting soil mixture (3:1). These pots were subsequently placed in an insect-rearing room maintained at a temperature of (25 ± 1) °C and a photoperiod of 16:8 h (L:D), with a light intensity of 250 $\mu mol\ m^{-2}\ s^{-1}$ and a relative humidity of 65 ± 5%. The *S. frugiperda* larvae were reared on maize plants at the six-leaf stages.

4.3. Host Strain Identification and Sequence Alignment

In order to identify the genotype of collected *S. frugiperda* larvae, DNA extraction from a single larva (five larvae for repeated processing) using the E.Z.N.A. Insect DNA Kit (Omega, Norcross, GA, USA). Briefly, samples were homogenized and lysed in a high-salt buffer containing CTAB and extracted with chloroform to remove polysaccharides. Subsequently, a rapid alcohol precipitation step, binding conditions were adjusted, and DNA was further purified using HiBind® DNA Mini Columns. PCR amplification was performed using a 25 µL reaction mix containing 12.5 µL 2× Hieff PCR Master Mix (Yeasen, Shanghai, China), 2 µL of a 10 µM primer mix, 1 µL DNA template, and the remaining volume filled with water. The thermocycling program consisted of an initial denaturation step at 94 °C for 5 min, followed by 31 cycles of denaturation at 94 °C for 30 s, annealing at 55 °C for 30 s, extension at 72 °C for 30 s, and a final extension at 72 °C for 5 min. Amplification of *COI* used the primer pair 101F (5′-TTCGAGCTGAATTAGGGACTC-3′) and 911R (5′-GATGTAAAATATGCTCGTGT-3′), Amplification of the *Tpi* region with the primers 282F (5′-GGTGAAATCTCCCCTGCTATG-3′) and 850R (5′-AATTTTATTACCTGCTGTGG-3′) (synthesized by Tsingke Biotech Co., Ltd., Xi'an, China) [72]. The PCR products were subjected to 1% agarose gel electrophoresis, followed by Sanger sequencing provided by Tsingke Biotechnology Co., Ltd. (Beijing, China). Newly obtained sequences were submitted to the GenBank database. Newly obtained sequences were submitted to the National

Center for Biotechnology Information (http://blast.ncbi.nlm.nih.gov/Blast.cgi, accessed on 13 December 2023). It was compared with sequences of other *S. frugiperda* species in GenBank using the nucleotide BLAST program. Multiple alignments were analyzed using the DNAMAN version 6.0.3.99 (Lynnon Corporation, Vaudreuil-Dorion, QC, Canada), published *COI* sequences (The corn biotype HM136586 and rice biotype HM136593) [73], and *Tpi* sequences (The corn biotype KT336237 and rice biotype KT336230) [74] from GenBank were included in our analyses.

4.4. Assessment of Nutritional Indices of 3rd Instar Larvae of S. frugiperda on Various Maize Varieties

Prior to nutritional assessments, third-instar larvae were subjected to a 12-h starvation period to ensure complete defecation, and then the weights of single larvae were recorded. Subsequently, the larvae were provided with sufficient leaves from corresponding maize seedings, with the initial fresh weight of the leaves meticulously documented. After a 48-h feeding period, larvae were isolated, again subjected to a 12-h starvation period, and then weighed to determine their fresh post-feeding mass. Subsequent to feeding, the residual leaves, larvae, and feces were desiccated at 80 °C until a consistent weight was achieved and determined the dry weight. This process allowed for the calculation of the dry weight of both leaves and larvae pre-consumption, adjusted based on the determined drying rate of the host plant material. The experiment was set up with 25 larvae per trial, repeated 5 times. The nutritional indices of *S. frugiperda* larvae were computed following the protocol established by Waldbauer [75].

$$\text{Relative growth rate (RGR)} = \frac{G}{B \times T}$$

$$\text{Relative consumption rate (RCR)} = \frac{I}{B \times T}$$

$$\text{Efficiency of conversion of ingested food (ECI)} = \frac{G}{I} \times 100\%$$

$$\text{Efficiency of conversion of digested food (ECD)} = \frac{G}{I - F} \times 100\%$$

$$\text{Approximate digestibility (AD)} = \frac{I - F}{I} \times 100\%$$

In the given equation, various parameters are defined as follows: G represents the weight gain of larvae (calculated as G = dry weight of larvae after feeding—dry weight of larvae before feeding), B represents the mean larval body weight during the test period (B = (dry weight of larvae before feeding + dry weight of larvae after feeding)/2), I represents the weight of food eaten during the test period (I = dry weight of leaf before feeding—the dry weight of leaf after feeding), F represents the dry weight of fecal matter during the test period, and T represents the duration of the test period (measured in days).

4.5. Measurement of Digestive Enzyme Activity

4.5.1. Midgut Dissection and Enzyme Extraction

The sample was prepared with minor modifications based on the methods described by Wang and Qin [76]. Fifth-instar larvae of *S. frugiperda* were collected and subjected to a 6-h period of starvation. Midguts were then dissected on ice and rinsed with chilled physiological saline to remove food remnants. Subsequently, tissues were homogenized in a pre-cooled homogenizer with distilled water. The homogenate was centrifuged at 15,000 rpm for 10 min at 4 °C, and the resultant supernatant was reserved as the enzyme source. This procedure was conducted on groups of 10 larvae, with the entire assay replicated thrice for consistency.

4.5.2. Amylase Activity Assay

The quantification of amylase activity was conducted using the dinitrosalicylic acid (DNS) method [77]. The assay mixture consisted of 20 μL enzyme extract, 100 μL of phosphate buffer (0.02 M, pH 7.1), and 40 μL of 1% soluble starch. The mixture was incubated at 37 °C for 30 min. The enzymatic reaction was then halted by adding 100 μL of DNS reagent, followed by a 10-min boiling step. After cooling, the absorbance was measured at 540 nm to determine amylase activity (BioTek Instruments, Inc., Winooski, VT, USA). One enzyme activity unit is defined as catalyzing the production of 1 mg of reducing sugar per gram of tissue per minute in 30 min at 37 °C.

4.5.3. Lipase Activity Assay

Lipase activity was assessed by a modified version of the method from Bai and Wang [78], using an olive oil emulsion as the substrate. The assay involved pre-mixing 0.5 mL of phosphate buffer (0.025 mol/L, pH 7.5) with 0.4 mL of olive oil, followed by a 10-min incubation at 40 °C. To this, 0.5 mL of the enzyme suspension solution was added, and the incubation continued for another 20 min at the same temperature. The reaction was terminated by the addition of 1.5 mL of 95% ethanol and a drop of 1% phenolphthalein indicator. Followed by titration with 0.05 mol/L NaOH until a reddish color persisted. The volume of NaOH dispensed was recorded, adjusting for the control, which included 95% ethanol added prior to the lipase activity initiation step. The specific activity of lipase is defined as the amount of enzyme required to hydrolyze 1 μmol of fatty acid per minute under pH 7.5 and 40 °C conditions per gram of lipase.

4.6. Determination of Detoxification Enzyme Activity

4.6.1. Enzyme Solution Preparation

Triplicates of five fifth-instar larvae on six maize varieties were homogenized in pre-cooled phosphate buffer (0.1 M, pH = 7.5). The supernatant, extracted post-centrifugation at 12,000 rpm at 4 °C for 10 min, was utilized as the source of enzymes.

4.6.2. Glutathione-S-Transferase (GST) Activity Assay

GST catalyzes the conjugation of 1-chloro-2,4-dinitrobenzene (CDNB) with reduced glutathione (GSH), yielding a dinitrophenyl thioether (GS-DNB) with a high extinction coefficient at 340 nm, facilitating its detection. The GST activity was quantified employing an established protocol [79]. Larvae were homogenized in 1 mL of pre-cooled sodium phosphate buffer (0.1 M, pH 7.2). The homogenate was centrifuged at 10,000 rpm for 30 min at 4 °C, and the resultant supernatant was reserved as the enzyme solution. For the enzymatic reaction, 810 μL of sodium phosphate buffer was mixed with 30 μL of 1-chloro-2,4-dinitrobenzene (CDNB), 50 μL of enzyme solution, and 30 μL of glutathione (GSH). The mixture was then incubated for 5 min at 25 °C. The absorbance was recorded at 340 nm over 5 min with a read 30-s interval using an ultraviolet spectrophotometer (BioTek Instruments, Inc., Winooski, VT, USA). For the control group, an equivalent volume of phosphate buffer is added instead of an enzyme solution. At 25 °C, one enzyme activity unit is defined as the catalysis of 1 μmol of CDNB binding with GSH per gram of sample per minute.

4.6.3. CarE Activity Determination

CarE activity was determined following a previously described method [80]. Briefly, the homogenization of samples occurred in an ice bath using 1 mL of sodium phosphate buffer (0.04 M, pH 7.0), followed by centrifugation at 11,000 rpm for 15 min at 4 °C. The clear supernatant was then prepared for the assay by combining it with 1.8 mL of substrate solution (containing 3×10^{-4} M α-NA and 3×10^{-4} eserine), 0.45 mL of sodium phosphate buffer (0.04 M, pH 7.0), and 0.05 mL of enzyme solution, incubated for 15 min at 30 °C. The reaction was terminated with 0.9 mL of dye reagent (1% of Fast blue B salt: 5% of sodium dodecyl sulfate solution = 2:5). The optical density (OD) for the reaction mixtures was measured at 600 nm (BioTek Instruments, Inc., Winooski, VT, USA). One enzyme activity unit (U) is defined as the increase of 1 absorbance unit per mL reaction system per gram of tissue per minute, and enzyme activity was expressed as U/g.

4.6.4. CYP450 Activity Determination

The procedure for CYP450 activity measurement was adapted from a previously reported method [81]. The reaction mixture consisted of 705 μL of Tris-HCl buffer, 25 μL of 2 mmol/L of 7-ethoxycoumarin (7-EC), 20 μL of 6 mmol/L NADPH, and 250 μL of enzymatic suspension, incubated for 30 min at 30 °C. The reaction was halted using 300 μL of trichloroacetic acid and subsequently centrifugation at 12,000 rpm for 3 min. The fluorescence of the resultant 7-hydroxycourmarin was measured with excitation at 368 nm and emission at 456 nm (BioTek Instruments, Inc., Winooski, VT, USA). As the control, the same amount of enzyme was added after introducing trichloroacetic acid.

4.7. Statistical Analysis

Data from these experiments were processed and analyzed using IBM SPSS Statistics version 23.0 (Chicago, IL, USA). The results are expressed as means ± standard error (SE). The effects of various maize cultivars on the nutritional indices, as well as digestive and detoxification enzyme activities, were assessed through a one-way analysis of variance (ANOVA). Post-hoc comparisons were made using Duncan's multiple-range test to identify specific differences between groups. A correlation analysis was performed to explore the relationship between the chemical substances of maize leaves, as measured in our previous study [39] (Tables S1 and S2 of the Supplementary Materials), and the nutritional and enzymatic parameters observed in the current study. These analyses employed the Pearson correlation coefficient to quantify the degree of association, with an accompanying *p*-value indicating its statistical significance. Thresholds for significance were set at $p \leq 0.05$ for significant differences and $p \leq 0.01$ for extremely significant differences. The figures related to nutritional indices and enzyme activities were generated using GraphPad Prism 6 (GraphPad Software, Boston, MA, USA).

5. Conclusions

This comprehensive study provided valuable insights into the complex interactions between *S. frugiperda* and various maize varieties. Our analysis revealed that both the *COI* and *Tpi* genes identified *S. frugiperda* as the corn strain. Our findings also reveal the significant impact of host plant diversity on the nutritional indices and digestive and detoxification enzyme activities in *S. frugiperda* larvae. The research revealed that special maize varieties facilitate higher nutrient assimilation efficiency in *S. frugiperda* compared with common maize varieties. This indicates a potential adaptation strategy of the pest to different host plant characteristics. The elevated activities of amylase and lipase enzymes in larvae reared on common maize varieties suggest a nuanced physiological adaptation to optimize nutrient extraction from these host plants. The correlation analysis further established a strong link between the maize plants' nutritional content and secondary metabolites and the physiological responses of *S. frugiperda*, highlighting the pest's ability to adjust its metabolic processes in response to the chemical composition of its food source. These findings contribute significantly to the understanding of *S. frugiperda*'s adaptability and resilience against different maize cultivars. They also offer a crucial theoretical basis for developing sustainable pest management strategies. By considering the diversity of maize crops and understanding the adaptive mechanisms of *S. frugiperda*, more effective and environmentally sustainable approaches to pest control can be devised. This study, therefore, not only enhances our knowledge of insect-plant interactions but also provides practical insights for agronomists and farmers in managing *S. frugiperda* infestations more effectively.

Supplementary Materials: The following supporting information can be downloaded at: https://www.mdpi.com/article/10.3390/plants13050597/s1, Table S1: Content of nutrients and secondary substances in six maize varieties; Table S2: The moisture and chlorophyll content in six maize varieties.

Author Contributions: Conceptualization, Q.Z., Y.Z., C.L. (Chunchun Li) and C.L. (Changzhong Liu); methodology, Q.Z. and Y.Z.; software, Q.Z.; formal analysis, Q.Z. and Y.Z.; data curation, Q.Z., Y.Z., K.Z., H.L. and Y.G.; writing-original draft, Q.Z., C.L. (Chunchun Li) and C.L. (Changzhong Liu); writing-review and editing, Q.Z., P.Q., I.U.H., C.L. (Chunchun Li) and C.L. (Changzhong Liu). All authors have read and agreed to the published version of the manuscript.

Funding: This research was funded by the National Key Research and Development Program of China, Grant number 2018YFD0200405, the Science Project of Agriculture and Rural Department of Gansu Province, Grant number GZB20191105, and Gansu Province excellent graduate "Innovation Star" project of China (2022CXZX-659) to Q.Z.

Data Availability Statement: The datasets in this study are available from the corresponding author upon reasonable request.

Acknowledgments: The authors would like to thank Wenhao Li for revising the manuscript.

Conflicts of Interest: The authors declare no conflict of interest.

References

1. Goergen, G.; Kumar, P.L.; Sankung, S.B.; Togola, A.; Tamò, M. First report of outbreaks of the fall armyworm *Spodoptera frugiperda* (JE Smith) (Lepidoptera, Noctuidae), a new alien invasive pest in West and Central Africa. *PLoS ONE* **2016**, *11*, e165632. [CrossRef]
2. Otuka, A. Prediction of the Overseas Migration of the Fall Armyworm, *Spodoptera frugiperda*, to Japan. *Insects* **2023**, *14*, 804. [CrossRef]
3. Yadava, P.; Abhishek, A.; Singh, R.; Singh, I.; Kaul, T.; Pattanayak, A.; Agrawal, P.K. Advances in maize transformation technologies and development of transgenic maize. *Front. Plant Sci.* **2017**, *7*, 1949. [CrossRef]
4. Scott, P.; Pratt, R.C.; Hoffman, N.; Montgomery, R. Specialty corns. In *Corn*; Elsevier: Amsterdam, The Netherlands, 2019; pp. 289–303.
5. Montezano, D.G.; Sosa-Gómez, D.R.; Specht, A.; Roque-Specht, V.F.; Sousa-Silva, J.C.; Paula-Moraes, S.D.; Peterson, J.A.; Hunt, T.E. Host plants of *Spodoptera frugiperda* (Lepidoptera: Noctuidae) in the Americas. *Afr. Entomol.* **2018**, *26*, 286–300. [CrossRef]
6. Prasanna, B.M.; Bruce, A.; Beyene, Y.; Makumbi, D.; Gowda, M.; Asim, M.; Martinelli, S.; Head, G.P.; Parimi, S. Host plant resistance for fall armyworm management in maize: Relevance, status and prospects in Africa and Asia. *Theor. Appl. Genet.* **2022**, *135*, 3897–3916. [CrossRef]
7. Guo, J.; Zhao, J.; He, K.; Zhang, F.; Wang, Z. Potential invasion of the crop-devastating insect pest fall armyworm *Spodoptera frugiperda* to China. *Plant Prot.* **2018**, *44*, 1–10.
8. Cui, L.; Rui, C.H.; Li, Y.P.; Wang, Q.Q.; Yang, D.B.; Yan, X.J.; Guo, Y.W.; Yuan, H.Z. Research and application of chemical control technology against *Spodoptera frugiperda* (Lepidoptera: Noctuidae) in foreign countries. *Plant Prot.* **2019**, *45*, 7–13.
9. Wu, P.; Wu, F.; Fan, J.; Zhang, R. Potential economic impact of invasive fall armyworm on mainly affected crops in China. *J. Pest. Sci.* **2021**, *94*, 1065–1073. [CrossRef]
10. Nagoshi, R.N.; Meagher, R.L. Behavior and distribution of the two fall armyworm host strains in Florida. *Fla. Entomol.* **2004**, *87*, 440–449. [CrossRef]
11. Nagoshi, R.N. The fall armyworm triose phosphate isomerase (*Tpi*) gene as a marker of strain identity and inter strain mating. *Ann. Entomol. Soc. Am.* **2010**, *103*, 283–292. [CrossRef]
12. Lu, Y.J.; Kochert, G.D.; Isenhour, D.J.; Adang, M.J. Molecular characterization of a strain-specific repeated DNA sequence in the fall armyworm *Spodoptera frugiperda* (Lepidoptera: Noctuidae). *Insect. Mol. Biol.* **1994**, *3*, 123–130. [CrossRef] [PubMed]
13. Dumas, P.; Legeai, F.; Lemaitre, C.; Scaon, E.; Orsucci, M.; Labadie, K.; Gimenez, S.; Clamens, A.; Henri, H.; Vavre, F. *Spodoptera frugiperda* (Lepidoptera: Noctuidae) host-plant variants: Two host strains or two distinct species? *Genetica* **2015**, *143*, 305–316. [CrossRef] [PubMed]
14. Nagoshi, R.N.; Murúa, M.G.; Hay-Roe, M.; Juárez, M.L.; Willink, E.; Meagher, R.L. Genetic characterization of fall armyworm (Lepidoptera: Noctuidae) host strains in Argentina. *J. Econ. Entomol.* **2012**, *105*, 418–428. [CrossRef] [PubMed]
15. Zhang, L.; Jin, M.H.; Zhang, D.D.; Jiang, Y.Y.; Liu, J.; Wu, K.M.; Xiao, Y.T. Molecular identification of invasive fall armyworm *Spodoptera frugiperda* in Yunnan Province. *Plant Prot.* **2019**, *45*, 19–24.
16. Li, X.; Garvey, M.; Kaplan, I.; Li, B.; Carrillo, J. Domestication of tomato has reduced the attraction of herbivore natural enemies to pest-damaged plants. *Agr. Forest. Entomol.* **2018**, *20*, 390–401. [CrossRef]
17. Zhang, Y.; Lou, Y. Research progress in chemical interactions between plants and phytophagous insects. *J. Appl. Ecol.* **2020**, *31*, 2151–2160.
18. Carrasco, D.; Larsson, M.C.; Anderson, P. Insect host plant selection in complex environments. *Curr. Opin. Insect. Sci.* **2015**, *8*, 1–7. [CrossRef]

19. Hwang, S.Y.; Liu, C.H.; Shen, T.C. Effects of plant nutrient availability and host plant species on the performance of two *Pieris* butterflies (Lepidoptera: Pieridae). *Biochem. Syst. Ecol.* **2008**, *36*, 505–513. [CrossRef]
20. Karlsson Green, K. The effects of host plant species and larval density on immune function in the polyphagous moth *Spodoptera littoralis*. *Ecol. Evol.* **2021**, *11*, 10090–10097. [CrossRef]
21. Gong, B.; Zhang, G. Interactions between plants and herbivores: A review of plant defense. *Acta Phytoecol. Sin.* **2014**, *34*, 325–336. [CrossRef]
22. Wang, W.; He, P.; Liu, T.; Jing, X.; Zhang, S. Comparative studies of ovipositional preference, larval feeding selectivity, and nutritional indices of *Spodoptera frugiperda* (Lepidoptera: Noctuidae) on 6 crops. *J. Econ. Entomol.* **2023**, *116*, 790–797. [CrossRef] [PubMed]
23. Guo, Z.M.; Jin, R.H.; Guo, Z.P.; Cai, T.W.; Zhang, Y.H.; Gao, J.Y.; Huang, G.Y.; Wan, H.; He, S.; Xie, Y.L. Insecticide susceptibility and mechanism of *Spodoptera frugiperda* on different host plants. *J. Agr. Food. Chem.* **2022**, *70*, 11367–11376. [CrossRef]
24. Xie, W.; Zhi, J.R.; Ye, J.Q.; Zhou, Y.M.; Li, C.; Liang, Y.J.; Yue, W.B.; Li, D.Y.; Zeng, G.; Hu, C.X. Age-stage, two-sex life table analysis of *Spodoptera frugiperda* (JE Smith) (Lepidoptera: Noctuidae) reared on maize and kidney bean. *Chem. Biol. Technol. Agr.* **2021**, *8*, 44. [CrossRef]
25. Scriber, J.M.; Slansky Jr, F. The nutritional ecology of immature insects. *Annu. Rev. Entomol.* **1981**, *26*, 183–211. [CrossRef]
26. Lu, Z.H.; He, S.Q.; Guo, J.; Chen, Y.P.; Hu, X.Q.; Li, H.; Xie, Q.; Gui, F.R. Effects of host plants on activities of three groups of enzymes in *Spodoptera frugiperda* (Smith)larvae. *J. South. Agric.* **2020**, *51*, 2461–2469.
27. Heidel-Fischer, H.M.; Vogel, H. Molecular mechanisms of insect adaptation to plant secondary compounds. *Curr. Opin. Insect. Sci.* **2015**, *8*, 8–14. [CrossRef] [PubMed]
28. Li, X.C.; Schuler, M.A.; Berenbaum, M.R. Molecular mechanisms of metabolic resistance to synthetic and natural xenobiotics. *Annu. Rev. Entomol.* **2007**, *52*, 231–253. [CrossRef]
29. Moreau, J.; Benrey, B.; Thiéry, D. Grape variety affects larval performance and also female reproductive performance of the European grapevine moth *Lobesia botrana* (Lepidoptera: Tortricidae). *Bull. Entomol. Res.* **2006**, *96*, 205–212. [CrossRef]
30. Awmack, C.S.; Leather, S.R. Host plant quality and fecundity in herbivorous insects. *Annu. Rev. Entomol.* **2002**, *47*, 817–844. [CrossRef]
31. Altaf, N.; Idrees, A.; Ullah, M.I.; Arshad, M.; Afzal, A.; Afzal, M.; Rizwan, M.; Li, J. Biotic potential induced by different host plants in the fall armyworm, *Spodoptera frugiperda* (Lepidoptera: Noctuidae). *Insects* **2022**, *13*, 921. [CrossRef]
32. Chiriboga Morales, X.; Tamiru, A.; Sobhy, I.S.; Bruce, T.J.; Midega, C.A.; Khan, Z. Evaluation of African maize cultivars for resistance to fall armyworm *Spodoptera frugiperda* (JE Smith) (Lepidoptera: Noctuidae) larvae. *Plants* **2021**, *10*, 392. [CrossRef] [PubMed]
33. He, L.M.; Wang, T.L.; Chen, Y.C.; Ge, S.S.; Wyckhuys, K.A.G.; Wu, K.M. Larval diet affects development and reproduction of East Asian strain of the fall armyworm, *Spodoptera frugiperda*. *J. Integr. Agr.* **2021**, *20*, 736–744. [CrossRef]
34. Sun, Y.; Liu, X.G.; Lv, G.Q.; Hao, X.Z.; Li, S.H.; Li, G.P.; Feng, H.Q. Comparison of population fitness of *Spodoptera frugiperda* (Lepidoptera: Noctuidae) feeding on wheat and different varieties of maize. *Plant Prot.* **2020**, *46*, 126–131.
35. De La Rosa-Cancino, W.; Rojas, J.C.; Cruz-Lopez, L.; Castillo, A.; Malo, E.A. Attraction, feeding preference, and performance of *Spodoptera frugiperda* larvae (Lepidoptera: Noctuidae) reared on two varieties of maize. *Environ. Entomol.* **2016**, *45*, 384–389. [CrossRef] [PubMed]
36. Omoto, C.; Bernardi, O.; Salmeron, E.; Sorgatto, R.J.; Dourado, P.M.; Crivellari, A.; Carvalho, R.A.; Willse, A.; Martinelli, S.; Head, G.P. Field-evolved resistance to Cry1Ab maize by *Spodoptera frugiperda* in Brazil. *Pest Manag. Sci.* **2016**, *72*, 1727–1736. [CrossRef] [PubMed]
37. Tang, Q.F.; Fang, M.; Yao, L.; Qiu, K.; Zheng, Z.Y.; Jin, T.; Li, G.T. Effects of feeding different corn organizations on growth, development and nutritional indexes of *Spodoptera frugiperda*. *Plant Prot.* **2020**, *46*, 24–27.
38. Zhang, Q.; Zhang, Y.; Quandahor, P.; Gou, Y.; Li, C.; Zhang, K.; Liu, C. Oviposition Preference and Age-Stage, Two-Sex Life Table Analysis of *Spodoptera frugiperda* (Lepidoptera: Noctuidae) on Different Maize Varieties. *Insects* **2023**, *14*, 413. [CrossRef]
39. Zhang, Y.L.; Zhang, K.X.; Ma, Y.; Zhang, Q.Y.; Liu, W.H.; Jiang, H.X.; Yang, L.K.; Liu, C.Z. Feeding selectivity of *Spodoptera frugiperda* on different maize cultivars and its relationship with chemical substances in maize leaves. *Chi Plant Prot.* **2021**, *41*, 10–16.
40. Yang, P.Y.; Zhu, X.M.; Guo, J.F.; Wang, Z.Y. Strategy and advice for managing the fall armyworm in China. *Plant Prot.* **2019**, *45*, 1–6.
41. Guan, F.; Zhang, J.; Shen, H.; Wang, X.; Padovan, A.; Walsh, T.K.; Tay, W.T.; Gordon, K.H.J.; James, W.; Czepak, C. Whole-genome sequencing to detect mutations associated with resistance to insecticides and Bt proteins in *Spodoptera frugiperda*. *Insect Sci.* **2021**, *28*, 627–638. [CrossRef]
42. Jiang, Y.Y.; Liu, J.; Xie, M.M.; Li, Y.H.; Yang, J.J.; Zhang, M.L.; Qiu, K. Observation on law of diffusion damage of *Spodoptera frugiperda* in China in 2019. *Plant Prot.* **2019**, *45*, 10–19.
43. Li, X.J.; Wu, M.F.; Ma, J.; Gao, B.Y.; Wu, Q.L.; Chen, A.D.; Liu, J.; Jiang, Y.Y.; Zhai, B.P.; Early, R. Prediction of migratory routes of the invasive fall armyworm in eastern China using a trajectory analytical approach. *Pest Manag. Sci.* **2020**, *76*, 454–463. [CrossRef] [PubMed]
44. Razmjou, J.; Naseri, B.; Hemati, S.A. Comparative performance of the cotton bollworm, *Helicoverpa armigera* (Hübner) (Lepidoptera: Noctuidae) on various host plants. *J. Pest. Sci.* **2014**, *87*, 29–37. [CrossRef]

45. Hoo, C.F.S.; Fraenkel, G. The consumption, digestion, and utilization of food plants by a polyphagous insect, *Prodenia eridania* (Cramer). *J. Insect. Physiol.* **1966**, *12*, 711–730. [CrossRef]
46. Behmer, S.T. Insect herbivore nutrient regulation. *Annu. Rev. Entomol.* **2009**, *54*, 165–187. [CrossRef] [PubMed]
47. Silva, D.M.D.; Bueno, A.D.F.; Andrade, K.; Stecca, C.D.S.; Neves, P.M.O.J.; Oliveira, M.C.N.D. Biology and nutrition of *Spodoptera frugiperda* (Lepidoptera: Noctuidae) fed on different food sources. *Sci. Agr.* **2017**, *74*, 18–31. [CrossRef]
48. Hemmati, S.A.; Shishehbor, P.; Stelinski, L.L. Life table parameters and digestive enzyme activity of *Spodoptera littoralis* (Boisd) (Lepidoptera: Noctuidae) on selected legume cultivars. *Insects* **2022**, *13*, 661. [CrossRef]
49. Kang, L. The chemical defenses of plants to phytophagous insects. *Chin. Bull. Bot.* **1995**, *12*, 22.
50. Borzoui, E.; Naseri, B.; Nouri-Ganbalani, G. Effects of food quality on biology and physiological traits of *Sitotroga cerealella* (Lepidoptera: Gelechiidae). *J. Econ. Entomol.* **2017**, *110*, 266–273. [PubMed]
51. Truzi, C.C.; Holzhausen, H.G.; Álvaro, J.C.; De Laurentis, V.L.; Vieira, N.F.; Vacari, A.M.; De Bortoli, S.A. Food consumption utilization, and life history parameters of *Helicoverpa armigera* (Lepidoptera: Noctuidae) reared on diets of varying protein level. *J. Insect. Sci.* **2019**, *19*, 12. [CrossRef]
52. Wu, L.; Li, L.S.; Wang, L.Y.; Yuan, Y.F.; Chen, M. Effects of gallic acid on the nutritional efficiency and detoxification enzymes in *Hyphantria cunea* larvae. *J. Environ. Entomol.* **2020**, *42*, 471–479.
53. Xue, M.; Pang, Y.H.; Wang, H.T.; Li, Q.L.; Liu, T.X. Effects of four host plants on biology and food utilization of the cutworm, *Spodoptera litura*. *J. Insect. Sci.* **2010**, *10*, 22. [CrossRef] [PubMed]
54. Sarate, P.J.; Tamhane, V.A.; Kotkar, H.M.; Ratnakaran, N.; Susan, N.; Gupta, V.S.; Giri, A.P. Developmental and digestive flexibilities in the midgut of a polyphagous pest, the cotton bollworm, *Helicoverpa armigera*. *J. Insect. Sci.* **2012**. *12*, 42. [CrossRef] [PubMed]
55. Shahbaz, M.; Nouri Ganbalani, G.; Naseri, B. Comparative damage and digestive enzyme activity of *Tuta absoluta* (Meyrick) (Lepidoptera: Gelechiidae) on 12 tomato cultivars. *Entomol. Res.* **2019**, *49*, 401–408. [CrossRef]
56. Zhao, X.D.; Sun, Y.H.; Chen, C.Y.; Tian, S.; Tao, R.; Hao, D.J. Cloning and spatio-temporal expression of serine protease gene HcSP1 and its expression in response to feeding on different host plants in *Hyphantria cunea* (Lepidoptera: Arctiidae). *Acta Entomol. Sin.* **2019**, *62*, 160–169.
57. Lu, Z.; Chen, Y.; Zhou, A.; He, S.; Li, H.; Bao, Y.; Gui, F. Effects of multi-generation feeding with different host plants on activities of enzyme in *Spodoptera frugiperda* larvae. *J. Environ. Entomol.* **2020**, *42*, 1361–1368.
58. Kotkar, H.M.; Sarate, P.J.; Tamhane, V.A.; Gupta, V.S.; Giri, A.P. Responses of midgut amylases of *Helicoverpa armigera* to feeding on various host plants. *J. Insect. Physiol.* **2009**, *55*, 663–670. [CrossRef]
59. Alfonso, J.; Ortego, F.; Sanchez-Monge, R.; Garcia-Casado, G.; Pujol, M.; Castañera, P.; Salcedo, G. Wheat and barley inhibitors active towards α-amylase and trypsin-like activities from *Spodoptera frugiperda*. *J. Chem. Ecol.* **1997**, *23*, 1729–1741. [CrossRef]
60. Namin, F.R.; Naseri, B.; Razmjou, J.; Cohen, A. Nutritional performance and activity of some digestive enzymes of the cotton bollworm, *Helicoverpa armigera*, in response to seven tested bean cultivars. *J. Insect. Sci.* **2014**, *14*, 93. [CrossRef]
61. Lazarevic, J.; Peric-Mataruga, V.; Vlahovic, M.; Mrdakovic, M.; Cvetanovic, D. Effects of rearing density on larval growth and activity of digestive enzymes in *Lymantria dispar* L. (Lepidoptera: Lymantriidae). *Folia. Biol.* **2004**, *52*, 105–112.
62. Broadway, R.M. Dietary regulation of serine proteinases that are resistant to serine proteinase inhibitors. *J. Insect. Physiol.* **1997**, *43*, 855–874. [CrossRef]
63. Jongsma, M.A.; Bolter, C. The adaptation of insects to plant protease inhibitors. *J. Insect. Physiol.* **1997**, *43*, 885–895. [CrossRef]
64. Howe, G.A.; Jander, G. Plant immunity to insect herbivores. *Annu. Rev. Plant Biol.* **2008**, *59*, 41–66. [CrossRef]
65. Huang, Y.F.; Xu, Z.B.; Lin, X.Y.; Feng, Q.L.; Zheng, S.C. Structure and expression of glutathione S-transferase genes from the midgut of the Common cutworm, *Spodoptera litura* (Noctuidae) and their response to xenobiotic compounds and bacteria. *J. Insect. Physiol.* **2011**, *57*, 1033–1044. [CrossRef]
66. Sun, X.Q.; Zhang, M.X.; Yu, J.Y.; Jin, Y.; Ling, B.; Du, J.P.; Li, G.H.; Qin, Q.M.; Cai, Q.N. Glutathione S-transferase of brown planthoppers (*Nilaparvata lugens*) is essential for their adaptation to gramine-containing host plants. *PLoS ONE* **2013**, *8*, e64026. [CrossRef]
67. Tao, X.Y.; Xue, X.Y.; Huang, Y.P.; Chen, X.Y.; Mao, Y.B. Gossypol-enhanced P450 gene pool contributes to cotton bollworm tolerance to a pyrethroid insecticide. *Mol. Ecol.* **2012**, *21*, 4371–4385. [CrossRef]
68. Zou, X.P.; Xu, Z.B.; Zou, H.W.; Liu, J.S.; Chen, S.N.; Feng, Q.L.; Zheng, S.C. Glutathione S-transferase SlGSTE1 in *Spodoptera litura* may be associated with feeding adaptation of host plants. *Insect Biochem. Mol. Biol.* **2016**, *70*, 32–43. [CrossRef]
69. Snyder, M.J.; Glendinning, J.I. Causal connection between detoxification enzyme activity and consumption of a toxic plant compound. *J. Comp. Physiol.* **1996**, *179*, 255–261. [CrossRef]
70. Xiang, Y.Y.; Niu, H.H.; Jin, B.L.; Zhang, Y.C.; Yin, P.F. Effects of Honeysuckle Varieties on Protective and Detoxifying Enzyme Activities in *Heterolocha Jinyinhuaphaga* Chu (Lepidoptera: Geometridae) Larvae. *J. Chem. Ecol.* **2023**, *49*, 205–213. [CrossRef]
71. Wang, S.Y.; Zhu, Q.Z.; Tan, Y.T.; Ma, Q.L.; Wang, R.F.; Zhang, M.F.; Xu, H.H.; Zhang, Z.X. Artificial diets and rearing technique of *Spodoptera frugiperda* (JE Smith) in laboratory. *J. Environ. Entomol.* **2019**, *41*, 742–747.
72. Nagoshi, R.N.; Koffi, D.; Agboka, K.; Tounou, K.A.; Banerjee, R.; Jurat-Fuentes, J.L.; Meagher, R.L. Comparative molecular analyses of invasive fall armyworm in Togo reveal strong similarities to populations from the eastern United States and the Greater Antilles. *PLoS ONE* **2017**, *12*, e181982. [CrossRef]

73. Fahmi, F.; Kusumah, R.Y.M.; Buchori, D. Genetic variation of pest fall armyworm *Spodoptera frugiperda* (JE Smith) (Lepidoptera: Noctuidae) in different landscapes in Bogor. *J. Entomol. Indones.* **2023**, *20*, 1. [CrossRef]
74. Jiang, J.; Huang, B.; Wu, Q.; Li, S.; Gu, J.; Huang, L. Identification of *Spodoptera frugiperda* (Lepidoptera: Noctuidae) and its two host strains in China by PCR-RFLP. *J. Econ. Entomol.* **2023**, *116*, 983–992. [CrossRef]
75. Waldbauer, G.P. The consumption and utilization of food by insects. *Adv. Insect Physiol.* **1968**, *5*, 229–288.
76. Wang, C.Z.; Qin, J.D. Identification of major protease activities in the midgut of *Helicoverpa armigera* larvae. *Acta Entomol. Sin.* **1996**, *39*, 7–11.
77. Bernfeld, P. Amylases, α and β. *Methods Enzymol.* **1955**, *1*, 149–158.
78. Bai, Y.; Wang, W.X. Methods for determination of protease, amylase, lipase and cellulase activities in the gut of *Apostichopus japonicas*. *Feed Ind.* **2012**, *33*, 28–32.
79. Han, Y.; Yu, W.; Zhang, W.; Yang, Y.; Walsh, T.; Oakeshott, J.G.; Wu, Y. Variation in P450-mediated fenvalerate resistance levels is not correlated with CYP337B3 genotype in Chinese populations of *Helicoverpa armigera*. *Pest. Biochem. Physiol.* **2015**, *121*, 129–135. [CrossRef]
80. Van Asperen, K. A study of housefly esterases by means of a sensitive colorimetric method. *J. Insect Physiol.* **1962**, *8*, 401–416. [CrossRef]
81. Aitio, A. A simple and sensitive assay of 7-ethoxycoumarin deethylation. *Anal. Biochem.* **1978**, *85*, 488–491. [CrossRef]

plants

MDPI

Article

Ontogenetic Changes in the Feeding Behaviour of *Helicoverpa armigera* Larvae on Pigeonpea (*Cajanus cajan*) Flowers and Pods

Trevor M. Volp [1,2,*], Myron P. Zalucki [2] and Michael J. Furlong [2,*]

[1] Agri-Science Queensland, Department of Agriculture and Fisheries, Toowoomba, QLD 4350, Australia
[2] School of the Environment, The University of Queensland, St. Lucia, QLD 4072, Australia; m.zalucki@uq.edu.au
* Correspondence: trevor.volp@daf.qld.gov.au (T.M.V.); m.furlong@uq.edu.au (M.J.F.)

Abstract: Despite substantial research examining caterpillar–plant interactions, changes in the feeding behaviour of lepidopteran larvae as they develop are poorly understood. In this study, we investigated ontogenetic changes in the behaviour of *Helicoverpa armigera* larvae feeding on reproductive structures of pigeonpea (*Cajanus cajan*). Specifically, we examined the preference for and avoidance of pigeonpea flowers and pods of first, second, third, and fourth instar *H. armigera* larvae. We also conducted a no-choice assay to compare the ability of third and fourth instar larvae to penetrate pigeonpea pod walls, which act as a physical defence against herbivory. When presented with a choice between pigeonpea pods and flowers, different instars behaved differently. First and second instar larvae largely avoided pigeonpea pods, instead feeding on flowers; third instar larvae initially avoided pods, but by 24 h, did not strongly discriminate between the structures; and fourth instars demonstrated a preference for pods. When initially placed on pods, first instars were slower than other instars to leave these structures, despite pods being suboptimal feeding sites for small caterpillars. We identified a clear instar-specific ability to penetrate through the pod wall to reach the seeds. Most third instar larvae were unable to penetrate the pod wall, whereas most fourth instars succeeded. Third instars suffered a physiological cost (measured by relative growth rate) when boring through the pod wall, which was not observed in fourth instars. Our study further illuminates the insect–plant interactions of the *H. armigera*–pigeonpea system and provides evidence for the significant changes in feeding behaviour that may occur during lepidopteran larval development.

Keywords: plasticity; foraging; host-plant resistance; plant defence; herbivory; herbivore

1. Introduction

Holometabolous herbivorous insects may substantially change their feeding behaviour throughout immature development [1]. Some 17.6% of 1137 species of British Lepidoptera make a single marked change in their feeding habit during their larval stages [2]. Such changes may occur due to the quantity and quality of food available, to reduce the risk of mortality (from either lower or higher trophic levels), or because of a larger mandible/body size that enables feeding in different locations and/or on different plant parts.

Lepidopteran larvae are exposed to heterogenous environments (e.g., Figure 1) wherein they should select feeding sites that increase performance and survival, while avoiding sites that decrease performance and increase mortality risk. Whether a feeding site is optimal depends on context—what feeding sites are available—and the instar stage of the larva. As larvae develop, they may switch to feeding in different locations [3], on different types of plant structures [4–6], or even to different species of plant [7,8].

In this study, we examine the larval behaviour of *Helicoverpa armigera* (Lepidoptera: Noctuidae), a polyphagous pestiferous moth that attacks numerous plant species throughout its nearly global distribution [9–11]. The behaviour of early instar *H. armigera* larvae has been examined in detail—neonates respond to gravity and light, moving up plants and preferring to feed at plant reproductive structures [3,12,13]. As larvae develop, their feeding behaviour changes. On mungbean (*Vigna radiata*), *H. armigera* neonates spend more time searching and located at the plant apex, whereas third instars spend more time feeding and less time at the plant apex [3]. On an artificial diet, third instars move more frequently among diet cubes than neonates [14]. As *H. armigera* larvae develop into larger instars and approach pupation, their food consumption drastically increases [15].

Helicoverpa armigera is the most significant insect pest of the major legume crop pigeonpea (*Cajanus cajan*) (Figure 1) [16,17]. Moths are highly attracted to flowering pigeonpea [18,19]. Larvae may feed on pigeonpea plants throughout vegetative and reproductive plant phenological stages [13], but when provided with a choice, early instars avoid leaves and prefer to feed on flowers [18–21]. As larvae develop into larger instars (Figure 2), they 'switch' to feeding on pods [20,22].

In this study, we investigate the ontogenetic changes in the feeding behaviour of *H. armigera* larvae on reproductive structures of pigeonpea and examine the purported 'switch' in feeding behaviour of *H. armigera* from flowers to pods. Specifically, we investigate the following questions: (i) do *H. armigera* larvae of different instars prefer to feed on different plant structures, (ii) do early instar *H. armigera* larvae avoid feeding on pods, (iii) does the pod wall limit the ability of early instar larvae to feed on pigeonpea seeds, and (iv) does boring through the pod wall impose a physiological 'cost' to larvae?

Figure 1. Heterogenous environment of flowering-podding pigeonpea plants (cv. ICPL 86012), where *H. armigera* larvae must make foraging 'decisions', see [21] for details on pigeonpea reproductive structures.

Figure 2. First, second, third, and fourth instar *H. armigera* larvae on the same pigeonpea leaf. After feeding on pigeonpea flowers in the laboratory at 25 °C, it takes 6–7 days for larvae to transition from the first instar to the fourth instar. The red circle indicates the first instar larva and the scale bar = 1 cm.

2. Results

2.1. Larval Preference for Pods or Flowers

In the larval preference experiment, larval instars differed in their distribution between plant structures at 6 h ($\chi^2 = 20.44$, df = 3, $p < 0.001$) (Figure 3). By 6 h, the majority of second (93%) and third instar (90%) larvae were on flowers. However, first instars were evenly divided between flowers (56%) and pods (43%), as were fourth instars (flowers 50%; pods 43%). All larvae were located on a plant structure at 6 h, except for two fourth instars which were located on the filter paper.

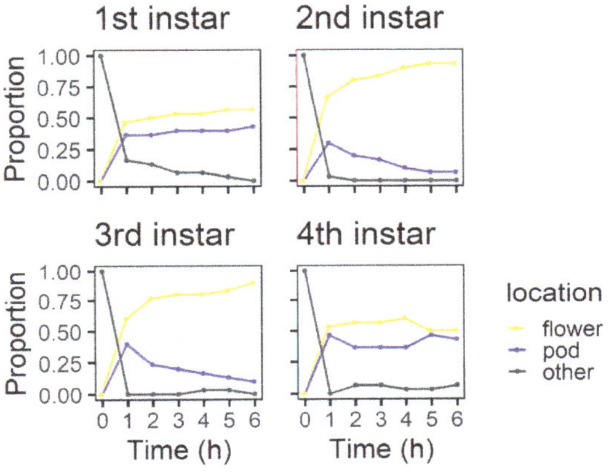

Figure 3. Proportion of larvae at different locations at each of the six hourly observations in the larval preference for pods or flowers experiment; "other" indicates larvae were on the filter paper or the internal walls of the Petri dish (n = 30 for each larval instar).

At 24 h, the larval instar again affected location ($\chi^2 = 56.04$, df = 3, $p < 0.001$) (Figure 4), with 80% of first instars and all second instars located on flowers. Third instars were evenly distributed between flowers (43%) and pods (53%), but most (83%) fourth instars were found on pods. At 24 h, only three larvae (two fourth instars and a single third instar) were located on the filter paper rather than on the plant structures.

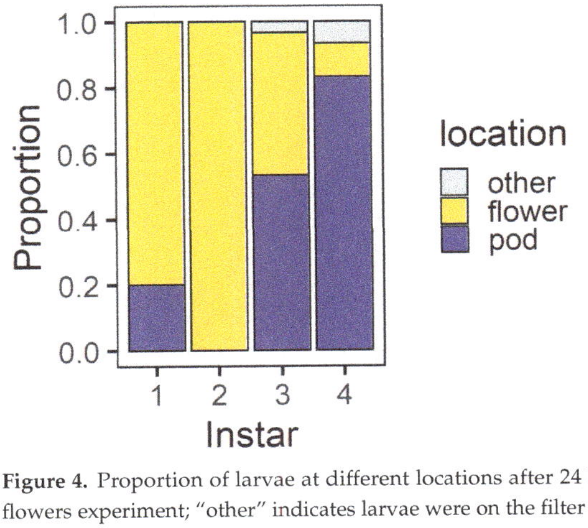

Figure 4. Proportion of larvae at different locations after 24 h in the larval preference for pods or flowers experiment; "other" indicates larvae were on the filter paper or the internal walls of the Petri dish (n = 30 for each larval instar).

Forty percent of the fourth instar larva had eaten >50% of the flower material presented. Only one third instar consumed this level of flower tissue, but none of the younger instars did. However, when these replicates were removed from the dataset, there was still a significant effect of larval instar on the final larval location ($\chi^2 = 46.96$, df = 3, $p < 0.001$).

We recorded the presence of visible feeding damage on plant structures to determine which plant structures that larvae fed on. For first instars, it was difficult to determine signs of feeding compared to the older instars; therefore, we excluded them from the analysis. Larval instar influenced whether feeding damage was recorded on flowers only, pods only, or both structures ($\chi^2 = 52.28$, df = 4, $p < 0.001$). For second instars, 83% of larvae fed only on flowers and 16% fed on both structures. For third instars, 73% of larvae fed on both structures, 16% fed on flowers only, and 10% only fed on pods. For fourth instars, 70% of larvae fed on both structures, 26% fed on pods only, and only a single larva fed on the flower only.

2.2. *Behavioural Avoidance of Pods*

In the pod avoidance experiment, larval instars differed in their distribution between plant structures at 6 h ($\chi^2 = 18.53$, df = 3, $p < 0.001$) (Figure 5). At 6 h, most first instars (80%) and most fourth instars (63%) were still on the pods, but only 46% of second instars and 26% third instars remained on the pods. At 6 h, two larvae (one third instar and one fourth instar) had left the pod and were found on the filter paper.

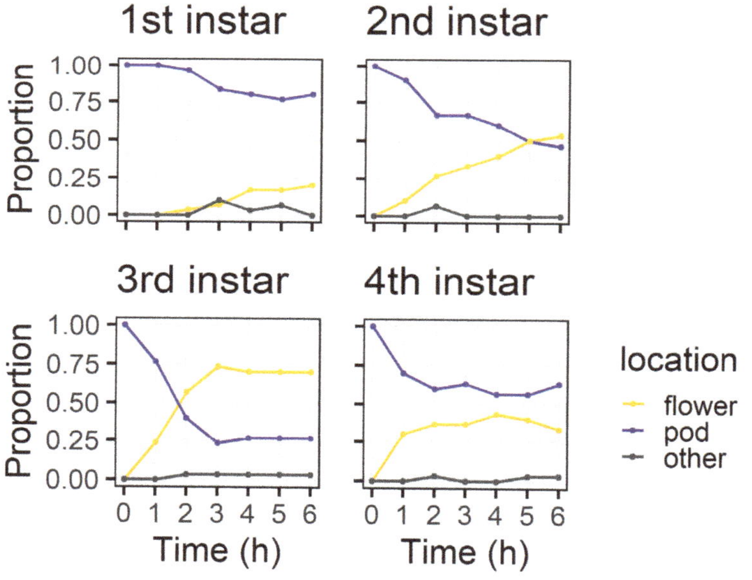

Figure 5. Proportion of larvae at different locations at each of the six hourly observations in the pod avoidance experiment; "other" indicates larvae were on the filter paper or the internal walls of the Petri dish (n = 30 for each larval instar).

At 24 h, larval location was again influenced by larval instar ($\chi^2 = 31.40$, df = 3, $p < 0.001$) (Figure 6). First instars were evenly distributed between flowers (43%) and pods (43%), with the remainder (14%) located on neither structure. Second instars showed a strong preference for flowers (80%). Third instars were evenly distributed between pods (46%) and flowers (53%), while most fourth instars (86%) were found on pods. There were also two fourth instar larvae located on neither a flower nor a pod. Thirty percent of fourth instar larvae had eaten >50% of flower tissue; when these data are excluded, the distribution of larvae as a function of instar remained significant ($\chi^2 = 22.59$, df = 3, $p < 0.001$). In this experiment, there was a single dead larva—a first instar that was dead in the pod exudate.

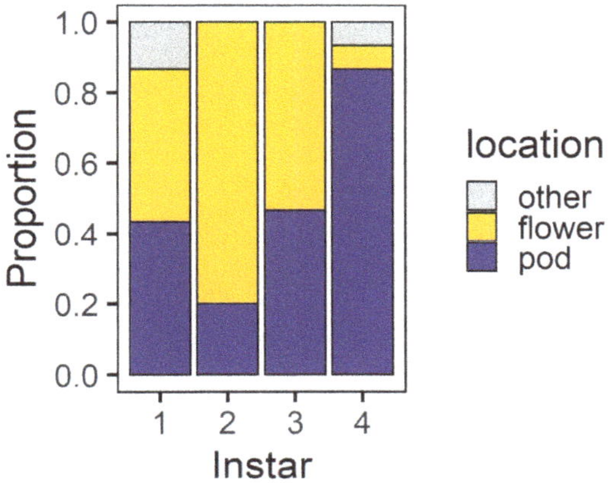

Figure 6. Proportion of larvae at different locations after 24 h in the pod avoidance experiment. Other includes larvae on the filter paper or the internal walls of the Petri dish.

As in the first experiment, we recorded the presence of visible feeding damage on plant structures and excluded first instars due to the difficulty of detecting their feeding damage. Larval instar influenced whether feeding damage was recorded on flowers only, pods only, or both structures ($\chi^2 = 22.84$, df = 4, $p < 0.001$). For second instars, most larvae fed on flowers only (50%), followed by both flowers and pods (27%), and then pods only (23%). For third instars, most larvae fed on both flowers and pods (60%), followed by pods only (20%) and flowers only (20%). For fourth instars, more larvae fed on both flowers and pods (70%) than pods only (30%).

In the pod avoidance experiment, larval instar affected the probability that a larva penetrated the pod through to the seed ($\chi^2 = 78.9$, df = 3, $p < 0.001$) (Figure 7). Most fourth instars (96%) but only a minority of third (30%) and second instars (6%) penetrated through to the seed. No first instars penetrated the pod through to the seed in this assay.

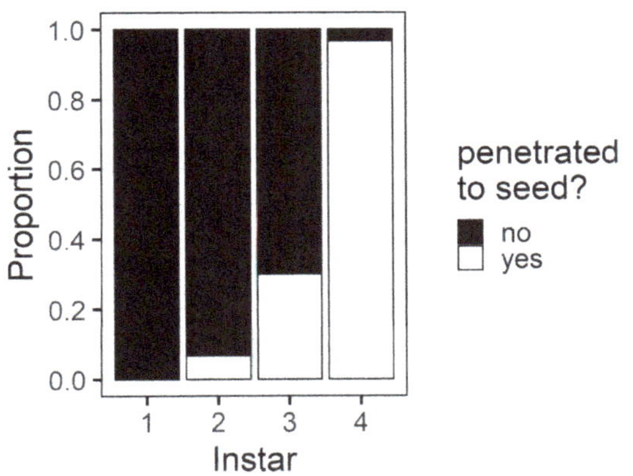

Figure 7. Proportion of larvae of different instars that penetrated the pod wall through to the seed within 24 h in the pod avoidance experiment. The white bars indicate the proportion of larvae that did penetrate through to the seed, and the black bars indicate the proportion of larvae that did not penetrate through to the seed.

2.3. Pod Feeding of Older Instars

Larval instar and pod treatment influenced whether larvae fed on pigeonpea seeds ($\chi^2 = 63.69$, df = 3, $p < 0.001$) (Figure 8). All larvae in the open pod treatment fed on seeds, but in the intact pod treatment, only 26% of third instars chewed their way through to the seed, compared with 83% of fourth instars. This result was reflected in the counts of the different types of holes on pods. When analysing the intact pod treatment only, the larval instar influenced the number of wall holes ($F_{1,58} = 10.92$, $p = 0.0016$) and seed holes ($F_{1,58} = 28.06$, $p < 0.001$) (Figure 9). Third instar larvae created more wall holes but fewer seed holes than fourth instars. Again, when only analysing the intact treatment, fourth instar larvae created larger seed holes than third instars ($F_{1,30} = 26.88$, $p < 0.001$; Figure S1), but wall hole size did not differ between instars ($F_{1,37} = 0.037$, $p = 0.85$; Figure S1).

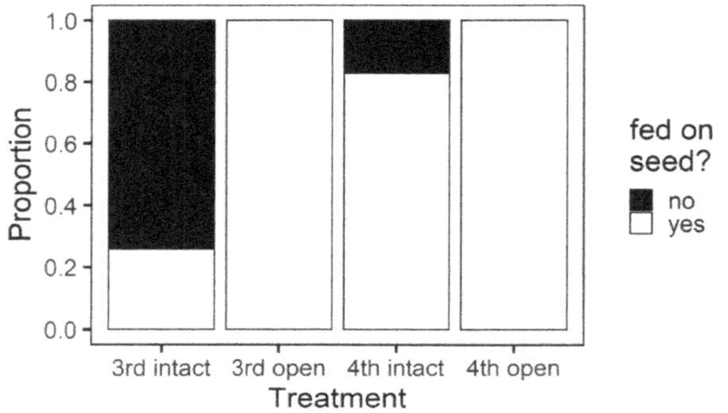

Figure 8. Proportion of larvae of different instars that fed on pigeonpea seeds during the older instar pod feeding experiment. All larvae in the open treatment fed on the seed as they did not have to bore through the pod wall. However, larvae in the intact treatment had to chew through the pod wall to feed on seeds.

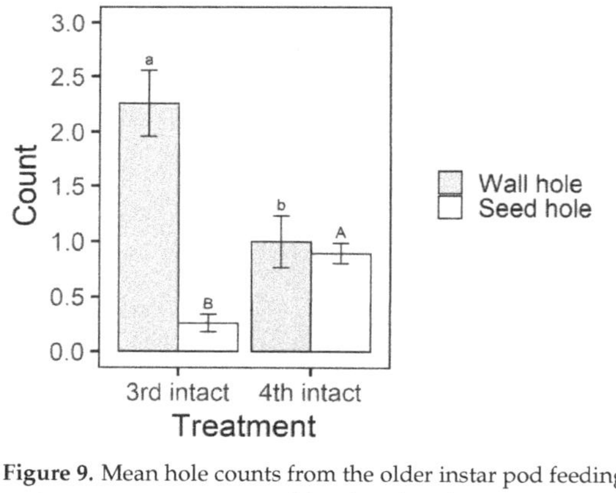

Figure 9. Mean hole counts from the older instar pod feeding experiment. Only data from the intact pod treatment are presented/analysed, as larvae in the open treatment had ready access to seeds. "Wall hole" indicates a larva has fed on the pod wall but not penetrated through to the seed and "seed hole" indicates that the larva has penetrated through to the seed. Bars are the means, and error bars are standard errors. Different letters indicate a significant difference between instars for the same hole type (wall hole = lower case letters and seed hole = upper case) according to Fisher's LSD test.

When analysing only the intact treatment larvae (i.e., those which had to chew through the pod wall to feed on seed), the initial larval weight affected the probability that third instars bored through the pod wall to feed on seeds (Z = 2.82, df = 30, $p = 0.0047$), but the larval weight did not affect the probability that fourth instars were able to access the seed (Z = 1.53, df = 28, $p = 0.13$) (Figure 10).

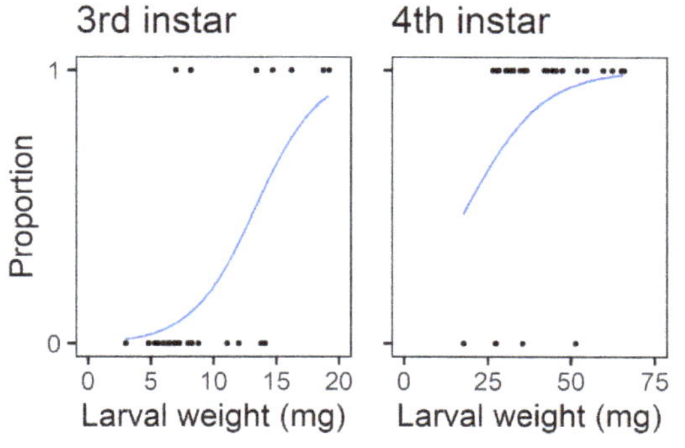

Figure 10. Logistic regressions for feeding on seed in the older instar pod feeding experiment, only for larvae in the control treatment where they had to chew through the pod wall to feed on seed. The y-axis represents success/failure whether a larva managed to feed on seed, 1 = fed on seed and 0 = did not feed on seed. The x-axis is the larval weight at the start of the experiment; note the different scale between third and fourth instars on the x-axis.

Larval RGR was not affected by instar ($F_{1,116} = 1.99$, $p = 0.16$), but it was affected by whether pods were intact or open ($F_{1,116} = 34.15$, $p < 0.001$), and there was also a significant interaction effect between larval instar and pod treatment ($F_{1,116} = 26.37$, $p < 0.001$).

We investigated if there was an instar-specific cost associated with larvae boring through the pod wall by separating larvae from the intact treatment into two groups—those that penetrated through to the seed (labelled 'P' in Figure 11) and those that did not (labelled 'NP' in Figure 11). Larval RGR was not affected by instar ($F_{1,114} = 2.44$, $p = 0.121$), but it was significantly affected by pod treatment ($F_{1,114} = 47.18$, $p < 0.001$) and there was a significant interaction effect between larval instar and pod treatment ($F_{2,114} = 4.05$, $p = 0.02$). The RGR of third instar larvae that had to bore through the pod wall (3rd intact *p*) was significantly lower that the RGR of those which did not (3rd open) (Fishers LSD, $p < 0.05$), but there was no difference between the larval RGR of fourth instars that had to bore through the pod wall (4th intact P) and those which did not (4th open) (Figure 11).

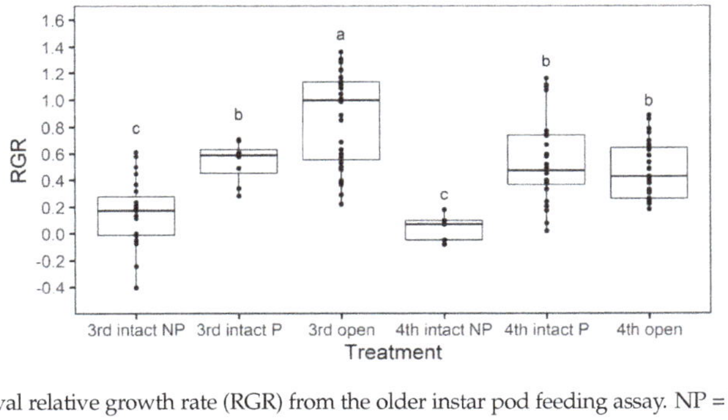

Figure 11. Larval relative growth rate (RGR) from the older instar pod feeding assay. NP = larvae did not penetrate the pod through to the seed, whereas P = larvae penetrated though the pod through to the seed. The box indicates the interquartile range, the horizontal line is the median, and individual data points are shown with black dots. Different letters indicate a significant difference according to Fisher's LSD test. Untransformed values are presented; however, square-root transformed values were used for analysis.

3. Discussion

In this study, we examined ontogenetic changes in the feeding behaviour of *H. armigera* larvae feeding on pigeonpea flowers and pods. Using simple laboratory assays, we identified clear differences among larval instars in their preference for, avoidance of, and ability to feed on different pigeonpea reproductive structures. When offered a choice of feeding sites; early instars preferred to feed on flowers and older instars preferred to feed on pods. Smaller instars are limited in their ability to penetrate though pigeonpea pod walls, preventing them from feeding on pigeonpea seeds. Finally, the process of boring through the pod wall imposes a physiological cost on third instar larvae, but not on fourth instars.

In our larval preference experiment, we found that, over a short time frame (6 h), second and third instar larvae overwhelmingly selected flowers as their feeding sites, whereas first and fourth instars were more evenly distributed between the plant parts offered (Figure 3). It was surprising that first instars were evenly divided between pods and flowers at 6 h, as first instars do not survive well on large pigeonpea pods compared to flowers [19]. However, after 24 h, most first and second instars were found on flowers (Figure 4), whereas third instars were evenly distributed between pods and flowers and most fourth instars were found on pods.

The objective of our pod avoidance experiment was to investigate what would happen if a larva was initially located on a pod (a suboptimal feeding site for smaller instars). First instars were slow to leave the pods compared to older instars (Figure 5). Our results from both assays indicate that second and third instar larvae are more mobile than first instars, likely due to their larger body size enabling a faster crawling speed. Therefore, larger instars may correct for 'mistakes' (i.e., choosing a suboptimal feeding site) more quickly than first instars. Pod wall trichomes may also slow the ability of first instars that initially select a pod (or are placed on a pod) to relocate to flowers.

The pod avoidance assay revealed the potential consequences of first instar larvae making an initial 'poor' selection of feeding site. At 24 h, only 43% of first instars had relocated to flowers (compared to 80% in the preference assay), and none of the first instars penetrated the pod wall through to the seed (Figure 7). Despite so many first instars remaining on the pods at 24 h, only a single first instar died in this assay after being trapped in pod exudate. First instar larvae may simply require longer than 24 h to relocate by crawling to a more suitable site. It is worth noting that the Petri dish assays prevented first instars dispersing by silking, which is an important dispersal mode for small larvae [23]. Identifying the mechanisms by which larvae of different instars disperse from poor feeding sites (crawling vs. silking) warrants further investigation.

Taken together, our Petri dish assay results indicate that first and second instars prefer feeding on flowers and tend to avoid pods, third instars are more evenly distributed between the structures, and fourth instar larvae prefer to feed on pods. This indicates that the third instar is the developmental stage at which *H. armigera* larvae may 'switch' to feeding on large filling pods.

The walls of pigeonpea pods function as a physical defence that prevents smaller instars from feeding on pigeonpea seeds (Figures 7 and 8). When restricted to pods in a no-choice assay, most fourth instars penetrated the pod through to the seed, but third instars struggled (Figure 8). Third instars repeatedly attempted to penetrate the pod wall, as evidenced by the larger number of 'wall holes' (Figure 9). Third instars also suffered a physiological cost of pod boring (evidenced by a reduced larval RGR) not documented for fourth instar larvae (Figure 11). This cost is likely due to the increased time and physical activity required for third instars to make their way through the pod wall. We suspect that a larger mandible size enables fourth instars to penetrate the pod wall more easily, resulting in the larger seed holes made by fourth instars (Figure S1). However, larval weight may also play a role (Figure 10), with a larger body mass potentially providing larvae with more energy reserves to persevere in chewing to penetrate the pod wall.

Ultimately, our experiments placed *H. armigera* larvae in artificial conditions to simplify the context in which they make foraging decisions. On whole pigeonpea plants, the foraging environment is more heterogenous (Figure 1) and it is important we place our results into the context of our knowledge of the *H. armigera*–pigeonpea system. As we have shown elsewhere [19,21], *H. armigera* moths are highly attracted to flowering pigeonpea plants and lay most of their eggs on flowers. Although eggs may be laid at other crop stages, floral structures are important for the establishment and development of early instar larvae, which preferentially feed at these sites. As larvae develop, flowering pigeonpea plants develop contemporaneously. Other studies have indicated that larvae 'switch' feeding from flowers to pods [20,22]. However, rather than a 'switch', this phenomenon might be better thought of as an increased capacity of *H. armigera* larvae to feed on a range of structures as they develop. Fourth instar larvae will still feed on flowers (Figures 3 and 5), although in our assays, after 24 h, they did prefer pods (Figures 4 and 6). Future work investigating the foraging of *H. armigera* larvae of different instars on whole pigeonpea plants would likely prove useful.

In a set of simple laboratory assays, we have demonstrated large changes between instars in the feeding behaviour of *H. armigera* larvae. In our experimental setup, it takes 6–7 days at 25 °C for a moderately susceptible *H. armigera* neonate to develop into a fourth instar larva that can feed on large pigeonpea pods 'cost-free' and cause seed damage which plants are likely unable to compensate for. Development from the third to fourth instar drastically changes the ability of larvae to feed inside large-filling pigeonpea pods. The challenge for pest management researchers, therefore, is to design strategies that increase the mortality of the cryptic small instar larvae before they quickly reach the damaging large instars (which are also more tolerant to insecticides and biopesticides).

Ontogenetic changes in the ecology of lepidopteran larvae have been underexplored due to understandable experimental difficulties. However, studying how ontogeny influences interactions among caterpillars and other trophic levels is a major frontier in lepidopteran ecology [24]. The interactions between larval instar, plant factors (phenology, structure availability, constitutive and induced defences), natural enemies (predators, parasitoids, and pathogens), and larval nutrient regulation are frighteningly complex. We refrain from speculating what methods will be more useful moving forward—the current reductionist approach disentangling the various factors to examine their effect (e.g., this study), or perhaps novel approaches using high-throughput data collection (e.g., videography, time lapse photography, etc.). However, if we desire to understand larval behaviour as a function of ontogeny, we must evaluate the best approach to generate and test hypotheses that include not only plant traits, but also nutritional geometry and natural enemies.

4. Materials and Methods

4.1. Plants

To obtain the relevant reproductive structures (flowers and large filling pods) for our assays, we grew pigeonpea plants of a short duration, determinate pigeonpea cultivar (ICPL 86012), in a controlled-temperature glasshouse (27 °C day, 25 °C night). Seeds were planted in 200 mm (4 L) ANOVA^{TM} pots using a 2:1 mix of commercial potting mix (Premium Potting Mix, Searles, Kilcoy, Australia) and sand. Plants were watered regularly, as required, and no additional fertiliser was provided. Plants were not treated with any insecticides, and any glasshouse pests were physically removed upon detection. Under these growing conditions, ICPL 86012 plants reached flowering in approximately 8 weeks. For our assays, we sourced flowers from plants that were 8–10 weeks old and large filling pods from plants that were 10–12 weeks old.

4.2. Insects

Helicoverpa armigera moths and larvae were obtained from a laboratory culture maintained at the Queensland Department of Agriculture and Fisheries laboratory in Toowoomba, Australia. The culture was established from insects collected from various field crops from south-east Queensland, Australia, in 2020 and the colony was regularly supplemented with field-collected insects to minimise inbreeding. Moths were kept in 5 litre plastic buckets and supplied with 10% sucrose solution using a cotton wick in a 70 mL plastic container. An 18 cm hole was cut in the bucket lid and the edges of the lid were used to secure nappy liner (bamboo rayon), which was used as an oviposition substrate. Eggs were removed daily, washed in 1% sodium hypochlorite solution, and collected onto filter paper using vacuum filtration. The filter paper was allowed to air dry, then placed in 90 mm Petri dishes with the edges sealed with parafilm until the neonates hatched. Upon hatching, neonate larvae were placed in groups onto a soybean-flour-based artificial diet (recipe modified from [25], ingredients provided in [21]) in 500 mL rectangular plastic containers. When larvae developed to the third instar, they were transferred to fresh diet in 32-well plastic trays, where they remained until pupation. Pupae were washed in 1% sodium hypochlorite, air-dried, and placed in 500 mL containers until eclosion. The colony and all experiments were maintained in a controlled temperature room (25 ± 2 °C, 12:12 L:D).

The larvae used in behavioural assays were fed on pigeonpea flowers, mimicking the natural feeding progression of larvae, to prevent any confounding effects of switching larvae from an artificial diet to plant parts [26]. Neonates were individually placed in a 90 mm Petri dish containing filter paper moistened with distilled water and provided with a single pigeonpea flower. To obtain larvae of different instars, old flowers were replaced with fresh flowers at days 4, 6, 7, and 8. Using this method, larvae typically reached the second instar in 2–3 days, third instar in 4–5 days, and fourth instar in 6–7 days. We established different larval cohorts on consecutive days to provide us with different instars to compare simultaneously in our assays. Larvae in the first instar (<4 h old) treatment group were not fed before their use in assays and all larger larvae were starved for 4 h before they were used in assays.

4.3. *Larval Preference for Pods or Flowers*

In the first experiment, we examined preference for flowers versus pods for first, second, third, and fourth instar *H. armigera* larvae. The experiment was conducted in 90 mm Petri dishes lined with water-moistened filter paper. In each dish, we placed one pod (large filling stage) and one flower on either side of the Petri dish (separated by approximately 2 cm). We randomised which side of the Petri dish either structure was placed. We then placed a single larva on the centre of filter paper, equidistant from either plant structure. Petri dishes were sealed with parafilm and placed in the controlled-temperature room. We recorded the location of larvae every hour for the first 6 h, and then again at 24 h. At 24 h, we recorded the presence of visible feeding damage on either plant structure. Thirty replicates were conducted for each instar.

4.4. *Behavioural Avoidance of Pods*

In the second experiment, we examined if different instar *H. armigera* larvae (first, second, third, and fourth instars) avoid feeding on pods in a Petri dish assay. Assays were conducted in a similar manner to the preference assay, except at the start of the assay, where we placed all larvae on pods (mimicking an initial feeding choice of pod). We monitored the location of larvae hourly for the first 6 h, and then again at 24 h. At 24 h, we recorded whether there was visible damage to either plant structure and if larvae had penetrated the pod through to the seed. Thirty replicates were conducted for each instar.

4.5. *Pod Feeding of Older Instars*

In the third experiment, guided by our results from the first two assays, we examined the pod feeding behaviour of third and fourth instar larvae. Specifically, we examined (i) if third and fourth instar larvae were able to feed inside pods under no-choice conditions, (ii) if the pod wall presents a barrier to larvae feeding on seeds, and (iii) if boring through the pod wall imposes a physiological cost on larvae.

Third and fourth instar *H. armigera* larvae were divided into two pod treatments. In the control treatment ('intact'), larvae were provided access to a single large filling pigeonpea pod. In the second treatment ('open'), pigeonpea pods were sliced along their ventral seam using a scalpel blade. The seam was then prised open slightly to provide larvae with access to all seeds within the pod without having to chew through the pod wall. Larvae and pods were placed inside 50 mL centrifuge tubes which were positioned vertically to allow larvae to access the entire surface of the pods. As in previous assays, larvae were starved for 4 h before placement into the tubes. After the starvation period, larvae were weighed on a microbalance (HR-250AZ; A&D, Tokyo, Japan) to obtain initial weights. After 24 h, larvae were removed from the tubes and re-weighed. Larval relative growth rates (RGRs) were calculated using Equation (1):

$$RGR = \ln(wt1) - \ln(wt0) \quad (1)$$

where $wt0$ is the initial larval weight after 4 h of starvation and $wt1$ is the larval weight after the 24 h assay. At 24 h, we recorded the number and diameter of holes on the pods. Holes were defined as either 'wall holes', where larvae had fed on the pod wall but not penetrated through to the seed, or 'seed holes', where larvae had fed through the wall and into the seed. For the hole diameter, we took two linear measurements with Vernier calipers (Protech, Shenzhen, China)—one measurement at 90° to the ventral pod edge and the second at 90° to the first measurement. In this experiment, the third instar treatments were replicated 31 times each and the fourth instar treatments were replicated 29 times each.

4.6. Statistical Analyses

We compared the location of larvae and evidence of feeding damage in our Petri dish choice assays using chi-square tests. We compared the pod hole counts and larval relative growth rates from the third experiment using ANOVAs, and post hoc comparisons were conducted using Fisher's LSD test. We square-root transformed (1 + sqrt(x)) our RGR data to meet ANOVA assumptions. We conducted logistic regression to examine how the larval weight influenced the probability of penetrating the pod walls. All analyses were performed in R version 3.6.2 [27]; for Fisher's LSD tests, we used the package 'agricolae' [28]; and graphs were made with the package 'ggplot2' [29].

Supplementary Materials: The following supporting information can be downloaded at: https://www.mdpi.com/article/10.3390/plants13050696/s1, Figure S1: Mean hole diameter from the older instar pod feeding experiment. Only data from the intact pod treatment are presented/analysed, as larvae in the 'open' treatment had ready access to seeds. "Wall hole" indicates a larva has fed on the pod wall but not penetrated through to seed, and "seed hole" indicates when the larva has penetrated through to the seed. Bars are the means and error bars are standard errors. Different letters indicate a significant difference between instars for the same hole type according to Fisher's LSD test.

Author Contributions: Conceptualisation, T.M.V., M.P.Z. and M.J.F.; methodology, T.M.V., M.P.Z. and M.J.F.; investigation, T.M.V.; resources, T.M.V.; data curation, T.M.V.; writing—original draft preparation, T.M.V.; writing—review and editing, T.M.V., M.P.Z. and M.J.F.; funding acquisition, T.M.V. All authors have read and agreed to the published version of the manuscript.

Funding: This research was funded by the Queensland Department of Agriculture and Fisheries' 'Pigeonpea Initiative' project.

Data Availability Statement: Data used in this study are available from the corresponding authors upon request.

Acknowledgments: We are grateful to Robyn Nobbs and Chloe Hockey who provided technical support.

Conflicts of Interest: The authors declare no conflict of interest.

References

1. Hochuli, D.F. Insect herbivory and ontogeny: How do growth and development influence feeding behaviour, morphology and host use? *Austral Ecol.* **2001**, *26*, 563–570. [CrossRef]

2. Gaston, K.J.; Reavey, D.; Valladares, G.R. Changes in feeding habit as caterpillars grow. *Ecol. Entomol.* **1991**, *16*, 339–344. [CrossRef]
3. Johnson, M.L.; Zalucki, M.P. Feeding and foraging behaviour of a generalist caterpillar: Are third instars just bigger versions of firsts? *Bull. Entomol. Res.* **2007**, *97*, 81–88. [CrossRef] [PubMed]
4. Suits, R.; Reisig, D.; Burrack, H. Feeding preference and performance of *Helicoverpa zea* (Lepidoptera: Noctuidae) larvae on various soybean tissue types. *Fla. Entomol.* **2017**, *100*, 162–167. [CrossRef]
5. Eckel, C.S.; Terry, L.I.; Bradley Jr, J.; Van Duyn, J.W. Changes in within-plant distribution of *Helicoverpa zea* (Boddie) (Lepidoptera: Noctuidae) on soybeans. *Environ. Entomol.* **1992**, *21*, 287–293. [CrossRef]
6. Smallegange, R.C.; Van Loon, J.J.A.; Blatt, S.E.; Harvey, J.A.; Agerbirk, N.; Dicke, M. Flower vs. Leaf Feeding by *Pieris brassicae*: Glucosinolate-Rich Flower Tissues are Preferred and Sustain Higher Growth Rate. *J. Chem. Ecol.* **2007**, *33*, 1831–1844. [CrossRef]
7. Mitra, S.; Firake, D.; Umesh, K.; Pandey, P.P.; Pandit, S. Polyphagous caterpillars of *Spodoptera litura* switch from a trap crop to the main crop, improve fitness, and shorten generation time. *J. Pest. Sci.* **2021**, *94*, 1091–1103. [CrossRef]
8. Dosdall, L. Do larvae of bertha armyworm, *Mamestra configurata* (Walker) (Lepidoptera: Noctuidae), express induced feeding responses? *J. Appl. Entomol.* **2010**, *134*, 531–541. [CrossRef]
9. Tay, W.T.; Soria, M.F.; Walsh, T.; Thomazoni, D.; Silvie, P.; Behere, G.T.; Anderson, C.; Downes, S. A brave new world for an old world pest: *Helicoverpa armigera* (Lepidoptera: Noctuidae) in Brazil. *PLoS ONE* **2013**, *8*, e80134. [CrossRef]
10. Zalucki, M.P.; Daglish, G.; Firempong, S.; Twine, P. The biology and ecology of *Heliothis armigera* (Hubner) and *Heliothis punctigera* Wallengren (Lepidoptera, Noctuidae) in Australia-What do we know. *Aust. J. Zool.* **1986**, *34*, 779–814. [CrossRef]
11. Fitt, G.P. The ecology of *Heliothis* species in relation to agroecosystems. *Annu. Rev. Entomol.* **1989**, *34*, 17–52. [CrossRef]
12. Perkins, L.E.; Cribb, B.W.; Hanan, J.; Glaze, E.; Beveridge, C.; Zalucki, M.P. Where to from here? The mechanisms enabling the movement of first instar caterpillars on whole plants using *Helicoverpa armigera* (Hübner). *Arthropod-Plant Interact.* **2008**, *2*, 197. [CrossRef]
13. Johnson, M.L.; Zalucki, M. Foraging behaviour of *Helicoverpa armigera* first instar larvae on crop plants of different developmental stages. *J. Appl. Entomol.* **2005**, *129*, 239–245. [CrossRef]
14. Wang, P.; Furlong, M.J.; Walsh, T.K.; Zalucki, M.P. Moving to keep fit: Feeding behavior and movement of *Helicoverpa armigera* (Lepidoptera: Noctuidae) on artificial diet with different protein: Carbohydrate ratios. *J. Insect. Sci.* **2019**, *19*, 20. [CrossRef]
15. Browne, L.B.; Raubenheimer, D. Ontogenetic changes in the rate of ingestion and estimates of food consumption in fourth and fifth instar *Helicoverpa armigera* caterpillars. *J. Insect Physiol.* **2003**, *49*, 63–71. [CrossRef]
16. Mula, M.G.; Saxena, K.B. *Lifting the Level of Awareness on Pigeonpea-a Global Perspective*; International Crops Research Institute for the Semi-Arid Tropics: Patancheru, India, 2010; 540p.
17. Shanower, T.; Romeis, J.; Minja, E. Insect pests of pigeonpea and their management. *Annu. Rev. Entomol.* **1999**, *44*, 77–96. [CrossRef]
18. Rajapakse, C.; Walter, G. Polyphagy and primary host plants: Oviposition preference versus larval performance in the lepidopteran pest *Helicoverpa armigera*. *Arthropod-Plant Interact.* **2007**, *1*, 17–26. [CrossRef]
19. Volp, T.M.; Zalucki, M.P.; Furlong, M.J. Pigeonpea (*Cajanus cajan*) crop stage strongly influences plant susceptibility to *Helicoverpa armigera* (Lepidoptera: Noctuidae). *J. Econ. Entomol.* **2024**, in press.
20. Green, P.; Stevenson, P.; Simmonds, M.; Sharma, H. Can larvae of the pod-borer, *Helicoverpa armigera* (Lepidoptera: Noctuidae), select between wild and cultivated pigeonpea *Cajanus sp.* (Fabaceae)? *Bull. Entomol. Res.* **2002**, *92*, 45–51. [CrossRef] [PubMed]
21. Volp, T.M.; Zalucki, M.P.; Furlong, M.J. *Helicoverpa armigera* preference and performance on three cultivars of short-duration pigeonpea (*Cajanus cajan*): The importance of whole plant assays. *Pest Manag. Sci.* **2023**, *79*, 627–637. [CrossRef] [PubMed]
22. Rajapakse, C.N.K. Host Plant Interactions of a Polyphagous Herbivorous Pest, Helicoverpa armigera, in Relation to a Primary Host Species. Ph. D. Thesis, The University of Queensland, Brisbane, Australia, 2007.
23. Zalucki, M.P.; Clarke, A.R.; Malcolm, S.B. Ecology and behavior of first instar larval Lepidoptera. *Annu. Rev. Entomol.* **2002**, *47*, 361–393. [CrossRef] [PubMed]
24. Marquis, R.J.; Koptur, S. (Eds.) Synopsis and the Future of Caterpillar Research. In *Caterpillars in the Middle: Tritrophic Interactions in a Changing World*; Springer Nature: Cham, Switzerland, 2022; pp. 609–622.
25. Teakle, R.; Jensen, J. Heliothis punctiger. In *Handbook of Insect Rearing*; Singh, P., Moore, R.F., Eds.; Elsevier: Amsterdam, The Netherlands, 1985; pp. 313–322.
26. Ennis, D.; Mader, B.; Burnside, K.; Bauce, E.; Despland, E. Is Feeding Behaviour on Foliage Affected by lab-Rearing on Artificial Diet? *J. Insect. Behav.* **2015**, *28*, 147–156. [CrossRef]
27. R Core Team. *R: A language and Environment for Statistical Computing*; R Core Team: Vienna, Austria, 2019.
28. De Mendiburu, F. *Package 'agricolae'. R Package, Version 2019*; R Core Team: Vienna, Austria, 2019.
29. Wickham, H.; Chang, W.; Wickham, M.H. R Package 'ggplot2'. *Version* **2016**, *2*, 1–189.

plants

MDPI

Article

Interactions of *Opuntia ficus-indica* with *Dactylopius coccus* and *D. opuntiae* (Hemiptera: Dactylopiidae) through the Study of Their Volatile Compounds

Esteban Rodríguez-Leyva [1], Esperanza García-Pascual [2], Marco M. González-Chávez [3,*], Santiago de J. Méndez-Gallegos [2,*], Juan A. Morales-Rueda [4], Juan C. Posadas-Hurtado [3], Ángel Bravo-Vinaja [2] and Avelina Franco-Vega [3]

1 Colegio de Postgraduados, Campus Montecillo, Texcoco C.P. 56264, Mexico; esteban@colpos.mx
2 Colegio de Postgraduados, Campus San Luis Potosí, Salinas de Hidalgo, San Luis Potosi C.P. 78622, Mexico; forestal.egp@gmail.com (E.G.-P.); abravo@colpos.mx (Á.B.-V.)
3 Facultad de Ciencias Químicas, Universidad Autónoma de San Luis Potosí, San Luis Potosi C.P. 78210, Mexico; j_c247@hotmail.com (J.C.P.-H.); avelina.franco@uaslp.mx (A.F.-V.)
4 Viscoelabs, Materials Research Center, Librado Rivera 390, San Luis Potosi C.P. 78200, Mexico; ja.morales@viscoelabs.com
* Correspondence: gcmm@uaslp.mx (M.M.G.-C.); jmendez@colpos.mx (S.d.J.M.-G.)

Abstract: *Opuntia ficus-indica* has always interacted with many phytophagous insects; two of them are *Dactylopius coccus* and *D. opuntiae*. Fine cochineal (*D. coccus*) is produced to extract carminic acid, and *D. opuntiae*, or wild cochineal, is an invasive pest of *O. ficus-indica* in more than 20 countries around the world. Despite the economic and environmental relevance of this cactus, *D. opuntiae*, and *D. coccus*, there are few studies that have explored volatile organic compounds (VOCs) derived from the plant–insect interaction. The aim of this work was to determine the VOCs produced by *D. coccus* and *D. opuntiae* and to identify different VOCs in cladodes infested by each *Dactylopius* species. The VOCs (essential oils) were obtained by hydrodistillation and identified by GC-MS. A total of 66 VOCs from both *Dactylopius* species were identified, and 125 from the Esmeralda and Rojo Pelón cultivars infested by *D. coccus* and *D. opuntiae*, respectively, were determined. Differential VOC production due to infestation by each *Dactylopius* species was also found. Some changes in methyl salicylate, terpenes such as linalool, or the alcohol *p*-vinylguaiacol were related to *Dactylopius* feeding on the cladodes of their respective cultivars. Changes in these VOCs and their probable role in plant defense mechanisms should receive more attention because this knowledge could improve *D. coccus* rearing or its inclusion in breeding programs for *D. opuntiae* control in regions where it is a key pest of *O. ficus-indica*.

Keywords: cactus pear; phytophagous insects; terpenes; methyl salicylate; *p*-vinylguaiacol

1. Introduction

Dactylopiidae, or cochineals, is a family of scale insects that includes only the genus *Dactylopius* and 11 recognized species [1] that are endemic to North and South America [2,3]. An important characteristic of these insects is that they produce carminic acid, probably as a defense mechanism against predation [4–6]. All the species of the genus are considered obligate parasites of Cactacea with high host specificity, particularly for the genera *Nopalea* Salm-Dyck and *Opuntia* Miller [7].

Because of the high carminic acid concentration (~20–25%) of *Dactylopius coccus* Costa, the true cochineal, it is the only species of commercial interest for production. It is reared on *Opuntia ficus-indica* (L.) Miller, the cactus pear. Carminic acid is recognized as a natural dye with cosmetic, food, pharmaceutical, textile, and plastic applications [8]. In addition, it is currently used in biomedicine [9] and as a photosynthesizing pigment in solar cells [10].

In contrast, *Dactylopius opuntiae* Cockerell, or wild cochineal, whose carminic acid content is less than 5%, is not considered useful for obtaining this substance. Rather, it is considered the key pest of *O. ficus-indica* in commercial plantations in Mexico [11,12], where plants and insects are native [7,13]. Additionally, *D. opuntiae* is an invasive pest in at least 20 countries in America, Europe, Africa, and Asia [14–16], where *O. ficus-indica* was adopted or naturalized and became one of the most important cultivated cactus species in the world because of its economic, environmental, and ecological benefits [13,14,17,18].

From a scientific perspective, most *D. coccus* research has focused on the basic biology of the species and the quest to understand the mechanisms of carminic acid production and its possible physiological or ecological functions [4,19,20]. On the other hand, research on *D. opuntiae* has focused on control tactics because it is a key pest of *O. ficus-indica* [14,15,21,22]. The different cultivars of *O. ficus-indica* used as hosts of both *Dactylopius* species are likely to have particular physical and chemical characteristics, as well as volatile organic compounds (VOCs) that influence the trophic plant–insect and plant–pest–natural enemy relationship, as has been shown in other models of tritrophic interactions where volatiles cause positive or negative responses in terms of attraction and establishment of insects of the same or different species [23,24].

Volatile organic compounds (VOCs) are synthesized as products of plant metabolism, and they are emitted into the environment [25] in response to biotic complexes or abiotic stresses [23,26]. These VOCs and essential oils are released from the leaves, flowers, and fruits into the atmosphere and from the roots into the soil [27,28]. This set of volatiles, essential in the defense mechanisms of plants against herbivores or in interspecific communication [23,24,27], is called the volatilome, and its analysis is carried out by gas chromatography-mass spectrometry (GC-MS) [29]. This is a field that is continuously growing with the development of analytical and data-processing methods [30]. In this regard, some research has been carried out on VOCs of *O. ficus-indica* emanating from cladodes, flowers, fruits, and the oils of its seeds [31–36]. These studies concluded that VOC composition is a function of the geographical area, species or cultivar, plant structure, state of development, and season, among other factors. However, none of these relatively recent papers included interaction with any of the important *Dactylopius* species, nor did they relate the production of VOCs to insect infestation. To our knowledge, there is only one study that evaluated VOCs in *O. ficus-indica* cladodes uninfested and infested by *D. coccus* [37]. This study reported eight types of compounds in uninfested cladodes and nine in cladodes infested by the insect. Furthermore, no other work is known to have explored VOCs of either *Dactylopius* species.

Because plant VOCs play an important role in interactions between insects and other organisms, e.g., pathogens or predators, and parasitoids [23,24,38], as well as in the plant's response to insect attacks [39], the objectives of this work were (1) to determine the VOCs of *D. coccus* and *D. opuntiae* feeding on *O. ficus-indica* and (2) to establish the changes in the composition and proportion of VOCs in cladodes of *O. ficus-indica* uninfested and infested with *Dactylopius*. This information could contribute to understanding the variation between cultivars of both species of insects and to exploring the potential of the biological functions that these compounds play in interspecies communication.

2. Results

Through essential oils, it was possible to recover and identify about 80% and 90% of the volatile organic compounds (VOCs) of *D. coccus* and *D. opuntiae*, respectively. The *Dactylopius* species had 20 VOCs in common. In addition, 12 and 34 VOCs were specifically produced by *D. coccus* and *D. opuntiae*, respectively (Figure 1). Thus, the volatilome of each species was 32 or 54 compounds, and the proportion of each compound varied greatly between species (Table 1). The VOCs belonged to eight chemical groups, of which three had the highest relative abundance. Carboxylic acids and their derivatives were the most important group, accounting for 59.28% and 78.29% of the VOC abundance for *D. coccus* and *D. opuntiae*, respectively. The second group was alcohols only for *D. coccus* (12.15%), and the third group was aldehydes with 5.8% and 7.68% of the relative abundance for *D. coccus* and *D. opuntiae*, respectively. The alkanes recovered were less than 2.5% for both species. The remaining four groups of recovered compounds (ether, terpenes, ketones, and alkenes) had less than 0.55% relative abundance per group (Table 1).

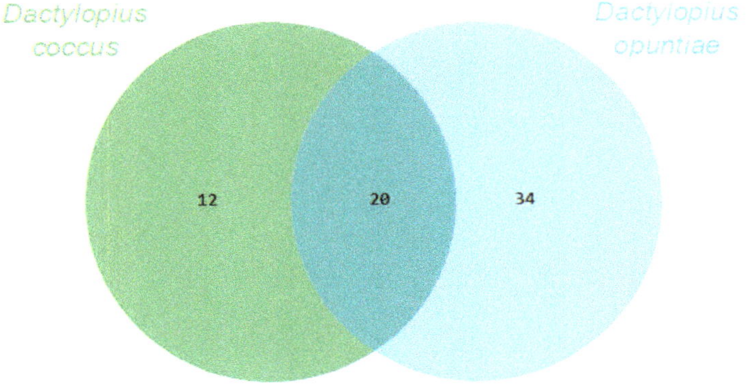

Figure 1. Comparison of the *Dactylopius* volatilomes using Venn diagrams based on the number of Volatile organic compounds (VOCs) obtained through essential oils for each *Dactylopius* species.

Table 1. Volatile organic compounds (VOCs) obtained through the essential oils of each *Dactylopius* species.

No.	Compounds	*D. coccus*		*D. opuntiae*		KI Ref
		RA (%)	KI Exp	RA (%)	KI Exp	
	Carboxylic acids and derivatives	*59.28%*		*78.29%*		
1	Hexanoic acid			1.09 ± 0.08		903
2	2-methylhexanoic acid			0.54 ± 0.01		950
3	Heptanoic acid	0.20 ± 0.05	1021	0.64 ± 0.16	1021	1005
4	2,4-dimethylhexanoic acid			0.37 ± 0.02		1015
5	2-Ethylhexanoic acid			10.90 ± 0.21	1036	1031
6	Lactic acid	5.58 ± 0.58	1061	0.86 ± 0.03	1052	1057
7	Glycolic acid			0.11 ± 0.03	1075	1072
8	2,6-dimethylheptanoic acid	0.61 ± 0.21	1087			
9	Octanoic acid	1.90 ± 0.74	1108	2.20 ± 0.24	1108	1108
10	2,3,4-Trimethylpentanoic acid			0.44 ± 0.02	1127	
11	Ethyl benzoate			0.12 ± 0.01	1153	1141
12	Ethyl octanoate			0.05 ± 1.77	1188	1175
13	Nonanoic acid	2.32 ± 0.64	1214	3.57 ± 0.37	1214	1205
14	2,4-dimethylnonanoic acid	0.31 ± 0.01	1234			
15	Benzoic acid	2.04 ± 0.00	1235	1.04 ± 0.01	1235	1232
16	Ethyl nonanoate	0.26 ± 0.03	1290			1282
17	2-Decenoic acid			0.42 ± 0.02	1310	1290
18	Decanoic acid	8.76 ± 2.53	1316	2.38 ± 0.12	1316	1309
19	Butanedioic acid			0.11 ± 0.03	1318	1314
20	(Z)-4-*tert*-butylcyclohexyl acetate	0.80 ± 0.02	1358	0.12 ± 0.03	1356	1346
21	Ethyl decanoate			0.18 ± 0.02	1388	1382
22	Undecanoic acid			0.63 ± 0.02	1414	1410
23	*cis*-5-Dodecenoic acid			3.18 ± 0.04	1504	
24	Dodecanoic acid	4.93 ± 0.90	1512			1509
25	Nonanedioic acid			0.40 ± 0.03	1535	1511
26	Ethyl dodecanoate	1.06 ± 0.07	1580			1566
27	Tridecanoic acid			0.49 ± 0.04	1606	1606
28	*p*-Hydroxybenzoic acid	0.24 ± 0.00	1615	0.05 ± 0.01	1616	1621
29	Hexyl salicylate	1.05 ± 0.01	1660	0.57 ± 0.02	1658	1684
30	(Z)-9-Tetradecenoic acid	0.42 ± 0.05	1702	0.22 ± 0.09	1707	1691

Table 1. *Cont.*

No.	Compounds	*D. coccus*		*D. opuntiae*		
		RA (%)	KI Exp	RA (%)	KI Exp	KI Ref
31	Tetradecanoic acid	21.25 ± 2.90	1717	30.15 ± 1.35	1718	1713
32	Ethyl tetradecanoate			0.40 ± 0.01	1793	1782
33	(Z)-9-Hexadecenoic acid			0.56 ± 0.12	1909	1885
34	Hexadecanoic acid	2.30 ± 1.83	1934	5.89 ± 0.42	1934	1909
35	Ethyl hexadecanoate			0.23 ± 0.01	1974	1968
36	Heptadecanoic acid			1.03 ± 0.06	2039	2009
37	(Z,Z) 9,12-Octadecadienoic acid	0.83 ± 0.73	2105	3.76 ± 0.67	2105	2087
38	(Z)-9-Octadecenoic acid	0.82 ± 1.11	2112	2.44 ± 0.33	2112	2088
39	Octadecanoic acid	1.69 ± 2.32	2139	2.58 ± 0.22	2140	2133
40	Ethyl octadecanoate	0.33 ± 0.03	2208			2181
41	Dehydroabietic acid	1.58 ± 0.00	2375	0.12 ± 0.02	2376	2385
Aldehydes		*5.80*		*7.68*		
42	Hexanal			1.79 ± 0.36		964
43	Heptanal			1.24 ± 0.11	1066	1069
44	Octanal			0.23 ± 0.07	1165	1162
45	Nonanal	0.34 ± 0.11	1235	2.57 ± 0.17	1268	1267
46	Decanal			0.28 ± 0.02	1367	1366
47	Dodecanal			0.28 ± 0.10	1663	
48	α-Hexylcinnamaldehyde	5.46 ± 0.08		0.11 ± 0.12	1719	1728
49	Heptadecanal			1.18 ± 0.33	2088	
Ether				*0.09*		
50	Benzyl methyl ether			0.09 ± 0.01		966
Terpene		0.54		0.08		
51	*p*-Cymene			0.08 ± 0.04	1018	1025
52	α-Ionone	0.36 ± 0.15	1415			1413
53	β-Ionone	0.18 ± 0.02	1472			1486
Ketones				0.54		
54	Benzophenone			0.09 ± 0.02	1600	1611
55	2-Nonadecanone			0.45 ± 0.08	2116	2087
Alcohols		*12.15*		*0.62*		
56	Phenol			0.31 ± 0.09	1045	1043
57	2-Ethylhexanol			0.22 ± 0.04	1095	
58	1-Dodecanol	4.31 ± 0.00	1559	0.09 ± 0.01	1559	1575
59	1-Tridecanol	0.42 ± 0.00	1659			1656
60	1-Tetradecanol	3.43 ± 0.00	1765			1770
61	1-Hexadecanol	3.47 ± 0.00	1977			1965
62	1-Octadecanol	0.52 ± 0.00	2175			2159
Alkene				*0.42*		
63	1-Tridecene			0.42 ± 0.00	1284	1287
Alkane		*1.80*		*2.41*		
64	Hexadecane	1.80 ± 0.00	1581	1.11 ± 0.29	1581	1600
65	Octadecane			0.25 ± 0.00	1797	1800
66	Heneicosane			1.05 ± 0.21	2309	
	Total	**79.57**		**90.13**		

RA, relative abundance; KI Exp, Kovats index experimental; KI Ref, Kovats index reference.

As mentioned above, the number and abundance of volatiles in each group of compounds also varied greatly in each *Dactylopius* species. For example, in the carboxylic acids and their derivatives, tetradecanoic acid was the most abundant in both species, but decanoic acid, lactic acid, and dodecanoic acid presented greater relative abundance in *D. coccus*. On the other hand, for *D. opuntiae*, 2-ethylhexanoic acid and cis-5-dodecenoic acid were detected only in this species in greater relative abundance. Hexadecanoic acid, (Z,Z)-9,12-octadecadienoic acid, (Z)-9-octadecenoic acid, and octadecanoic acid occurred in both species, but their abundance differed considerably between species; again, they were more abundant for *D. opuntiae* (Table 1).

The Esmeralda and Rojo Pelón cultivars had VOC production profiles that differed before and after *Dactylopius* infestation. In both cultivars, 28 VOCs were commonly produced and identified. In addition, 35 specific compounds were identified in Esmeralda and 19 in Rojo Pelón (Figure 2). After infestation by each *Dactylopius* species in the respective *O. ficus-indica* cultivar, a contrasting difference occurred between uninfested and infested cladodes of each cultivar (Table 2). The changes were not only in the number of VOCs but also in their abundance and variation. Sometimes they decreased, sometimes they increased, sometimes some VOCs were no longer detected, and of course there were also some *de novo* compounds (Table 2). After infestation by *D. coccus*, the Esmeralda cultivar increased the number of volatiles from 63 (uninfested) to 87, of which 48 were produced de novo and belonged to nine chemical groups. In the case of Rojo Pelón *D. opuntiae*, uninfested cladodes produced 47 VOCs, and after infestation, they decreased to 38, 13 of which were identified as de novo, belonging to seven chemical groups (Table 2, Figures S1 and S2).

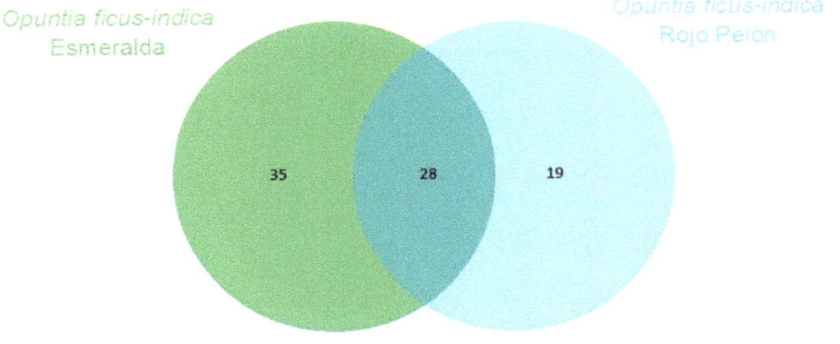

Figure 2. Comparison of volatilomes of the *Opuntia ficus-indica* uninfested cladodes of each cultivar using a Venn diagram, based on the number of VOCs obtained through essential oils.

Although there was an enormous variation between the number and proportion of VOCs before and after infestation, it was observed that four chemical groups maintained the highest abundance in both infested cultivars. These groups were (a) carboxylic acid and derivatives, (b) terpenes, (c) alcohols, and (d) aldehydes and their derivatives. Another group, the heterocycles, was only abundant for the uninfested Rojo Pelón cultivar (8.91%), but after *D. opuntiae* infestation, it decreased to less than 1.4%. The rest of the recovered chemical groups (ethers, ketones, aromatic derivatives, and alkanes) were less than 1.16% of the relative abundance per group in either cultivar infested by the respective *Dactylopius* species. Two of these groups (ethers and aromatic derivatives) were not detected in the infested Rojo Pelón cultivar (Table 2).

As indicated above, because of *Dactylopius* infestation in each cultivar, there were many changes in the relative abundance of compounds and the production of some de novo compounds. The de novo compounds were mostly of low relative abundance (equal to or less than 1.0%), except for some terpenes and alcohols. For example, in the uninfested Rojo Pelón cultivar, the relative abundance of terpenes was around 0.8%, but this relative abundance of terpenes changed to 15.5% after *D. opuntiae* infestation. On the other hand, the relative abundance of terpenes in the Esmeralda cultivar decreased from 18 to 13.9% due to *D. coccus* infestation (Table 2). The amount and type of terpenes were different between infested *O. ficus-indica* cultivars, but monoterpenes or their derivatives predominated in both cases (Figure 3).

Table 2. Volatilomes of *Opuntia ficus-indica* (OFI) cultivars before and after infestation by *Dactylopius* species.

No.	Compounds	OFI Esmeralda RA (%)	OFI Esmeralda Infested by *D. coccus* RA (%)	KIExp	OFI Rojo Pelón RA (%)	OFI Rojo Pelón Infested by *D. opuntiae* RA (%)	KI Exp	KI Ref
	Carboxilic acid and derivatives	48.79	44.28		31.78	20.05		
1	Hexanoic acid	0.83 ± 0.63	0.82 ± 0.00	942				904
67	3-Methyl-2-pentenoic acid	0.35 ± 0.32		959				926
68	2-Hexenoic acid	0.66 ± 0.01	0.36 ± 0.00	972				939
69	4-Oxopentanoic acid	0.22 ± 0.14		991				956
70	Heptanoic acid	0.59 ± 0.50	0.39 ± 0.05	1022	0.39 ± 0.11	0.25 ± 0.15	1030	1005
5	2-Ethylhexanoic acid				0.11 ± 0.07		1044	1031
71	4-Methylvaleric acid		0.15 ± 0.10	1033				1039
72	2-Methyl-4-pentenoic acid		0.11 ± 0.16	1062				
73	Lactic acid	3.22 ± 1.91	1.63 ± 0.16	1065				1057
15	Benzoic acid	1.49 ± 1.77	0.29 ± 0.03	1080	2.06 ± 0.13	0.50 ± 0.09	1080	1084
74	Methyl benzoate	0.41 ± 0.14		1081		0.91 ± 0.34		1084
9	Octanoic acid	1.23 ± 1.52		1112	1.21 ± 0.07	1.34 ± 0.07	1112	1109
11	Ethyl Benzoate	0.05 ± 0.01		1156	0.08 ± 0.04	0.24 ± 0.11	1152	1141
75	Benzeneacetic acid					0.17 ± 0.05	1160	1150
76	Salicylic acid					0.18 ± 0.25	1171	1176
77	Methyl salicylate	1.21 ± 0.37	6.96 ± 0.14	1181		0.32 ± 0.07	1172	1176
78	2-Nonenoic acid		0.86 ± 0.00	1179		0.33 ± 0.08	1184	
13	Nonanoic acid	1.09 ± 0.24	0.76 ± 0.07	1216	1.45 ± 0.12	1.72 ± 0.14	1212	1206
79	Ethyl salycilate		1.03 ± 0.01	1244				1241
18	Decanoic acid	0.77 ± 0.75	0.63 ± 0.00	1318	0.79 ± 0.08		1311	1309
19	Butanedioic acid	0.49 ± 0.19		1320				1314
80	Gliceric acid	0.73 ± 0.02		1346				1342
81	2-Methoxybenzoic acid		0.10 ± 0.01	1331				1362
82	Methyl 2-methoxy benzoate		0.45 ± 0.10	1319				1295
83	Glutaric acid	0.16 ± 0.03		1410				1400
22	Undecanoic acid	0.16 ± 0.02	0.14 ± 0.07	1417	0.08 ± 0.02	0.12 ± 0.08	1408	1410
24	Dodecanoic acid	5.70 ± 6.08	2.77 ± 0.06	1516	7.19 ± 0.50	2.35 ± 0.16	1505	1509
84	2,5-Dimethoxy benzenemethanol acetate				0.13 ± 0.04		1523	
26	Ethyl Dodecanoate	0.26 ± 1.27	0.17 ± 0.09	1582	0.33 ± 0.01		1571	1581
28	*p*-Hydroxybenzoic acid	0.94 ± 0.45		1620				1621
29	Hexylsalicylate	0.25 ± 3.04	0.46 ± 0.07	1662				1652
85	Methyl tetradecanoate	0.84 ± 0.16		1719				1714
27	Tridecanoic acid		0.06 ± 0.04	1611				1606
86	12-Methyltridecanoic acid		0.07 ± 0.00	1678				1680
31	Tetradecanoic acid	2.98 ± 3.09	1.36 ± 0.06	1720	1.56 ± 0.10	0.19 ± 0.07	1714	1714
87	Methyl benzoate		1.46 ± 0.03	1752				
88	Benzyl Benzoate		2.65 ± 0.51	1741	0.15 ± 0.04		1754	1765
32	Ethyl tetradecanoate		0.10 ± 0.13	1784				1782
89	Nonanedioic acid	0.28 ± 0.01		1808				1788
90	Pentadecanoic acid	0.42 ± 0.47	0.43 ± 0.02	1826				1807
91	Isopropyl tetradecanoate		0.05 ± 0.34	1820				1827
92	Benzyl salicylate		0.31 ± 0.12	1855				1860
34	Hexadecanoic acid	7.19 ± 2.16	4.88 ± 0.16	1935	9.35 ± 0.74	8.25 ± 0.45	1916	1909
93	15-Methylhexadecanoic acid	0.17 ± 0.03		2040				1974
37	(Z,Z)-9,12-Octadecadienoic acid	1.90± 0.44	1.03 ± 0.12	2106	0.88 ± 0.14	2.20 ± 1.43	2087	2087
38	(Z)-9-Octadecenoic acid	2.35 ± 1.66	1.94 ± 0.10	2113	2.47 ± 0.98		2090	2100
39	Octadecanoic acid	2.53 ± 0.59	1.76 ± 0.01	2141	2.76 ± 0.15	0.59 ± 0.15	2119	2133
94	Methyl octadecanoate				0.33 ± 0.04		1809	
40	Ethyl octadecanoate		0.11 ± 0.00	2199				2202
41	Dehydroabietic acid	9.32 ± 2.24	9.99 ± 0.04	2376	0.46 ± 0.05	0.39 ± 0.02	2344	2373

Table 2. *Cont.*

No.	Compounds	OFI Esmeralda RA (%)	OFI Esmeralda Infested by *D. coccus* RA (%)	KIExp	OFI Rojo Pelón RA (%)	OFI Rojo Pelón Infested by *D. opuntiae* RA (%)	KI Exp	KI Ref
	Aldehides and derivatives	2.15	6.25		4.3	4.82		
42	Hexanal	0.44 ± 0.44	0.47 ± 0.00	984				964
43	Heptanal		0.18 ± 0.02	1069				1068
95	Benzaldehyde		0.15 ± 0.07	1094	0.32 ± 0.03	0.55 ± 0.33	1094	1080
96	Diethyl acetal hexanal	0.25 ± 0.14	0.46 ± 0.10	1086				1082
97	5,5-Dimethyl-3-oxo-1-cyclohexene-1-carboxaldehyde		0.15 ± 0.03	1104				
44	Octanal		0.17 ± 0.01	1160	0.40 ± 0.05	0.09 ± 0.05	1165	1167
98	Phenylacetaldehyde	0.62 ± 0.61	0.56 ± 0.06	1198	0.82 ± 0.06	0.81 ± 0.17	1201	1208
45	Nonanal	0.53 ± 0.21	1.03 ± 0.01	1271	1.73 ± 0.12	1.32 ± 0.09	1265	1267
46	Decanal		0.14 ± 0.00	1370	0.14 ± 0.08		1366	1366
99	Nonanaldimethylacetal	0.21 ± 0.10	0.37 ± 0.05	1374				1379
100	3-(4-(*tert*-butyl)phenyl-2-methylpropanal		0.30 ± 0.04	1497				1500
101	4-Hydroxy-3-methoxybenzaldehyde	0.10 ± 0.04	0.62 ± 0.02	1524	0.89 ± 0.04	2.05 ± 0.01	1511	1544
102	3-Ethoxy-4-hydroxybenzaldehyde		0.11 ± 0.02	1554				1560
48	α-Hexylcinnamaldehyde		1.22 ± 0.74	1725				1726
103	Octadecanal		0.32 ± 0.13	2187				
	Heterocycles		0.67		8.91	1.38		
104	2-Isopropyl-3-metoxypirazina		0.25 ± 3.25	1070				1080
105	2-Methoxy-3-isopropylpyrazine				0.30 ± 7.22	1.05 ± 0.33	1083	1089
106	Ethyl 2-(5-methyl-5-vinyltetrahydrofuran-2-yl)propan-2-yl carbonate				8.18 ± 0.10		1064	1090
107	3-Isobutyl-2-methoxypyrazine				0.43 ± 0.03		1164	1170
108	3-Ethyl-4-methyl-1H-pyrrole-2,5-dione		0.35 ± 0.03	1209				1192
109	3-Hydroxy-2-methylpyran-4-one		0.07 ± 0.00	1266				1293
110	2,3-Dihydro-2,2,4,6-tetramethylbenzofuran					0.33 ± 0.01	1410	
	Ethers	0.31	0.16					
50	Benzylmethylether	0.31 ± 0.20		966				966
111	1,2-Dimethoxybenzene		0.16 ± 0.02	1111				1106
	Ketones	1.95	1.16		1.17	0.61		
112	5-Hexen-2-one	0.29 ± 0.01		1007				
113	2,2,6-Trimethylcyclohexanone				0.14 ± 0.09		1031	
114	Acetophenone		0.57 ± 0.08	1047		0.45 ± 0.05	1055	1049
115	Isophorone	0.24 ± 0.06	0.01 ± 0.22	1106	0.17 ± 0.03		1038	1094
116	Phenylacetone	0.48 ± 0.05	0.44 ± 0.00	1114	0.24 ± 0.02	0.16 ± 0.07	1110	1116
117	4-Oxoisophorone	0.13 ± 0.02		1131	0.07 ± 0.04		1125	1105
118	2-(1-Hydroxybut-2-enylidene)cyclohexanone				0.14 ± 0.03		1145	
119	1-(1-cyclohexen-1-yl)(-1-1-Butenone)	0.70 ± 0.34		1214				
54	Benzophenone	0.11 ± 0.07	0.14 ± 0.01	1607	0.41 ± 0.05		1584	1607

Table 2. *Cont.*

No.	Compounds	OFI Esmeralda RA (%)	OFI Esmeralda Infested by *D. coccus* RA (%)	KIExp	OFI Rojo Pelón RA (%)	OFI Rojo Pelón Infested by *D. opuntiae* RA (%)	KI Exp	KI Ref
	Terpenes	17.89	13.92		0.8	15.52		
120	Limonene	0.85 ± 0.23		1023				1020
121	Linalool oxide	8.48 ± 0.58	5.06 ± 0.42	1063				1064
122	*trans*-Linalool oxide		5.70 ± 0.03	1064				1068
123	1,5,5-Trimethyl-3-methylene cyclohexene	0.55 ± 0.63		1071				
124	β-Linalool	5.00 ± 0.58	0.26 ± 0.35	1088				1082
125	α-Terpineol	2.00 ± 0.36		1178				1172
126	Linalool	0.19 ± 0.22		1232		5.61 ± 0.10	1227	1227
127	Geraniol	0.44 ± 0.27	0.33 ± 0.04	1250		1.84 ± 0.06	1357	1238
128	Nerol		0.33 ± 0.06	1232		0.79 ± 0.04	1328	1260
129	β-Damascenone		0.18 ± 0.17	1362		0.55 ± 0.30	1360	1361
52	α-Ionona		0.09 ± 0.26	1404				1413
130	α-Isomethylionone		0.88 ± 0.15	1453				1478
53	β-Ionone		0.40 ± 0.06	1460		0.18 ± 0.04	1458	1486
131	Dihydroactinidiolide	0.38 ± 0.13	0.30 ± 0.02	1537				1532
132	Neophytadiene		0.39 ± 0.08	1832				1842
133	28-Nor-17β(H)-hopane				0.45 ± 0.01		2942	
134	β-Sitosterol				0.35 ± 0.03	6.55 ± 0.04	3244	3284
	Alcohols	12.72	9.91		29.37	30.78		
135	1,2-Dihydroxy-4-methylpentane		0.27 ± 0.02	990				
136	Hexanol		0.06 ± 0.01	9994				992
137	(*Z*)-2-Hexen-1-ol		0.36 ± 0.19	1010	7.65 ± 0.18	4.21 ± 0.08	1025	1001
56	Phenol	0.22 ± 0.09	0.16 ± 1.30	1045				1043
138	Heptanol		0.11 ± 0.23	1067				1092
57	2-Ethylhexanol	0.58 ± 0.23	0.22 ± 0.24	1099	2.13 ± 0.09	1.49 ± 0.05	1103	
139	Benzylalcohol	0.27 ± 0.11	0.56 ± 0.22	1143	0.90 ± 0.02	2.08 ± 0.06	1132	1156
140	1-Octanol		0.29 ± 0.14	1158	1.27 ± 0.04	1.51 ± 0.18	1177	1177
141	Guaiacol		0.35 ± 0.03	1209				1192
142	Nonanol		0.07 ± 0.01					
143	Glycerol	0.33 ± 0.52		1290	1.46 ± 0.09		1288	1292
144	*p*-Vinylguaiacol	10.62 ± 7.24	1.76 ± 0.34	1305	14.67 ± 0.93	17.03 ± 5.14	1294	1282
145	1-Methyl-1(4-methyl-3-cyclohexenyl)ethanol		0.63 ± 0.00	1318				1309
146	Isododecanol		0.09 ± 0.01	1479				
58	1-Dodecanol	0.47 ± 0.09	0.60 ± 0.01	1563	0.76 ± 0.04		1553	1575
60	1-Tetradecanol		0.66 ± 0.00	1756				1768
61	1-Hexadecanol	0.23 ± 0.07	0.98 ± 0.08	1978	0.53 ± 0.03		1960	1965
62	1-Octadecanol		0.92 ± 0.04	2177				2159
147	3,7,11,15-Tetramethyl-2-hexadecenol		1.82 ± 0.15	2198		3.52 ± 0.17	2173	2179
148	Octacosanol					0.94 ± 0.04	3125	3154
	Aromatic derivatives	0.77						
149	1,2-Dihydro-1,1,6-trimethyl naphthalene	0.18 ± 0.07		1338				1332
150	10,18,Bisnorabieta-8,11,13.triene	0.59 ± 0.03		2041				

Table 2. *Cont.*

No.	Compounds	OFI Esmeralda RA (%)	OFI Esmeralda Infested by *D. coccus* RA (%)	KIExp	OFI Rojo Pelón RA (%)	OFI Rojo Pelón Infested by *D. opuntiae* RA (%)	KI Exp	KI Ref
	Alkanes	0.69	0.55		0.78			
64	Hexadecane	0.69 ± 0.19	0.07 ± 0.01	1585				1600
151	Heptadecane		0.12 ± 0.05	1692				1700
152	Nonadecane		0.36 ± 0.06	1906				1900
153	Eicosane				0.19 ± 2.08		1992	2000
66	Heneicosane				0.23 ± 0.42		2092	2100
154	Docosane				0.36 ± 0.07		2188	2200
Total		**85.27**	**76.9**		**77.11**	**73.16**		

RA, relative abundance; KI Exp, Kovats index experimental; KI Ref, Kovats index reference.

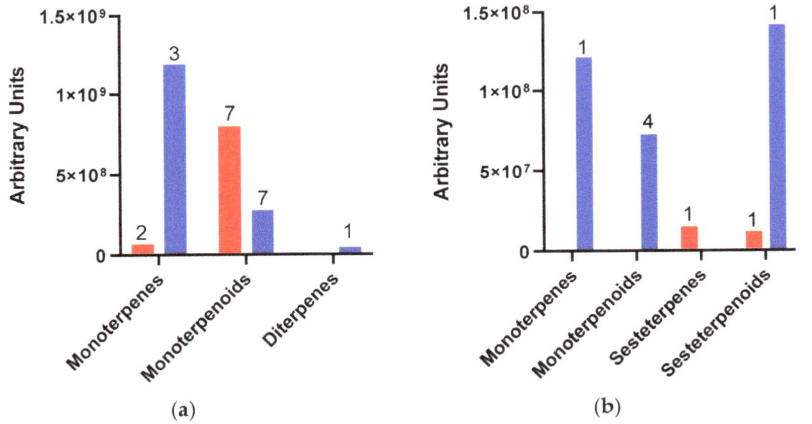

Figure 3. Amount and type of terpenes released by *Opuntia ficus-indica* (OFI) after *Dactylopius* infestation. (**a**) OFI Esmeralda-*D. coccus*; (**b**) OFI Rojo Pelón-*D. opuntiae*. The red columns represent uninfested cladodes, and the blue columns represent cladodes infested by each *Dactylopius* species. Data are presented as means of the peak area of each terpene (grouped by type and number of compounds).

The terpenes linalool oxide, trans-linalool oxide, and the alcohol 3,7,11,15-tetramethyl-2-hexadecenol reached a relative abundance of 5.06%, 5.7%, and 1.8% in the Esmeralda cultivar infested by *D. coccus*. On the other hand, the terpenes linalool, geraniol, and the alcohol 3,7,11,15-tetramethyl-2-hexadecenol registered 5.6%, 1.84%, and 3.5% of the relative abundance in the Rojo Pelón cultivar infested by *D. opuntiae*, respectively. Also, *p*-vinylguaiacol increased 2.3% in relative abundance after *D. opuntiae* infestations (Table 2).

3. Discussion

Previous assays of *Dactylopius* VOCs extraction, such as Headspace (HS-SPME) and extraction by Autosampler Headspace coupled to CG-MS (HS-CG-MS), did not provide the results expected for GC-MS analysis. Thus, to identify the volatiles from *Dactylopius* and its cultivar hosts, we preferred to do so using their essential oils. Essential oils were obtained by the hydrodistillation method (Table S1), which is frequently used to obtain essential oils from plants that contain low-vapor pressure compounds or low-volatile compounds. This technique is also used for concentrating compounds with lower concentrations in the essential oil and allows working with a larger sample mass than microextraction techniques, which can potentially improve the characterization of insect VOCs [29].

In the volatilome of *D. coccus* and *D. opuntiae*, 32 and 54 VOCs were identified for each species, respectively. To our knowledge, neither of these volatilomes had been reported previously, and this may be the first contribution to this work. By their composition, these VOCs corresponded to eight different chemical groups, but there were three groups of greater abundance. These were (a) carboxylic acids and their derivatives, 59.28% and 78.29% abundance for *D. coccus* and *D. opuntiae*, respectively; (b) alcohols, which were abundant only for *D. coccus* (12%); and (c) aldehydes, 5.8% and 7.68% abundance for *D. coccus* and *D. opuntiae*, respectively (Figures S3 and S4). This composition could be one of the reasons that results were not obtained with the HS-SPME and HS-CG-MS techniques. The VOCs of *Dactylopius* species are mostly fatty acids, some of which may be part of the fat content of the insects or of the complexity of their waxy coat [40,41]. In fact, each VOC in those groups may have more than one role in structure, function, metabolism, and probably in intra- or interspecific communication. For example, *D. coccus* produces a sex pheromone [42], and *D. opuntiae* is suspected to do so as well [43]. Regarding tetradecanoic acid, which is one of the most abundant VOCs for both species of *Dactylopius*, and hexadecanoid acid, relevant to *D. opuntiae*, they have many functions in insect metabolism. One of these is to participate in the metabolic pathways of sex pheromones of some *Lepidoptera*, such as *Spodoptera lottoralis* Boisduval and *Plodia interpunctella* Hubner [44,45], but none of these compounds appear to have relevance in the pheromones of Coccoidea [46], which is the superfamily to which the Dactylopiidae belong. The methodology for identifying insect pheromones begins with live females at a particular moment of maturity and sexual behavior, and so much work remains to be carried out in order to decipher the main functions of the VOCs that turned out to be more abundant, which could lead to novel acids with shorter chains and perhaps more specific for each *Dactylopius* species.

The volatilomes of the Esmeralda and Rojo Pelón cultivars were different before and after *Dactylopius* infestation. The variation in compound production in cladodes of both cultivars prior to infestation (by *Dactylopius*) may be specific to each cultivar, as variations of other bioactive and volatile compounds have been reported in different cultivars of *O. ficus-indica* [31,35,47]. However, variation in the number and abundance of VOCs within each cultivar after infestation can be attributed to *D. coccus* or *D. opuntiae* feeding on its corresponding cultivar host, as has been demonstrated in other plants where the change in production of VOCs, particularly terpenes and sesquiterpenes, was directly associated with phytophagous insect feeding [23,24,48,49].

In the volatilomes of the Esmeralda and Rojo Pelón, before or after *Dactylopius* infestation, four chemical groups were identified as the most abundant: (a) carboxylic acid and derivatives, (b) terpenes, (c) alcohols, and (d) aldehydes and derivatives (Figures S5 and S6). The structural composition of the host, particularly the quantity of waxes, could be related to the abundance of some of these VOCs in both cultivars [47,50]. This suggestion is related to the anatomical and physiological adaptations of cacti to develop in arid environments, such as a thick and impermeable epidermis covered by a layer of waxy cuticle, a hypodermis with chollenchyma, plenty of cells with mucilage distributed in the parenchyma, and crassulaceae acid metabolism (CAM), among other characteristics [50]. Of the first and most abundant chemical groups (a), it is probable that we should mention methyl salicylate, which increased in abundance after infestation by *D. coccus* (about 5%) in the Esmeralda cultivar. The same compound was identified de novo in the Rojo Pelón cultivar infested by *D. opuntiae*, although it was low in abundance (0.3%). Methyl salicylate is a phenolic compound that has been reported to be an herbivore-induced plant volatile (HIPV) [49,51,52]. Some of these HIPVs can induce direct defense against the phytophagous insect and indirect defense by attracting their natural enemies. It is also useful for communication among plants damaged by phytophagy and others that are not yet damaged. For example, methyl salicylate emitted by plants with phytophagous mite damage was attractive to *Phytoseiulus persimilis* Athias-Henriot (Phytoseiidae) [51,52]. In the same way, it was observed that emission of this compound, after damage by psyllids in pear trees, was attractive to the predatory bug *Anthocoris nemoralis* F. (Hemiptera: Anthocoridae) [53].

In general, a slight decrease in terpene abundance (18 to 14%) was observed after *D. coccus* infestation, but a considerable increase (0.8 to 15%) occurred after *D. opuntiae* infestation. In the Esmeralda cultivar, β-linalool abundance decreased from 5.0 to 0.3%, but linalool oxide and trans-linalool oxide increased to 5.0 and 5.7%, respectively. On the other hand, in the cultivar Rojo Pelón infested by *D. opuntiae*, five de novo terpenes were identified, of which the most abundant was linalool (5.6%). Terpenes are one of the most studied groups of HIPVs, and it has been shown that some of them have a relevant role in the direct defense system against phytophages, and some volatile terpenes constitute indirect defenses of plants as they attract natural enemies such as predators and parasitoids [23,24,27,48,49].

Linalool is a monoterpene that occurs naturally in flowers and aromatic plants, but it is also produced in response to feeding by phytophagous insects, and it is part of the indirect defenses of plants [54]. For example, an increase in linalool production in tobacco plants caused by feeding Lepidoptera larvae increased the rate of egg predation and decreased the oviposition of another Lepidoptera [55]. Linalool also increased due to phytophages feeding on corn, bean, cotton, and potato plants [23], or by a zoophytophagous mirid feeding on pepper plants, and favored the action of natural enemies of their pests [49]. This can suggest that significant changes in the abundance of methyl salicylate from the above group and terpenes, particularly linalool, are probably related to each *Dactylopius* species feeding on its corresponding cultivar host.

The alcohol of greatest abundance and change was *p*-vinylguaiacol. This compound is common in plants and is part of many essential oils. In addition, it can be found in the guts of some insects, probably through the process of lignin degradation [56]. Regarding secondary plant defenses due to damage by phytophagous insects, *p*-vinylguaiacol stimulated the ovipositional behavior of the natural enemy *Coleomegilla maculata* [57], and it was also a deterrent to the oviposition of the cerambycid *Monochamus alternatus* [56]. Therefore, it is suggested that some changes in *p*-vinylguaiacol abundance may be a consequence of *Dactylopius* feeding.

In this work, 66 VOCs of both *Dactylopius* species were identified, and 125 of the Esmeralda and Rojo Pelón cladodes were infested by *D. coccus* and *D. opuntiae*, respectively. A proportion of VOCs were commonly produced in both insect species or cultivars, but others were specific to each species or cultivar (Figure 4). This is a first approach to the diversity of VOCs produced by *O. ficus-indica* and the changes that occur due to *D. coccus* and *D. opuntiae* feeding on cultivars suitable for the development of each *Dactylopius* species. More time and work will now be needed to understand the functions performed by the most relevant compounds in these interactions.

If knowledge of the interaction is improved, for example, if it is confirmed that some terpenoids favor the direct or indirect defenses of *O. ficus-indica* against *D. coccus* or *D. opuntiae*, this information could be considered in breeding programs. These programs could be aimed at improving the rearing of *D. coccus* or inducing resistance to *D. opuntiae*. In this regard, breeding programs for *O. ficus-indica* resistant to *D. opuntiae* have already been developed in Brazil and Morocco, and these have focused on physical and biochemical defense mechanisms [15,21,58]. For example, selecting cultivars with high concentrations of calcium oxalates can physically and biochemically limit phytophagous insects [59,60]. However, there are no known breeding programs for *O. ficus-indica* that consider the abundance of terpenes in cultivars and the response this can trigger in the plant's direct or indirect defenses. This mechanism would be classified as biochemical defense, and measuring terpenes in different cultivars could improve the direction and understanding of the response.

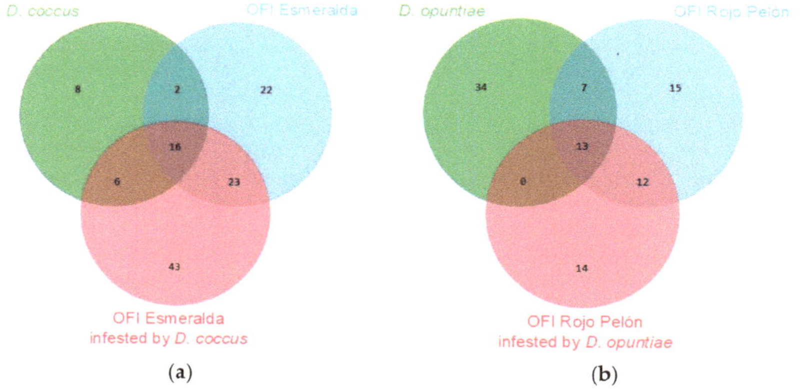

Figure 4. Number of VOCs, obtained through its essential, identified as common or de novo compounds between uninfested and infested *O. ficus-indica* (OFI) cultivars, (**a**) by *Dactylopius coccus*, (**b**) by *Dactylopius opuntiae*.

Besides, SIMPER analysis (Tables S2–S5) showed the components that are typical of each *Dactylopius* species and its hosts; these contribute a low percentage of each sample, so their contribution to the dissimilarity is low. This observation highlights the need to better understand the interaction between *O. ficus-indica* and *Dactylopius* because it can increase the possibilities of making proposals for sustainable management in the production of *D. coccus* or in the control of *D. opuntiae*.

4. Materials and Methods

4.1. Chemicals

The reagents used in this study were *N*, *O*-bis(trimethylsilyl) trifluoroacetamide (BSTFA), trimethylsilyl chloride (TMCS), boron trifluoride methanol solution (Sigma-Aldrich, St. Louis, MO, USA), and ethylic ether (JT Baker, Deventer, Holland).

4.2. Insects and Uninfested and Infested O. ficus-indica Cultivars

Dactylopius coccus and *Opuntia ficus-indica* Esmeralda cultivars (infested and uninfested) were originally obtained from a local provider in Jerez, Zacatecas, Mexico. *Dactylopius opuntiae* and *O. ficus-indica* Rojo Pelón cultivars (infested and uninfested) were collected from an experimental field at Colegio de Postgraduados, Campus San Luis Potosí (Salinas, SLP). These cactus pear cultivars were selected with the knowledge that each one is favorable for the development of the respective *Dactylopius* species [58]. The taxonomic identity of *Dactylopuis* species was corroborated by S. J. Méndez-Gallegos using De Lotto (1974) [40] and Ferris (1955) keys [61]. To increase material for the samples and analyses, *D. coccus* and *D. opuntiae* colonies were reared on the respective cultivars mentioned under greenhouse conditions (15 ± 2 °C, 22 ± 2 °C, and 50% RH).

4.3. Essential Oil of Dactylopius Species and Hosts

One hundred grams of adult females previous to the reproduction stage of *D. coccus* (80 to 85 d old) and *D. opuntiae* (30 to 35 d old) with their protective coverings (secretion substances) and 1000 g of infested and uninfested *O. ficus-indica* cladodes were used independently to obtain their essential oils by hydrodistillation. The *Dactylopius* species were manually separated from their hosts just before hydrodestillation, and the *O. ficus-indica* cladodes were cut into cubes just before hydrodestillation. The VOCs, which are components of essential oils, were obtained at boiling water temperature and extracted from the condensed water by liquid–liquid extraction with ethyl ether. The solvent was distillated, and the residual water was removed from the organic phase with anhydrous sodium sulfate. Each sample was then concentrated (to 1 mL) at 40 °C under vacuum, and the residual solvent was eliminated from each sample at atmospheric pressure at 0 °C.

4.4. Derivatization for Alcohol Detection

Essential oils were diluted to 2% in 500 μL heptane and introduced into a 10 mL microwave reaction tube with a gasket. Then, 100 μL of BSTFA/TMCS solution (9:1 *v*/*v*) was added to the same tube as a silanizing agent. The mixture was reacted at 90 °C under microwave irradiation (250 W microwave power) for 10 min using the Discover System 908,005 (CEM Corporation, Matthews, NC, USA) with autogenous pressure.

4.5. Derivatization for Aldehydes and Carboxylic Acid Detection

Essential oils were diluted to 2% in 500 μL heptane and introduced into a 10 mL microwave reaction tube with a gasket. Then, 500 μL of boron trifluoride (14% in methanol solution) was added to the same tube. The mixture reacted at 90 °C under microwave irradiation (250 W microwave power) for 10 min using the Discover System 908,005 with autogenous pressure.

4.6. Essential Oil GS-MS Analysis

Samples without derivatization were diluted to 2% in heptane, using 1 μL of each sample for the analysis, and each sample was analyzed in triplicate. GC-MS analysis was performed using a 7802A Network GC System coupled to a 5977E Network mass selective detector (MSD).

The separation was performed using an HP-5 capillary column (0.25 mm i.d., 30 mm, 0.25 mm film thickness) (J&W, Folsom, CA, USA). The injector was operated in splitless mode at 300 °C, with a flow of 1.0 mL/min, and the oven temperature was programmed to 40 °C for 3 min, and then heated at 3 °C/min to 300 °C with a holding time of 5 min at the final temperature. The MSD was operated at 70 eV; the ion source was set at 150 °C and the transfer line at 300 °C. VOCs were identified by interpreting their mass spectra as fragmentation in the mass range of 15 to 800 atomic mass units. The software MassHunter (Agilent B.07.01.1805, Santa Clara, CA, USA) was used for data recording. The compounds were identified by comparing the obtained mass spectra with those of reference compounds from the National Institute of Standards and Technology (NIST11) and Wiley 09. The identities of the compounds were confirmed by the Kovats retention index calculated for each peak with reference to the n-alkane standards (C7–C38) running under the same conditions.

4.7. Statistical Analysis

The relative percentage of each metabolite was calculated considering the peak area obtained by GC-MS of each metabolite in relation to the total area of peaks analyzed. The data represent the mean of the relative percentage of three repeats ± SD. Metabolites grouped by type for each essential oil were compared with the Mann–Whitney U test, considering the peak area of each metabolite and a $p \leq 0.05$. The data in the graphics were expressed as the median and range of each group. GraphPad Prism 5 was used to perform the analysis. Venn diagrams were constructed using an online tool (http://jvenn.toulouse.inra.fr/app/example.html, accessed on 23 November 2023) [62]. PAST statistical software (version 4.09) was used to perform the SIMPER analysis [63].

5. Conclusions

This work presents an approach to better understanding the interaction between *O. ficus-indica*, *D. coccus*, and *D. opuntiae* by identifying volatile compounds in their essential oils. The abundance and proportion of VOCs of *D. coccus* and *D. opuntiae* were determined in the Esmeralda and Rojo Pelón cultivars, viable for the development of each insect species, respectively. Differential VOC production due to infestation by each *Dactylopius* species in each cultivar was also identified. Changes in methyl salicylate, terpenes, and *p*-vinylguaiacol and their likely role in plant defense mechanisms should receive more attention because they could contribute to the development of proposals to improve *D. coccus* rearing or for the control of *D. opuntiae* in those regions of the world where it is a key pest of *O. ficus-indica*.

Supplementary Materials: The following supporting information can be downloaded at: https://www.mdpi.com/article/10.3390/plants13070963/s1, Table S1 shows the yields of essential oils of *Dactylopius* species and *Opuntias* varieties. Tables S2–S5 show the SIMPER analysis. Figures S1 and S2 show *O. ficus-indica* with *Dactylopius* species relationships by Venn diagrams. Figures S3–S6 show compound groups of *Dactylopius* species and *O. ficus-indica* cultivars.

Author Contributions: Conceptualization, M.M.G.-C. and S.d.J.M.-G.; methodology, E.G.-P., M.M.G.-C. and S.d.J.M.-G.; formal analysis, E.G.-P., E.R.-L., M.M.G.-C. and S.d.J.M.-G.; investigation, E.G.-P., E.R.-L., M.M.G.-C., S.d.J.M.-G., J.A.M.-R., J.C.P.-H., Á.B.-V. and A.F.-V.; writing—original draft preparation, E.R.-L. and S.d.J.M.-G.; writing—review and editing, E.G.-P., M.M.G.-C., J.A.M.-R., J.C.P.-H., Á.B.-V. and A.F.-V.; supervision, M.M.G.-C. and S.d.J.M.-G.; funding acquisition, M.M.G.-C. All authors have read and agreed to the published version of the manuscript.

Funding: The present study was financially supported by CONAHCYT from México (Ciencia de Frontera, funding number: 320298).

Data Availability Statement: The datasets used and analyzed during the current study are available from the corresponding author upon reasonable request.

Acknowledgments: We thank CONAHCYT for a graduate fellowship for Esperanza García-Pascual (CVU: 1146133). We also thank Vicente Rodríguez González (IPICYT) for access to GC-MS equipment and Brayan Arias-Alvarez, Juan J. Rodríguez-Silva, and María G. Ortega Salazar for the technical assistance.

Conflicts of Interest: The authors declare no conflicts of interest.

References

1. García-Morales, M.; Denno, B.D.; Miller, D.R.; Miller, G.L.; Ben-Dov, Y.; Hardy, N.B. ScaleNet: A literature-based model of scale insect biology and systematics. *Database* **2016**, *2016*, bav118. [CrossRef] [PubMed]
2. Van-Dam, A.R.; Martinez, L.P.; Chavez, A.J.; May, B.P. Range wide phylogeography of *Dactylopius coccus* (Hemiptera: Dactylopiidae). *Ann. Entomol. Soc. Am.* **2015**, *108*, 299–310. [CrossRef]
3. Campana, M.G.; Robles, N.M.; Tuross, N. America's red gold: Multiple lineages of cultivated cochineal in Mexico. *Ecol. Evol.* **2015**, *5*, 607–617. [CrossRef] [PubMed]
4. Barreto-García, O.A.; Rodríguez-Leyva, E.; Lomeli-Flores, J.R.; Vanegas-Rico, J.M.; Vigueras, A.L.; Portillo, L. *Laetilia coccidivora* feeding on two cochineal insect species, Does the prey affect the fitness of the predator? *BioControl* **2020**, *65*, 727–736. [CrossRef]
5. Eisner, T.; Nowicki, S.; Goetz, M.; Meinwald, J. Red Cochineal Dye (Caminic Acid): Its Role in Nature. *Science* **1980**, *208*, 1039–1042. [CrossRef] [PubMed]
6. Martínez-Martínez, S.; Rodríguez-Leyva, E.; Aranda-Ocampo, S.; Santillán-Galicia, M.T.; Hernández-López, A.; Guzmán-Franco, A.W. Bacteria associated with carminic acid metabolism in the intestinal tract of three predators of *Dactylopius opuntiae*. *Entomol. Exp. Appl.* **2023**, *172*, 183–192. [CrossRef]
7. Chávez-Moreno, C.K.; Tecante, A.; Casas, A. The *Opuntia* (Cactaceae) and *Dactylopius* (Hemiptera: Dactylopiidae) in Mexico: A historical perspective of use, interaction and distribution. *Biodivers. Conserv.* **2009**, *18*, 3337–3355. [CrossRef]
8. Borges, M.E.; Tejera, R.L.; Díaz, L.; Esparza, P.; Ibáñez, E. Natural dyes extraction from cochineal (*Dactylopius coccus*). New extraction methods. *Food Chem.* **2012**, *132*, 1855–1860. [CrossRef]
9. Müller-Maatsch, J.; Gras, C. The "Carmine Problem" and Potential Alternatives. In *Handbook on Natural Pigments in Food and Beverages*; Woodhead Publishing: Cambridge, UK, 2016; pp. 385–428. [CrossRef]
10. Sánchez-García, M.A.; Bokhimi, X.; Velázquez, S.; Jiménez-González, A.E. Dye-Sensitized solar cells prepared with Mexican pre-hispanic dyes. *J. Nanotechnol.* **2018**, *2018*, 1236878. [CrossRef]
11. Vanegas-Rico, J.M.; Rodríguez-Leyva, E.; Lomeli-Flores, J.R.; González-Hernández, H.; Pérez-Panduro, A.; Mora-Aguilera, G. Biology and life history of *Hyperaspis trifurcata* feeding on *Dactylopius opuntiae*. *BioControl* **2016**, *61*, 691–701. [CrossRef]
12. Vanegas-Rico, J.M.; Lomeli-Flores, J.R.; Rodríguez-Leyva, E.; Mora-Aguilera, G.; Valdez, J.M. Natural enemies of *Dactylopius opuntiae* (Cockerell) on *Opuntia ficus-indica* (L.) Miller in Central Mexico. *Acta Zool. Mex.* **2010**, *26*, 415–433. [CrossRef]
13. Griffith, M.P. The origins of an important cactus crop, *Opuntia ficus-indica* (Cactaceae): New molecular evidence. *Am. J. Bot.* **2004**, *91*, 1915–1921. [CrossRef] [PubMed]
14. Mazzeo, G.; Nucifora, S.; Russo, A.; Suma, P. *Dactylopius opuntiae*, a new prickly pear cactus pest in the Mediterranean: An overview. *Entomol. Exp. Appl.* **2019**, *167*, 59–72. [CrossRef]

15. Torres, J.B.; Giorgi, J.A. Management of the false carmine cochineal *Dactylopius opuntiae* (Cockerell): Perspective from Pernambuco state, Brazil. *Phytoparasitica* **2018**, *46*, 331–340. [CrossRef]
16. Mendel, Z.; Protasov, A.; Vanegas-Rico, J.M.; Lomeli-Flores, J.R.; Suma, P.; Rodríguez-Leyva, E. Classical and fortuitous biological control of the prickly pear cochineal, *Dactylopius opuntiae*. *Biol. Control* **2020**, *142*, 104157. [CrossRef]
17. Silva, M.A.; Albuquerque, T.G.; Pereira, P.; Ramalho, R.; Vicente, F.; Oliveira, M.B.; Costa, H.S. *Opuntia ficus-indica* (L.) mill.: A multi-benefit potential to be exploited. *Molecules* **2021**, *26*, 951. [CrossRef] [PubMed]
18. Kiesling, R. Origen, domesticación y distribución de *Opuntia ficus-indica*. *J. Prof. Assoc. Cactus Dev.* **1998**, *3*, 50–60. [CrossRef]
19. Hernandez-Hernandez, F.C.; García-Gil, F.; Rojas-Martínez, A.; Hernández-Martínez, S.; Lanz-Mendoza, H Carminic acid dye from the homopteran *Dactylopius coccus* hemolymph is consumed during treatment with different microbial elicitors. *Arch. Insect Biochem. Physiol.* **2003**, *54*, 37–45. [CrossRef]
20. Bravo-Vinaja, A.; Méndez-Gallegos, J. Emerging trends in scientific research on *Dactylopius coccus* Costa (Hemiptera: Dactylopiidae), carminic acid and its derivatives: A bibliometric analysis. *Agric. Soc. Desarro.* **2023**, *20*, 139–165. [CrossRef]
21. Sabbahi, R.; Hock, V. Control of the prickly pear cochineal, *Dactylopius opuntiae* (Cockerell), in Morocco: An overview. *J. Plant Dis. Prot.* **2022**, *129*, 1323–1330. [CrossRef]
22. García-Pascual, E.; González-Chávez, M.M.; Franco, A.; Rodríguez-Leyva, E.; Méndez-Gallegos, S.J.; Morales Rueda, J.A.; Bravo, A. *Dactylopius opuntiae* (Hemiptera: Dactylopiidae) control tactics: A bibliometric analysis. *Investig. Bibl.* **2023**, *38*, 13–29. [CrossRef]
23. Paré, P.W.; Tumlinson, J.H. Update on plant-insect interactions plant volatiles as a defense against insect herbivores. *Plant Physiol.* **1999**, *121*, 325–332. [CrossRef] [PubMed]
24. Turlings, T.C.J.; Erb, M. Tritrophic interactions mediated by herbivore-induced plant volatiles: Mechanisms, ecological relevance, and application potential. *Annu. Rev. Entomol.* **2017**, *63*, 433–452. [CrossRef] [PubMed]
25. Pophof, B.; Stange, G.; Abrell, L. Volatile organic compounds as signals in a plant-herbivore system: Electrophysiological responses in olfactory sensilla of the moth *Cactoblastis cactorum*. *Chem. Senses* **2005**, *30*, 51–68. [CrossRef] [PubMed]
26. Beck, J.J.; Baig, N.; Cook, D.; Mahoney, N.E.; Marsico, T.D. Semiochemicals from ex situ abiotically stressed cactus tissue: A contributing role of fungal spores? *J. Agric. Food Chem.* **2014**, *62*, 12273–12276. [CrossRef] [PubMed]
27. Rasmann, S.; Köllner, T.; Degenhardt, J.; Hiltpold, I.; Toepfer, S.; Kuhlmann, U.; Gershenzon, J.; Turlings, T. Recruitment of entomopathogenic nematodes by insect-damaged maize roots. *Nature* **2005**, *434*, 732–737. [CrossRef] [PubMed]
28. Maffei, M.E. Sites of synthesis, biochemistry and functional role of plant volatiles. *S. Afr. J. Bot.* **2010**, *76*, 612–631. [CrossRef]
29. Cázares-Samaniego, P.J.; Castillo, C.G.; Ramos-López, M.A.; González-Chávez, M.M. Volatilome and essential oil of *Ulomoides dermestoides*: A broad-spectrum medical insect. *Molecules* **2021**, *26*, 6311. [CrossRef]
30. Kasote, D.; Lee, J.; Sreenivasulu, N. Editorial: Volatilomics in plant and agricultural research: Recent trends. *Front. Plant Sci.* **2023**, *14*, 1289998. [CrossRef]
31. Flath, R.A.; Takahashi, J.M. Volatile Constituents of Prickly Pear (*Opuntia ficus indica* Mill., de *Castilla* Variety). *J. Agric. Food Chem.* **1971**, *26*, 835–837. [CrossRef]
32. Chahdoura, H.; Barreira, J.C.M.; Fernández-Ruiz, V.; Morales, P.; Calhelha, R.C.; Flamini, G.; Soković, M.; Ferreira, I.C.F.R.; Achour, L. Bioactivity, proximate, mineral and volatile profiles along the flowering stages of *Opuntia microdasys* (Lehm.): Defining potential applications. *Food Funct.* **2016**, *7*, 1458–1467. [CrossRef] [PubMed]
33. Nounah, I.; Chbani, M.; Matthäus, B.; Charrouf, Z.; Hajib, A.; Willenberg, I. Profile of volatile aroma-active compounds of cactus seed oil (*Opuntia ficus-indica*) from different locations in morocco and their fate during seed roasting. *Foods* **2020**, *9*, 1280. [CrossRef] [PubMed]
34. Wright, C.R.; Setzer, W.N. Volatile compositions of two cactus species growing in the sonoran desert of Southern Arizona. *Am. J. Essent. Oils Nat. Prod.* **2013**, *1*, 41–47. Available online: https://www.researchgate.net/publication/270572666 (accessed on 23 March 2024).
35. Apablaza, E.; Sáenz, C.; Prat, L.; Ubeda, C. Comprehensive characterization and volatile profile of cactus pear fruits from six different species and cultivars. *ACS Food Sci. Technol.* **2021**, *1*, 928–936. [CrossRef]
36. El-Hawary, S.S.; El-tantawy, M.E.; Rabeh, M.A.; Badr, W.K. Chemical composition and antimicrobial activity of volatile constituents of cladodes, fruits peel and fruits pulp from *Opuntia ficus indica* (L.) Mill. (Prickly pear) growing in Egypt. *Egypt J. Chem.* **2020**, *64*, 437–444. [CrossRef]
37. Zhang, W.W.; Zheng, H.; Zhang, H.; Chen, X.M.; Guo, Y.H. Volatile changes of *Opuntia ficus-indica* Mill before and after parasitizing by *Dactylopius coccus* Costa. *Chin. J. Appl. Entomol.* **2012**, *49*, 986–991.
38. Stierlin, É.; Nicolè, F.; Costes, T.; Fernandez, X.; Michel, T. Metabolomic study of volatile compounds emitted by lavender grown under open-field conditions: A potential approach to investigate the yellow decline disease. *Metabolomics* **2020**, *16*, 31. [CrossRef] [PubMed]
39. Cui, S.; Ling, P.; Zhu, H.; Keener, H.M. Plant pest detection using an artificial nose system: A review. *Sensors* **2018**, *18*, 378. [CrossRef]
40. De Lotto, G. On the status and identity of the cochineal insects (Homoptera: Coccoidea: Dactylopiidae). *J. Entomol. Soc. South. Afr.* **1974**, *37*, 167–193. Available online: https://hdl.handle.net/10520/AJA00128789_2848 (accessed on 1 June 1974).
41. Waku, Y.; Foldi, I. The fine structure of insect glands secreting waxy substances. In *Insect Ultrastructure*; King, R.C., Akai, H., Eds.; Springer: Boston, MA, USA, 1984; pp. 303–322. [CrossRef]

42. Rodríguez, L.C.; Faúndez, E.H.; Niemeyer, H.M. Mate searching in the scale insect, *Dactylopius coccus* (Hemiptera: Coccoidea: Dactylopiidae). *Eur. J. Entomol.* **2005**, *102*, 305–306. [CrossRef]
43. Palafox-Luna, J.A.; Rodríguez-Leyva, E.; Lomeli-Flores, J.R.; Vigueras-Guzmán, A.L.; Vanegas-Rico J., M. Ciclo de vida y fecundidad de Dactylopius opuntiae (Hemiptera: Dactylopiidae) en *Opuntia ficus-indica* (Caryophyllales: Cactaceae). *Agrociencia* **2018**, *52*, 103–114.
44. Martinez, T.; Fabriás, G.; Camps, F. Sex pheromone biosynthetic pathway in *Spodoptera littoralis* and its activation by a neurohormone. *J. Biol. Chem.* **1990**, *265*, 1381–1387. [CrossRef] [PubMed]
45. Tsfadia, O.; Azrielli, A.; Falach, L.; Zada, A.; Roelofs, W.; Rafaeli, A. Pheromone biosynthetic pathways: PBAN-regulated rate-limiting steps and differential expression of desaturase genes in moth species. *Insect Biochem. Mol. Biol.* **2008**, *38*, 552–567. [CrossRef] [PubMed]
46. Zou, Y.; Millar, J.G. Chemistry of the pheromones of mealybug and scale insects. *Nat. Prod. Rep.* **2015**, *32*, 1067–1113. [CrossRef] [PubMed]
47. Rodrigues, C.; de Paula, C.D.; Lahbouki, S.; Meddich, A.; Outzourhit, A.; Rashad, M.; Pari, L.; Coelhoso, I.; Fernando, A.L.; Souza, V.G.L. *Opuntia* spp.: An overview of the bioactive profile and food applications of this versatile crop adapted to arid lands. *Foods* **2023**, *12*, 1465. [CrossRef]
48. Holopainen, J.K.; Blande, J.D. Where do herbivore-induced plant volatiles go? *Front. Plant Sci.* **2013**, *4*, 48322. [CrossRef] [PubMed]
49. Pérez-Hedo, M.; Bouagga, S.; Zhang, N.X.; Moerkens, R.; Messelink, G.; Jaques, J.A.; Flors, V.; Broufas, G.; Urbaneja, A.; Pappas, M.L. Induction of plant defenses: The added value of zoophytophagous predators. *J. Pest Sci.* **2022**, *95*, 1501–1517. [CrossRef]
50. Terrazas, T.; Mauseth, J.D. Shoot Anatomy and Morphology. In *Cacti: Biology and Uses*; Nobel, P., Ed.; University of California Press: Oakland, CA, USA, 2002. [CrossRef]
51. Dicke, M.; Sabelis, M.W.; Takabayashi, J.; Bruin, J.; Posthumus, M.A. Plant strategies of manipulating predator-prey interactions through allelochemicals: Prospects for application in pest control. *J. Chem. Ecol.* **1990**, *16*, 3091–3118. [CrossRef]
52. Ozawa, R.; Arimura, G.I.; Takabayashi, J.; Shimoda, T.; Nishioka, T. Involvement of jasmonate-and salicylate-related signaling pathways for the production of specific herbivore-induced volatiles in plants. *Plant Cell Physiol.* **2000**, *41*, 381–398. [CrossRef] [PubMed]
53. Drukker, B.; Bruin, J.; Sabelis, M.W. Anthocorid predators learn to associate herbivore-induced plant volatiles with presence or absence of prey. *Physiol. Entomol.* **2001**, *25*, 260–265. [CrossRef]
54. Clavijo-McCormick, A.; Unsicker, S.B.; Gershenzon, J. The specificity of herbivore-induced plant volatiles in attracting herbivore enemies. *Trends Plant Sci.* **2012**, *17*, 303–310. [CrossRef]
55. Kessler, A.; Baldwin, I.T. Plant responses to insect herbivory: The emerging molecular analysis. *Annu. Rev. Plant Biol.* **2002**, *53*, 299–328. [CrossRef] [PubMed]
56. Li, S.Q.; Zhang, Z.N. The influence of volatiles from the hindgut of the pine sawyer, *Monochamus alternatus* (Coleoptera: Cerambycidae), on its oviposition behavior. *Zool. Stud.* **2007**, *46*, 726–733.
57. Smith, B.C.; Williams, R.R. Temperature relations of adult *Coleomegilla maculata lengi* and *C. m. medialzs* (Coleoptera: Coccinellidae) and responses to ovipositional stimulants. *Can. Entomol.* **1976**, *108*, 925–930. [CrossRef]
58. Berhe, Y.K.; Portillo, L.; Vigueras, A.L. Resistance of *Opuntia ficus-indica* cv 'Rojo Pelon' to *Dactylopius coccus* (Hemiptera: Dactylopiidae) under greenhouse condition. *J. Prof. Assoc. Cactus Dev.* **2022**, *24*, 293–309. [CrossRef]
59. Nakata, P.A. An assessment of engineered calcium oxalate crystal formation on plant growth and development as a step toward evaluating its use to enhance plant defense. *PLoS ONE* **2015**, *10*, e0141982. [CrossRef] [PubMed]
60. War, A.R.; Buhroo, A.A.; Hussain, B.; Ahmad, T.; Nair, R.M.; Sharma, H.C. Plant defense and insect adaptation with reference to secondary metabolites. In *Co-Evolution of Secondary Metabolites*; Springer: Berlin/Heidelberg, Germany, 2020; pp. 795–822. [CrossRef]
61. Ferris, G.F. Atlas of the Scale Insects of North America. *Nature* **1943**, *151*, 657. [CrossRef]
62. Bardou, P.; Mariette, J.; Escudié, F.; Djemiel, C.; Klopp, C. Jvenn: An interactive Venn diagram viewer. *BMC Bioinform.* **2014**, *15*, 293. [CrossRef]
63. Clarke, K.R. Non-parametric multivariate analyses of change in community structure. *Aust. J. Ecol.* **1993**, *18*, 117–143. [CrossRef]

plants

MDPI

Article

Using Age-Stage Two-Sex Life Tables to Assess the Suitability of Three Solanaceous Host Plants for the Invasive Cotton Mealybug *Phenacoccus solenopsis* Tinsley

Khaled Abbes [1], Ahlem Harbi [1], Emilio Guerrieri [2] and Brahim Chermiti [1,*]

[1] Department of Biological Sciences and Plant Protection, High Agronomic Institute of Chott-Mariem, University of Sousse, Sousse 4042, Tunisia; abbes.kaled@live.fr (K.A.); harbi.ahlem@hotmail.fr (A.H.)
[2] Institute for Sustainable Plant Protection, National Research Council of Italy, URT IPSP-DISIT, 15121 Alessandria, Italy; emilio.guerrieri@cnr.it
* Correspondence: chermiti54@yahoo.fr; Tel.: +216-73-327-542

Abstract: *Phenacoccus solenopsis* Tinsley (Hemiptera: Coccomorpha: Pseudococcidae), the cotton mealybug, is an invasive polyphagous species that has been extending its geographic range, posing a conspicuous threat to many Mediterranean crops of economic importance. These include three species of Solanaceae, namely *Solanum lycopersicum* L. (tomato), *Solanum tuberosum* L. (potato) and *Solanum melongena* L. (eggplant) all of which are economically important worldwide. In this study, we used age-stage two-sex life tables to investigate the suitability of these three plant species as hosts for *P. solenopsis* and to calculate pest fitness, life history parameters and population projection parameters. All tested host plants that were suitable for the pest and eggplant host plant induced a higher fecundity (276.50 ± 10.78 eggs/female), net reproductive rate (R_0) (243.32 ± 15.83 offspring/female) and finite rate of increase (λ) (1.18 ± 0.0043 day^{-1}) and more extended adult longevity (males: 6.50 ± 0.34 days and females: 24.15 ± 0.50 days). Population growth predictions over a period of 90 days of infestation, commencing with an initial population of 10 eggs showed that adult population size was 674,551 on tomato, 826,717 on potato and 355,139 on eggplant. Our data on plant host preference of *P. solenopsis* will aid the development of appropriate management strategies and achieve successful control of this invasive pest in key Mediterranean crop systems.

Keywords: tomato; potato; eggplant; polyphagous pest

1. Introduction

Invasive insect pests remain a constant threat to agricultural production systems worldwide. Their invasion benefits from both climate change and the rapid global trade of agricultural products including ornamentals, fruits and seedlings. In most cases, their establishment outside their native range can cause significant economic and ecological losses by disturbing ecosystem balance and impairing already implemented Integrated Pest Management (IPM) strategies, often leading to the overuse of synthetic insecticides [1–3].

Among invasive agricultural pests, mealybugs (Hemiptera: Coccomorpha: Pseudococcidae) are a major threat due to their biological features including cryptic behavior, high reproductive capacity, ecological plasticity and, in many cases, the ability to develop resistance to insecticides [4,5]. The cotton mealybug *Phenacoccus solenopsis* Tinsley has expanded its geographical distribution over the last decades [6]. Thus far, it has been reported in more than 70 countries worldwide and it is considered one of the most devastating pests of cotton in Asia (e.g., China, India, Iran and Pakistan) [7–9]. More recently, this pest has invaded several Mediterranean countries including Algeria, CyprusEgypt, France, Greece, Israel, Italy, Morocco, Tunisia and Turkey [10–17].

Phaenacoccus solenopsis is a highly polyphagous sap-feeding insect attacking about 300 host plant species belonging to 65 families, in particular species of Amaranthaceae,

Asteraceae, Cucurbitaceae, Euphorbiaceae, Fabaceae, Lamiaceae, Malvaceae and Solanaceae. It primarily feeds on the aerial parts of the plants, but can attack also roots and collars, producing abundant wax and honeydew, the latter promoting the development of sooty mould. Mealybug infestations cause distorted and bushy shoots, crinkled and/or twisted and bunchy leaves and ultimately death in the absence of efficient control measures [8,9,14,18].

Currently, cotton remains the preferred host crop of *P. solenopsis* with yield losses in Pakistan, China and India of 0.48, 1.4 and 1.12 million tons, respectively, during 2008–2009 [18]. Hence, an intensive and irrational use of insecticides for its management was adopted, leading to the emergence of resistance to many insecticides in several field populations [19–22]. In other invaded regions where cotton is not a key crop, *P. solenopsis* spread adapting to wild and ornamental hosts such as *Lantana camara* L. (Verbenaceae) and *Hibiscus rosa-sinensis* L. (Malvaceae) or has become a serious pest of Solanaceous crops, as has happened in Israel and Egypt where it has been reported on bell pepper, eggplant, tomato and potato [11,23,24]. Owing to the recent invasion of *P. solenopsis* in European and Mediterranean countries, it is therefore critical to assess the suitability of most relevant Solanaceous vegetables as host plants to reduce the possible yield losses and to develop efficient monitoring and control protocols.

Life table analyses including life history parameters and population fluctuations are an essential tool for studying insect ecology and fitness. They provide helpful information about insect biology and reproductive capacity allowing prediction of their performance and mortality patterns in relation to different host plants and environmental conditions [25]. However, for biparental species, conventional age-specific life tables often omit the contribution of males and differentiation by stage [26]. To fill this gap, Chi and Liu [27], Chi [28] and Chi et al. [29] developed the age-stage two-sex life table which has been successfully applied to study various ecological aspects of insect pests and their natural enemies [30–32].

In this study, we investigated the biology, survival, reproduction and life table parameters of *P. solenopsis* when reared on three economically important Solanaceous host plants, *S. lycopersicum* (tomato), *S. tuberosum* (potato) and *S. melongena* (eggplant), based on the age-stage two-sex life table using the TWOSEX-MSChart® software. We also assessed population-projection parameters using the TIMING-MSChart® program.

2. Results

Values for the developmental time of each stage, longevity and total duration of the life cycle of males and females on different hosts are presented in Table 1.

The egg-incubation time was the same on all host plants (1 d). For females, the duration of the first instar was longer on eggplant (6.18 ± 0.12 days) than on potato (5.58 ± 0.11 days) and tomato (4.88 ± 0.13 days) (P_{TP} = 0.0001; P_{TE} = 0; P_{PE} = 0.0005). The duration of the second instar did not differ among host plants ranging from 4.70 ± 0.15 to 4.84 ± 0.18 days (P_{TP} = 0.701; P_{TE} = 0.8721; P_{PE}= 0.1412). The duration of the third instar was greater on eggplant (7.09 ± 0.24 days) than on tomato (6.30 ± 0.26 days) and potato (6.28 ± 0.20 days) (P_{TP} = 0.0202; P_{TE} = 0.0287; P_{PE} = 0.0111). For males, the duration of the first instar was greater on eggplant (6.50 ± 0.43 days) than on potato (5.36 ± 0.15 days) and tomato (4.71 ± 0.36 days) (P_{TP} = 0.1045; P_{TE} = 0.0035; P_{PE} = 0.0185). The duration of the second instar was greater on potato (6.54 ± 0.41 days) than on tomato (5.28 ± 0.47 days) and eggplant (5.16 ± 0.48 days) (P_{TP} = 0.047; P_{TE} = 0.8721; P_{PE} = 0.1412). The duration of the pupa was greater on potato (7.09 ± 0.54 days) and tomato (6.71 ± 0.47 days) than on eggplant (4.50 ± 0.22 days) (P_{TP} = 0.602; P_{TE}= 0.0001; P_{PE}= 0.0001). The total female pre-adult developmental duration was greater on eggplant than tomato and potato (P_{TP} = 0.0717; P_{TE} = 0; P_{PE} = 0.001) and that of male was greater on potato than on tomato and eggplant (P_{TP} = 0.0082; P_{TE} = 0.5937; P_{PE} = 0). The longevity of adults was greater on eggplant than on tomato and potato for both males (P_{TP} = 0.9248; P_{TE} = 0; P_{PE} = 0) and females (P_{TP} = 0.0008; P_{TE} = 0; P_{PE} = 0). The duration of the life cycle of females was greatest on eggplant and lowest on tomato and potato (P_{TP} = 0.3548; P_{TE} = 0; P_{PE} = 0) and that of males was the longest on potato and eggplant and the shortest on tomato (P_{TP} = 0.0158; P_{TE} = 0.0203; P_{PE} = 0.5512).

Table 1. Developmental duration and adult longevity of *Phenacoccus solenopsis* on different host plants (25 ± 1 °C, 60 ± 5% RH and 16:8 h (L:D)). Values are means ± standard errors. Means in a row followed by different letters are significantly different at $p < 0.05$ using paired bootstrap test.

Developmental Duration (Days)	**Host Plants**		
	Tomato	**Potato**	**Eggplant**
Egg incubation	1 ± 0.00 a (*n* = 50)	1 ± 0.00 a (*n* = 50)	1 ± 0.00 a (*n* = 50)
First-instar nymph			
Female	4.8837 ± 0.1323 c (*n* = 27)	5.5897 ± 0.1132 b (*n* = 28)	6.1818 ± 0.1216 a (*n* = 23)
Male	4.7142 ± 0.3619 b (*n* = 23)	5.3636 ± 0.1511 b (*n* = 22)	6.50 ± 0.4320 a (*n* = 27)
Second-instar nymph			
Female	4.7441 ± 0.1870 a (*n* = 26)	4.8461 ± 0.1835 a (*n* = 26)	4.7045 ± 0.1564 a (*n* = 23)
Male	5.2857 ± 0.4755 b (*n* = 23)	6.5454 ± 0.4103 a (*n* = 22)	5.1666 ± 0.4819 b (*n* = 27)
Third-instar			
Female nymph	6.3023 ± 0.2688 b (*n* = 26)	6.2820 ± 0.2084 b (*n* = 26)	7.0909± 0.2417 a (*n* = 22)
Male pupa	6.7142 ± 0.4751 a (*n* = 23)	7.0909 ± 0.5436 a (*n* = 22)	4.50 ± 0.2249 b (*n* = 27)
Total pre-adult			
Female	16.9302 ± 0.3591 b (*n* = 26)	17.7179 ± 0.2493 b (*n* = 26)	18.9772 ± 0.2868 a (*n* = 22)
Male	17.7142 ± 0.8108 b (*n* = 23)	20.0000 ± 0.1899 a (*n* = 22)	17.1666 ± 0.6044 b (*n* = 27)
Adult longevity			
Female	21.0697 ± 0.3421 b (*n* = 26)	19.8205 ± 0.1197 c (*n* = 26)	24.1590 ± 0.5029 a (*n* = 22)
Male	3.1428 ± 0.1437 b (*n* = 23)	3.1818 ± 0.2248 b (*n* = 22)	6.5000 ± 0.3457 a (*n* = 27)
Total life cycle			
Female	38.0000 ± 0.4319 b (*n* = 26)	37.5384 ± 0.2466 b (*n* = 26)	43.1363 ± 0.5683 a (*n* = 22)
Male	20.8571 ± 0.8877 b (*n* = 23)	23.1818 ± 0.3234 a (*n* = 22)	23.6666 ± 0.7670 a (*n* = 27)

Data on fecundity, adult preoviposition period (APOP), total preoviposition period (TPOP) and oviposition days on the different host plants tested are shown in Table 2.

Table 2. Fecundity, adult preoviposition period (APOP), total preoviposition period (TPOP) and oviposition days of *Phenacoccus solenopsis* on different host plants (25 ± 1 °C, 60 ± 5% RH and 16:8 h (L:D)). Values are means ± standard errors. Means in a row followed by different letters are significantly different at $p < 0.05$ using paired bootstrap test.

Parameters	**Host Plants**		
	Tomato	**Potato**	**Eggplant**
APOP (days)	6.3953 ± 0.2162 b	5.2051± 0.1964 c	10.0909 ± 0.5255 a
TPOP (days)	23.3256 ± 0.405 b	22.9222 ± 0.3091 b	29.0682 ± 0.5807 a
Fecundity (eggs)	155.6046 ± 11.9904 c	244.9230 ± 9.7554 b	276.5000 ± 10.7814 a
Oviposition (days)	12.7209 ± 0.2469 b	13.6410 ± 0.1725 a	10.8409 ± 0.1109 c

The greatest APOP was registered on eggplant followed by tomato and potato (P_{TP} = 0.0001; P_{TE} = 0; P_{PE} = 0). Similarly, TPOP was greater on eggplant than on tomato and potato (P_{TP} = 0.4267; P_{TE} = 0; P_{PE} = 0). Concerning fecundity, it was greater on eggplant

(276.50 ± 10.78 eggs/female) than on potato (244.92 ± 9.75 eggs/female) and on tomato (155.60 ± 11.99 eggs/female) (P_{TP} = 0; P_{TE} = 0; P_{PE} = 0.0298). The number of oviposition days was greatest on potato (13.64 ± 0.17 days) followed by tomato (12.72 ± 0.24 days) and eggplant (10.84 ± 0.11 days) (P_{TP} = 0.0022; P_{TE} = 0; P_{PE} = 0).

The mean net reproductive rate (R_0) of *P. solenopsis* was significantly different on the selected host plants, attaining its highest value on eggplant (243.32 ± 15.83 nymphs/female), followed by potato (191.04 ± 16.23 nymphs/female) and tomato (133.82 ± 12.78 nymphs/female) (P_{TP} = 0.0056; P_{TE} = 0; P_{PE} = 0.0213). The average intrinsic rate of increase (r) was greatest on potato (0.18 ± 0.00 d^{-1}) compared to tomato and eggplant (P_{TP} = 0.0245; P_{TE} = 0.5411; P_{PE} = 0.0024). The finite rate of increase (λ) was significantly greater on tomato and eggplant (1.18 ± 0.005 d^{-1} and 1.18 ± 0.004 d^{-1}, respectively) than on potato (1.20 ± 0.004 d^{-1}) (P_{TP} = 0.0244; P_{TE} = 0.5411; P_{PE} = 0.0024) (Table 3).

Table 3. Net reproductive rate (R_0), the intrinsic rate of increase (r), finite rate of increase (λ) and generation time (T) of *Phenacoccus solenopsis* on different host plants (25 ± 1 °C, 60 ± 5% RH and 16:8 h (L:D)). Values are means ± standard errors. Means in a row followed by different letters are significantly different at $p < 0.05$ using the paired bootstrap test.

Parameters	Host Plants		
	Tomato	Potato	Eggplant
R_0	133.82 ± 12.7829 c	191.04 ± 16.23395 b	243.32 ± 15.83068 a
r	0.1731 ± 0.00424 b	0.1860 ± 0.0039 a	0.1696 ± 0.00364 b
λ	1.1889 ± 0.005042 a	1.2044 ± 0.00469 b	1.1848 ± 0.00431 a
T	28.2854 ± 0.4532 b	28.2371 ± 0.3660 b	32.3899 ± 0.5958 a

The age-stage-specific survival rate (s_{xj}) of *P. solenopsis* on different host plants indicates the probability that a newborn will survive to age x and develop to stage j (Figure 1).

Due to variable developmental rates among individuals, significant overlap was observed between stages in the survival curves. The first females emerged on days 12, 14 and 15, while the first males appeared on days 14, 19 and 16 on tomato, potato and eggplant, respectively. The survival rates of preadult instars ranged between 92% and 100%, while those of females were 84%, 78% and 86% on tomato, potato and eggplant, respectively. The lowest survival rate was recorded for males not exceeding 20% on all studied host plants.

The single age-stage survival rate (l_x) predicts that an egg will survive to age x (Figure 2). On all host plants, the l_x curve was constantly around 100% during the early stages, indicating a relatively low mortality rate.

The age-stage-specific fecundity (f_x) curve peak on potato was greater than that recorded for eggplant and tomato. The curve of age-specific fecundity (m_x) showed that reproduction began at 19 days on tomato and potato and 3 days later on eggplant, and that the fecundity on potato and eggplant was greater than on tomato.

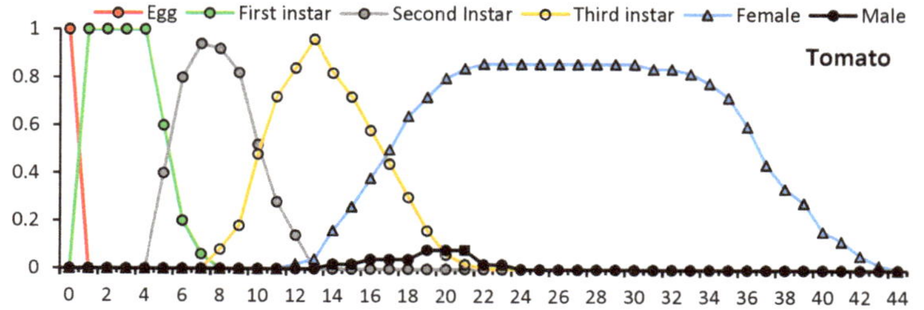

Figure 1. *Cont.*

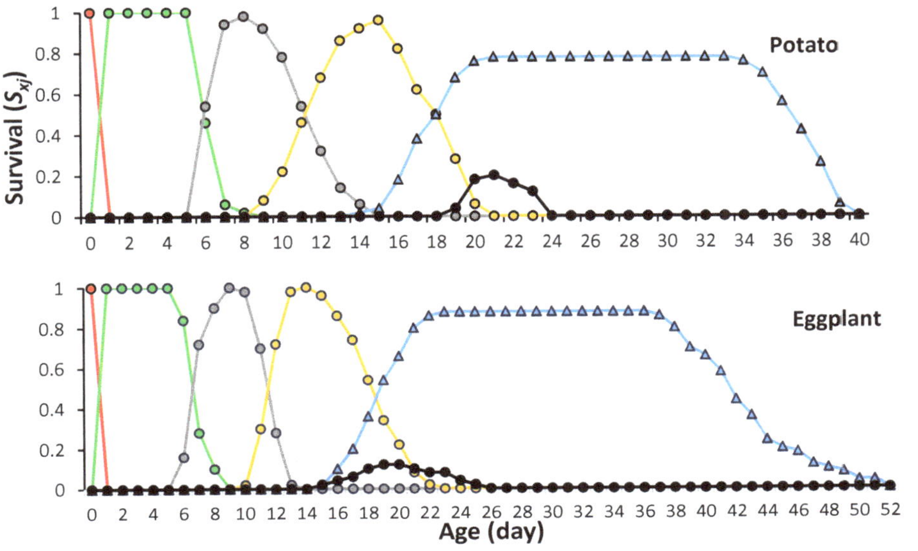

Figure 1. Survival rate of different development stages of *Phenacoccus solenopsis* on different host plants (25 ± 1 °C, 60 ± 5% RH and 16:8 h (L:D)).

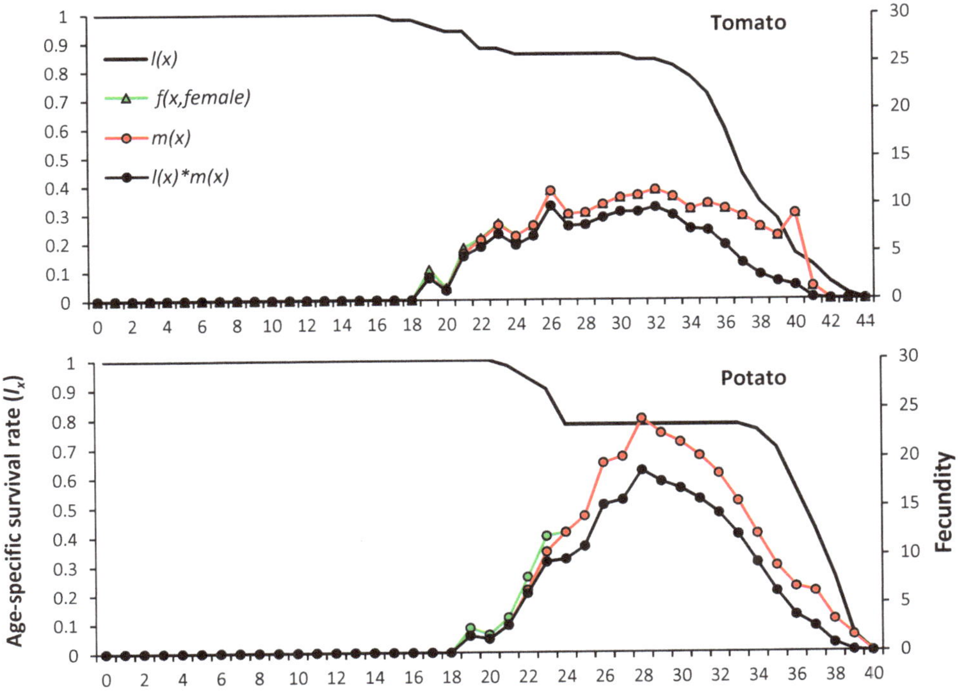

Figure 2. *Cont.*

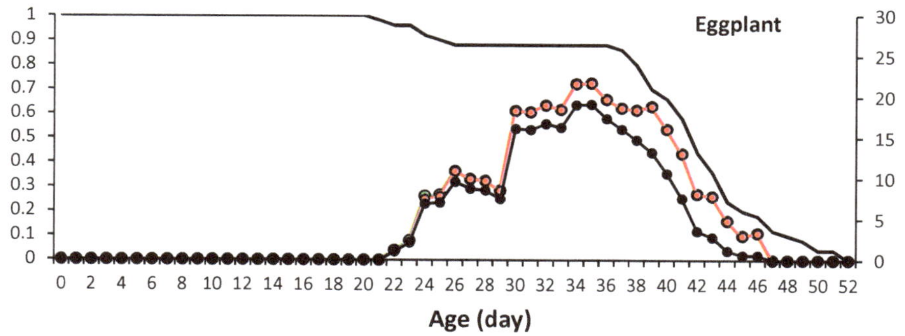

Figure 2. Age-specific survival rate (l_x), female age-specific fecundity (f_x), age-specific fecundity (m_x) and age-specific maternity (l_x*m_x) versus age of *Phenacoccus solenopsis* on different host plants (25 ± 1 °C, 60 ± 5% RH and 16:8 h (L:D)).

The age-stage life expectancy (e_{xj}) estimates the life duration of an individual of age x and stage j. The longevity of *P. solenopsis* at age zero (e_{01}) was 35.6 days on tomato, 34.38 days on potato and 40.80 days on eggplant (Figure 3).

The age-stage reproductive value (v_{xj}) shows the contribution of an individual from age x to stage j to the future population. The curves of reproductive value significantly increased when reproduction began, as shown in Figure 4. The value of v_{xj} peaked on day 26 for tomato and potato and on day 30 for eggplant with values of 54.87 d^{-1}, 99.57 d^{-1} and 103.59 d^{-1}, respectively.

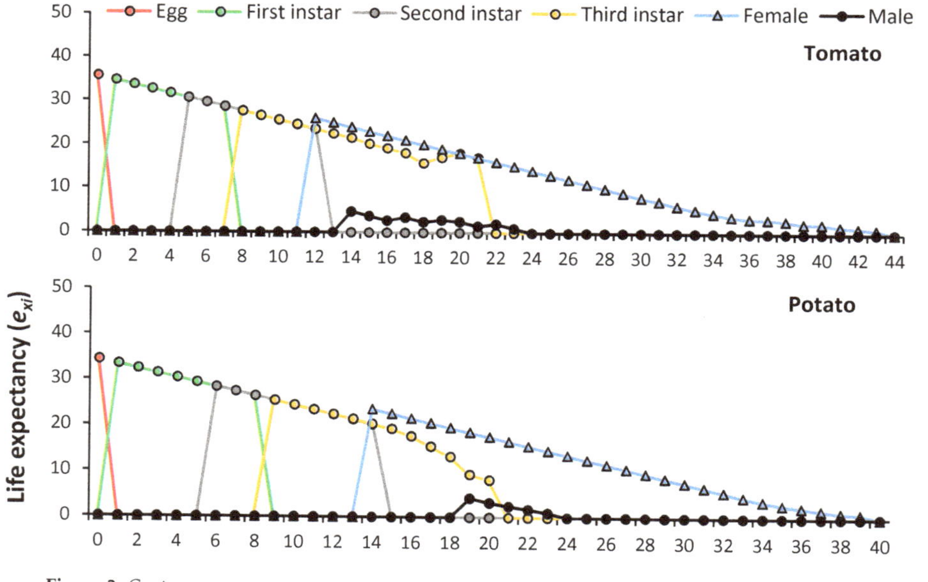

Figure 3. *Cont.*

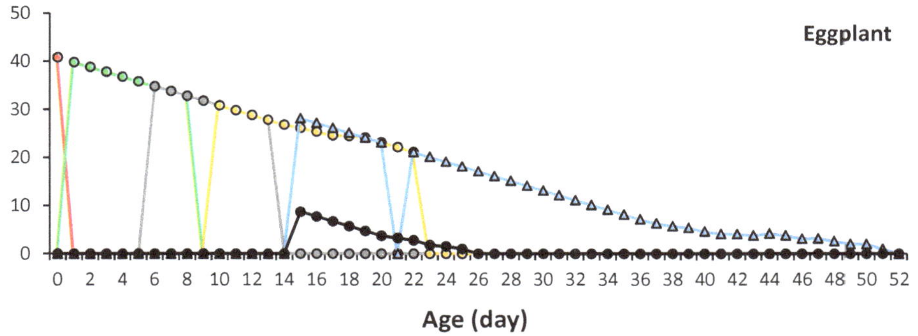

Figure 3. Age-stage life expectancy of *Phenacoccus solenopsis* on different host plants (25 ± 1 °C, 60 ± 5% RH and 16:8 h (L:D)).

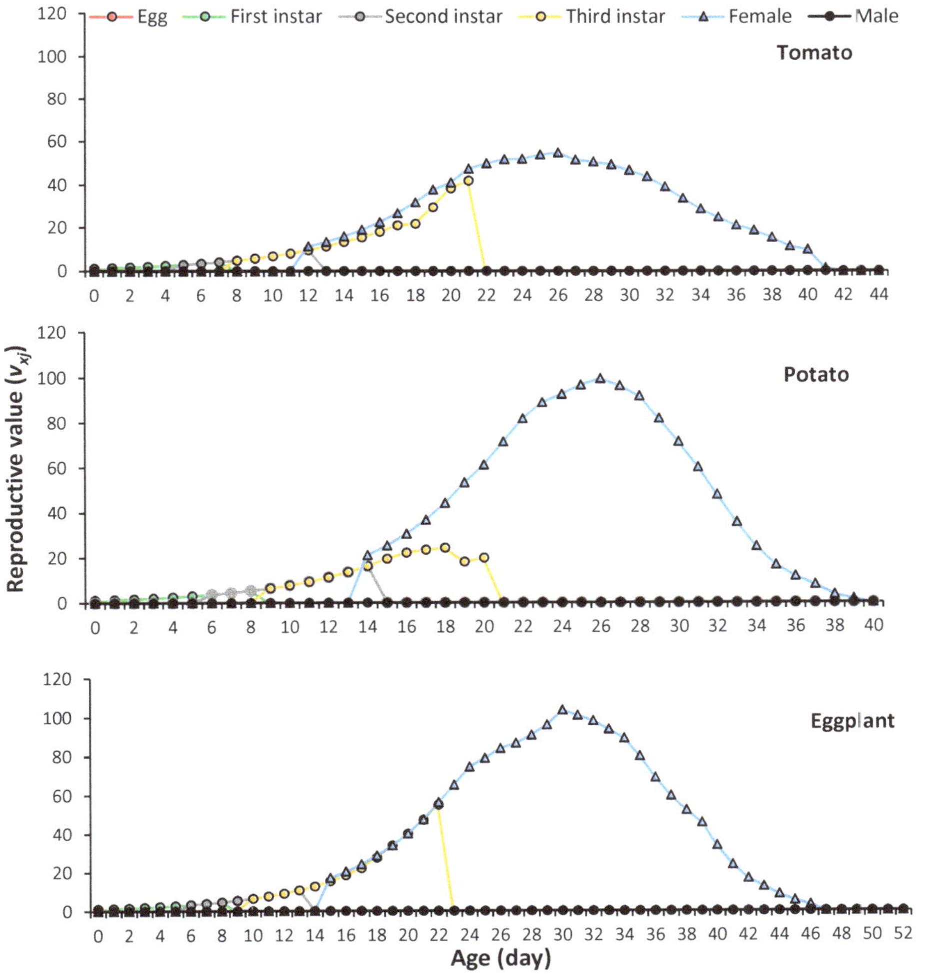

Figure 4. Age-stage reproductive value of *Phenacoccus solenopsis* on different host plants (25 ± 1 °C, 60 ± 5% RH and 16:8 h (L:D)).

Population growth predictions of *P. solenopsis* on considered host plants generated via the TIMING-MSChart® program are shown in Figure 5, which reveals considerable growth curves. Simulations suggest that they start to appear on the 12th, 14th and 15th days on eggplant, respectively (Figure 6A). The total predicted adult population size (N_t) after 90 days was 674,551 on tomato, 826,717 on potato and 355,139 on eggplant (Figure 6B,C).

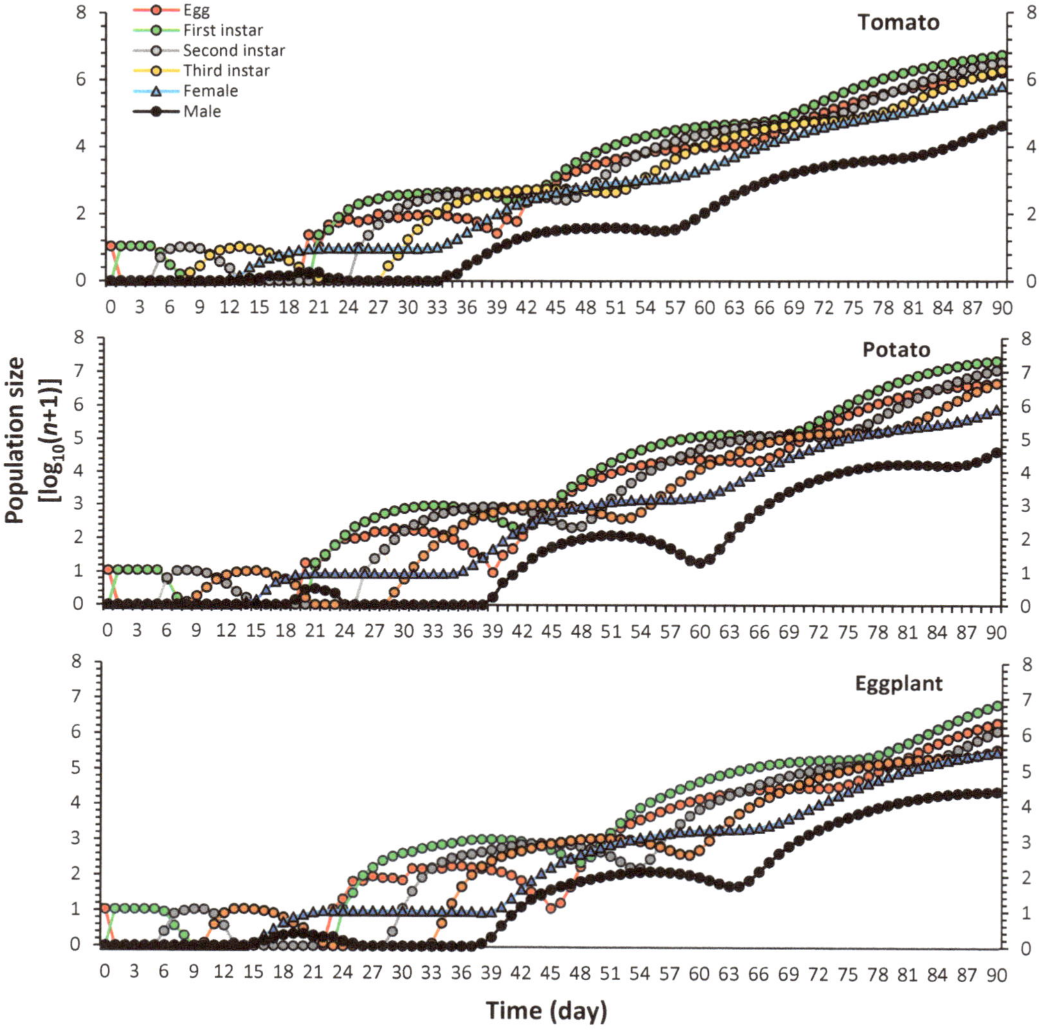

Figure 5. Population growth predictions of *Phenacoccus solenopsis* on different host plants.

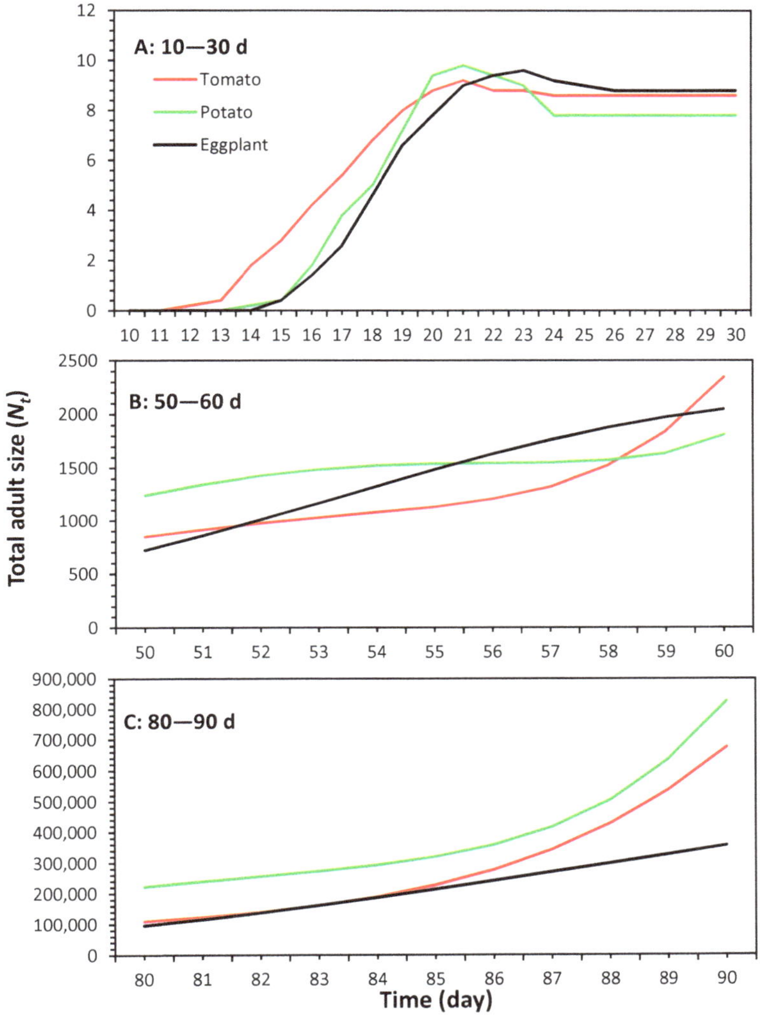

Figure 6. The total adult size (N_t) of *Phenacoccus solenopsis* on different host plants in time intervals (**A**) 10–30 d, (**B**) 50–60 d and (**C**) 80–90 d.

3. Discussion

Although cotton remains its preferred host, *P. solenopsis* is considered to be a potential economic pest of many other crops [33]. Its host range has been expanding over the last decade alongside its geographic spread largely due to favorable climate change. As a result, many Solanaceous crops including tomato, potato and eggplant, have become common hosts for this species in several newly colonized countries such as Algeria, Egypt, Israel, Iran, Italy and Tunisia, and in these newly invaded countries, where cotton is limited or absent, the damage to Solanaceous crops can be severe due, in particular, to its considerable reproductive capacity: *P. solenopsis* may go through many generations per cropping cycle. Furthermore, modelling studies predict that there will be a rise in the number of generations/year of this species, prompted by global warming [6]. In addition, it has been demonstrated that *P. solenopsis* produces more honeydew on tomato than on cotton [34], causing a greater amount of indirect damage, linked to the development of sooty mould which causes a reduction of photosynthesis and a depreciation of fruit quality. The production of honeydew also prompts a mutualism between ants and *P. solenopsis* that contributes to the rapid infestation by the mealybug, also hampering the possible action of natural enemies. For example, it has been demonstrated how ants successfully facilitated the invasion of another mealybug, *Delottococcus aberiae* De Lotto (Hemiptera: Pseudococcidae), on citrus in Spain [35]. In Mediterranean crops, the situation may be even more complicated due to the absence of coevolved specific natural enemies of this pest and the concomitant pressure posed by other invasive pests such as *Tuta absoluta* (Meyrick, 1917) (Lepidoptera: Gelechiidae) (the tomato leafminer) on tomato and eggplant and *Leptinotarsa decemlineata* (Say, 1824) (Coleoptera: Chrysomelidae) (the Colorado potato beetle) on potato [3,14].

Previous studies have shown that host plants can significantly affect the life history parameters and population dynamics of *P. solenopsis* [36–39] and have a significant impact on *Aenasius bambawalei* (Girault, 1915) (Hymenoptera: Encyrtidae), an efficient solitary endo-parasitoid of this pest, described from India and possibly accidentally introduced into that country along with its host pest [40].

In our study, tomato, potato and eggplant were all found to be suitable hosts for the pest. The theory suggests that the most appropriate host plant for a polyphagous insect pest should allow shorter APOP and TPOP and higher fecundity, net reproductive rate (R_0) and finite rate of increase (λ). Nabil [41] reported very high population densities of *P. solenopsis* in open field crops of eggplant in the Hihhya district, Sharqia Governorate in Egypt, reaching 328.75 individuals/leaf during September 2016. High densities of different instars of the pest reaching 150 individuals per plant were also recorded when reared on tomato in Israel by Spodek et al. [11]. In Egypt, natural infestation of the pest on tomato was also reported in Qalyubia governorate [42].

Significant differences in the selected parameters may be reasonably attributed to specific biochemical and morphological features and are in accordance with what has been reported in previous studies. For example, Shahid et al. [43] tested 25 different host plants for the development of *P. solenopsis*, correlating their morphological traits to the population dynamic of the pest. The authors concluded that eggplant is one of the most favorable host plants for the mealybug. Similarly, Nagrare et al. [44] found that the highest net reproductive rate was on cotton (284 females/female/generation) and the lowest was on tomato.

It is reasonable to hypothesize that the abundant glandular trichomes scattered on tomato leaves and stems play a key role in hampering the development of *P. solenopsis*. Nonetheless, it can conclude its cycle even on this plant, thus becoming another serious threat in the whole Mediterranean area.

The potential damage of *P. solenopsis* to eggplant and tomato can be particularly significant when they are grown in greenhouses, where it takes advantage of optimal climatic conditions and high host density. Indeed, Prasad et al. [45] observed the greatest fecundity and survival of crawlers at 30 °C and 32 °C, respectively, which are both common temperatures of protected crops in Mediterranean countries. This situation may lead to the overuse of chemicals in these crops while increasing plant protection costs and disrupting already implemented control schemes for other relevant pests. Furthermore, in Asia (in particular, Pakistan and India) *P. solenopsis* has already developed resistance to a wide range of insecticides including organophosphates, pyrethroids and neonicotinoids due to the continuous and severe use of these compounds on cotton cultivations [19–21]. This has severe consequences on efficient chemical control options of this pest on invaded host crops in the Mediterranean Basin since recent phylogenetic analyses performed in Tunisia and Italy revealed that the populations introduced in both countries most probably derive from Asian stock [16,17]. The data presented in this study may be a starting point for the development of suitable agroecological management strategies, such as those based on inter- and border cropping, with a view to progressive reduction of pest populations on cultivated crops and with the enhancement of biological control. These strategies are mostly needed in relation to reducing the application of synthetic insecticides as required by the European Union and by consumers.

4. Materials and Methods

4.1. Plants

Plants used in this study were tomato (Variety 'Dorra'), potato (Variety 'Spunta') and eggplant (Variety 'Tizona') grown in a glasshouse from seed (or tuber) in plastic trays (330 mm × 250 mm × 130 mm) in peat substrate, watered with tap water every two days and maintained under identical natural conditions without fertilizers and chemicals until they reached 150 mm height.

4.2. Insects

A laboratory rearing of *P. solenopsis* was initiated at the High Agronomic Institute of Chott-Mariem, Sousse, Tunisia, using specimens collected on *Lantana camara* L. (Verbenaceae) in Tunis, Tunisia, in spring 2022. The identity of specimens was confirmed using both morphological and molecular approaches [17]. Initially, the mealybug was reared for five generations on potato plants (Variety 'Spunta') obtained from tubers placed in plastic containers (330 mm × 250 mm × 130 mm) filled with fine sand. To avoid possible effects of host shifting during the experiments, two other colonies were set up using individuals reared on potato, on tomato and on eggplant. All colonies were maintained for five generations in a climatic chamber at 25 ± 2 °C, 60–70% RH and 16:8 h (L:D) photoperiod before their use in the bioassay.

4.3. Experimental Protocol

For each host plant (tomato, potato and eggplant), 50 plants were transplanted individually into plastic cups with 200 mL peat substrate. Peer cohorts of *P. solenopsis* were used to collect eggs and a single freshly laid egg (<1 h) was transferred to the central vein of an apical leaf using a fine soft paintbrush under a binocular microscope (Leica® MZ8, Leica Microsystems, Wetzlar, Germany). Plants bearing eggs were then incubated in a climatic cabinet (Scimmit, Shanghai Scimmit Technology, Shanghai, China) at 25 ± 1 °C temperature, 60 ± 5% relative humidity and 16:8 h (L:D) photoperiod. Plants were watered daily with tap water using a 50 mL volume syringe. The entire life cycle of each mealybug individual on each plant was monitored daily until its death. The moults during the larval stages were recorded by the presence of exuviae. Newly emerged adults were kept as couples to record the following parameters: egg incubation period, duration of each immature stage, pre-oviposition period, oviposition period, fecundity, post-oviposition period, adult sex and adult longevity.

4.4. Demographic Analyses

Collected data on the development and reproduction on each considered host plant were analyzed according to the age-stage two-sex life table theory as described by Chi and Liu [27] and Huang and Chi [46]. We calculated age-stage-specific survival rate (s_{xj}: the probability that a newly laid egg will survive to age x and stage j), age-stage-specific fecundity (f_{xj}: the mean fecundity of females at age x), age-specific survival rate (l_x: the probability that a newly laid egg will survive to age x), and age-specific fecundity (m_x: the mean fecundity of individuals at age x) (Supplementary Materials S1).

The means and standard errors of the life table parameters were estimated using the bootstrap method with a bootstrap number of m = 100,000 in order to ensure precise estimates [47]. TWOSEX-MSChart® [48] for Windows® (Version 2023.12.15) was used to construct and analyze age-stage two-sex life tables. A paired bootstrap test within TWOSEX-MSChart® was used to compare differences in developmental time, adult longevity, adult preoviposition period (APOP), total preoviposition period (TPOP), oviposition days and fecundity between treatments. The population parameters were also compared using the paired bootstrap test, based on the confidence interval of difference [47,49]. The p values of the paired bootstrap test were defined as follows: P_{TP}, tomato to potato; P_{TE}, tomato to eggplant; P_{PE}, potato to eggplant.

4.5. Population Projection

The TIMING-MSChart® [50] program (Version 04/18/2024) was used to simulate population growth rate and the structure of each age-stage of *P. solenopsis* over a period of 90 days with an initial population of 10 eggs and without control.

Supplementary Materials: The following supporting information can be downloaded at: https://www.mdpi.com/article/10.3390/plants13101381/s1. References [51–55] are cited in the supplementary materials.

Author Contributions: Conceptualization, K.A. and A.H.; methodology, K.A., B.C. and A.H.; software, K.A.; resources, B.C. and E.G.; data curation, K.A. and A.H; writing—original draft preparation, K.A.; writing—review and editing, A.H and E.G.; funding acquisition, E.G. and B.C. All authors have read and agreed to the published version of the manuscript.

Funding: This research was funded by financial support of the project "Agroecology-inspired Strategies and Tools to Enhance Resilience and ecosystem services in tomato crop, ASTER", Prima Section II 2021-MULTI-TOPIC, n°E73C22000240006. A.H. was the recipient of a postdoctoral grant from the Tunisian Ministry of Higher Education and Scientific Research (MHESR), through the Doctoral School "Agronomy and Environment" of the High Agronomic Institute of Chott-Mariem, University of Sousse.

Data Availability Statement: The datasets used and analyzed during the current study are available from the corresponding author upon reasonable request.

Acknowledgments: The authors are grateful to Hsin Chi from the National Chung Hsing University of Taiwan for his genuine help and guidance in analyzing data using TWOSEX-MSChart® and TIMING-MSChart® softwares. We also thank Mohamed El Hady Saleck for his help in rearing the mealybug and assistance in conducting the assays. Authors would like also to thank two anonymous reviewers for their constructive suggestions and remarks which substantially increased the quality of the paper.

Conflicts of Interest: The authors declare no conflicts of interest. The funders had no role in the design of the study; in the collection, analysis, or interpretation of data; in the writing of the manuscript; or in the decision to publish the results.

References

1. Charles, H.; Dukes, J.S. Impacts of Invasive Species on Ecosystem Services. In *Biological Invasions*; Springer: Berlin/Heidelberg, Germany, 2007; pp. 217–237.
2. Desneux, N.; Wajnberg, E.; Wyckhuys, K.A.G.; Burgio, G.; Arpaia, S.; Narváez-Vasquez, C.A.; González-Cabrera, J.; Catalán Ruescas, D.; Tabone, E.; Frandon, J.; et al. Biological invasion of European tomato crops by *Tuta absoluta*: Ecology, geographic expansion and prospects for biological control. *J. Pest. Sci.* **2010**, *83*, 197–215. [CrossRef]
3. Desneux, N.; Han, P.; Mansour, R.; Arnó, J.; Brévault, T.; Campos, M.R.; Chailleux, A.; Guedes, R.N.C.; Karimi, J.; Konan, K.A.J.; et al. Integrated pest management of *Tuta absoluta*: Practical implementations across different world regions. *J. Pest. Sci.* **2022**, *95*, 17–39. [CrossRef]
4. Pellizzari, G.; Germain, J.-F. Scales (Hemiptera, Superfamily Coccoidea). Chapter 9.3. *BioRisk* **2010**, *4*, 475–510. [CrossRef]
5. Miller, D.R.; Miller, G.L.; Watson, G.W. Invasive species of mealybugs (Hemiptera: Pseudococcidae). *Proc. Entomol. Soc. Wash.* **2002**, *104*, 825–836.
6. Fand, B.B.; Tonnang, H.E.Z.; Kumar, M.; Bal, S.K.; Singh, N.P.; Rao, D.V.K.N.; Kamble, A.L.; Nangare, D.D.; Minhas, P.S. Predicting the impact of climate change on regional and seasonal abundance of the mealybug *Phenacoccus solenopsis* Tinsley (Hemiptera: Pseudococcidae) using temperature-driven phenology model linked to GIS. *Ecol. Model.* **2014**, *288*, 62–78. [CrossRef]
7. Wang, Y.; Watson, G.W.; Zhang, R. The potential distribution of an invasive mealybug *Phenacoccus solenopsis* and its threat to cotton in Asia. *Agric. For. Entomol.* **2010**, *12*, 403–416. [CrossRef]
8. Tong, H.-J.; Ao, Y.; Li, Z.-H.; Wang, Y.; Jiang, M.-X. Invasion biology of the cotton mealybug, *Phenacoccus solenopsis* Tinsley: Current knowledge and future directions. *J. Integr. Agric.* **2019**, *18*, 758–770. [CrossRef]
9. Waqas, M.S.; Shi, Z.; Yi, T.-C.; Xiao, R.; Shoaib, A.A.; Elabasy, A.S.; Jin, D.-C. Biology, ecology, and management of cotton mealybug *Phenacoccus solenopsis* Tinsley (Hemiptera: Pseudococcidae). *Pest. Manag. Sci.* **2021**, *77*, 5321–5333. [CrossRef]

10. El-Zahi Saber, E.-Z.; Safwat Abd El-Salam, A.; Samy Kamal Mohammad, K. The cotton mealybug, *Phenacoccus solenopsis* Tinsley (Hemiptera: Pseudococcidae) as a new menace to cotton in Egypt and its chemical control. *J. Plant Prot. Res.* **2016**, *56*, 111–115. [CrossRef]
11. Spodek, M.; Ben-Dov, Y.; Mondaca, L.; Protasov, A.; Erel, E.; Mendel, Z. The cotton mealybug, *Phenacoccus solenopsis* Tinsley (Hemiptera: Pseudococcidae) in Israel: Pest status, host plants and natural enemies. *Phytoparasitica* **2018**, *46*, 45–55. [CrossRef]
12. Katbeh Bader, A.; Al-Jboory, I.J. First record of cotton mealybug, *Phenacoccus solenopsis* Tinsley 1898 (Hemiptera: Pseudococcidae), from Saudi Arabia. *EPPO Bull.* **2020**, *50*, 557–560. [CrossRef]
13. Aroua, K.; Kaydan, M.B.; Ercan, C.; Biche, M. First Record of *Phenacoccus solenopsis* Tinsley (Hemiptera: Coccoidea: Pseudococcidae) in Algeria. *Entomol. News* **2020**, *129*, 63–66. [CrossRef]
14. Bragard, C.; Dehnen-Schmutz, K.; Di Serio, F.; Gonthier, P.; Jacques, M.-A.; Jaques Miret, J.A.; Justesen, A.F.; Magnusson, C.S.; Milonas, P.; Navas-Cortes, J.A.; et al. Pest categorisation of *Leptinotarsa decemlineata*. *EFSA J.* **2020**, *18*, e06359. [CrossRef] [PubMed]
15. El Aalaoui, M.; Sbaghi, M. First record of the mealybug *Phenacoccus solenopsis* Tinsley (Hemiptera: Pseudococcidae) and its seven parasitoids and five predators in Morocco. *EPPO Bull.* **2021**, *51*, 299–304. [CrossRef]
16. Ricupero, M.; Biondi, A.; Russo, A.; Zappalà, L.; Mazzeo, G. The Cotton Mealybug Is Spreading along the Mediterranean: First Pest Detection in Italian Tomatoes. *Insects* **2021**, *12*, 675. [CrossRef] [PubMed]
17. Abbes, K.; Harbi, A.; Wanassi, T.; Ricupero, M.; Mazzeo, G.; Russo, A.; Biondi, A.; Zappalà, L.; Chermiti, B. The cotton mealybug *Phenacoccus solenopsis* Tinsley (Hemiptera: Pseudococcidae) is spreading in North Africa: First report in Tunisia. *Orient. Insects* **2024**, *58*, 172–186. [CrossRef]
18. Fand, B.B.; Suroshe, S.S. The invasive mealybug *Phenacoccus solenopsis* Tinsley, a threat to tropical and subtropical agricultural and horticultural production systems—A review. *Crop Prot.* **2015**, *69*, 34–43. [CrossRef]
19. Ahmad, M.; Akhtar, S. Development of resistance to insecticides in the invasive mealybug *Phenacoccus solenopsis* (Hemiptera: Pseudococcidae) in Pakistan. *Crop Prot.* **2016**, *88*, 96–102. [CrossRef]
20. Saddiq, B.; Shad, S.A.; Aslam, M.; Ijaz, M.; Abbas, N. Monitoring resistance of *Phenacoccus solenopsis* Tinsley (Homoptera: Pseudococcidae) to new chemical insecticides in Punjab, Pakistan. *Crop Prot.* **2015**, *74*, 24–29. [CrossRef]
21. Nagrare, V.S.; Fand, B.B.; Chinna Babu Naik, V.; Naikwadi, B.; Deshmukh, V.; Sinh, D. Resistance development in Cotton mealybug, *Phenacoccus solenopsis* Tinsley (Hemiptera: Pseudococcidae) to insecticides from Organophosphate, Thiadiazines and Thiourea derivatives. *Int. J. Trop. Insect Sci.* **2020**, *40*, 181–188. [CrossRef]
22. Shankarganesh, K.; Ricupero, M.; Sabtharishi, S. Field evolved insecticide resistance in the cotton mealybug *Phenacoccus solenopsis* and its direct and indirect impacts on the endoparasitoid *Aenasius arizonensis*. *Sci. Rep.* **2022**, *12*, 16764. [CrossRef] [PubMed]
23. Ibrahim, S.S. Study on cotton host plants of mealybug *Phenacoccus solenopsis* (Tinsley) and efficiency release the predator *Chrysoperla carnea* (Stephens) for its controlling on cotton plants in Egypt. *J. Plant Prot. Pathol.* **2018**, *9*, 247–252. [CrossRef]
24. Rezk, M.; Hassan, A.-N.T.; El-Deeb, M.F.; Shaarawy, N.; Dewer, Y. The impact of insecticides on the cotton mealybug, *Phenacoccus solenopsis* (Tinsley): Efficacy on potato, a new record of host plant in Egypt. *J. Plant Prot. Res.* **2019**, *59*, 50–59. [CrossRef]
25. Price, P.W. *Insect Ecology*; John Wiley & Sons: Hoboken, NJ, USA, 1997.
26. Birch, L.C. The intrinsic rate of natural increase of an insect population. *J. Anim. Ecol.* **1948**, *17*, 15–26. [CrossRef]
27. Chi, H.; Liu, H. Two new methods for the study of insect population ecology. *Bull. Inst. Zool. Acad. Sin.* **1985**, *24*, 225–240.
28. Chi, H. Life-Table Analysis Incorporating Both Sexes and Variable Development Rates Among Individuals. *Environ. Entomol.* **1988**, *17*, 26–34. [CrossRef]
29. Chi, H.; You, M.S.; Atlihan, R.; Smith, C.L.; Kavousi, A.; Özgökçe, M.S.; Güncan, A.; Tuan, S.J.; Fu, J.W.; Xu, Y.Y.; et al. Age-Stage, two-sex life table: An introduction to theory, data analysis, and application. *Entomol. Gen.* **2020**, *40*, 103–124. [CrossRef]
30. El Aalaoui, M.; Sbaghi, M. Life cycle and population growth parameter analysis of the mealybug *Phenacoccus solenopsis* on three new host plants. *Arthropod-Plant Interact.* **2022**, *16*, 437–448. [CrossRef]
31. El Aalaoui, M.; Sbaghi, M. Life table parameters and predation potential of the coccinellid *Hyperaspis trifurcata*, feeding on the invasive cactus scale *Dactylopius opuntiae*. *Int. J. Trop. Insect Sci.* **2023**, *43*, 2021–2031. [CrossRef]
32. Gharekhani, G.; Salekebrahimi, H.; Chi, H. Demography of *Tuta absoluta* (Meyrick) (Lepidoptera: Gelechiidae) reared on elicitor-treated tomato plants with an innovative comparison of projected population sizes and application of the multinomial theorem for population survival. *Pest. Manag. Sci.* **2023**, *79*, 4964–4976. [CrossRef]
33. Bragard, C.; Di Serio, F.; Gonthier, P.; Jaques Miret, J.A.; Justesen, A.F.; Magnusson, C.S.; Milonas, P.; Navas-Cortes, J.A.; Parnell, S.; Potting, R.; et al. Pest categorisation of *Phenacoccus solenopsis*. *EFSA J.* **2021**, *19*, e06801. [CrossRef] [PubMed]
34. Zhou, A.; Lu, Y.; Zeng, L.; Xu, Y.; Liang, G. Effect of Host Plants on Honeydew Production of an Invasive Mealybug, *Phenacoccus solenopsis* (Hemiptera: Pseudococcidae). *J. Insect Behav.* **2013**, *26*, 191–199. [CrossRef]
35. Plata, Á.; Gómez-Martínez, M.A.; Beitia, F.J.; Tena, A. Native ants facilitate the invasion by *Delottococcus aberiae* in Mediterranean citrus. *J. Pest. Sci.* **2024**, *97*, 255–267. [CrossRef]
36. Arif, M.J.; Shahid, M.R.; Gogi, M.D.; Arshad, M.; Khan, M.A. Studies on biological parameters of an invasive mealybug, Phenacoccus solenopsis Tinsely (Pseudococcidae: Hemiptera) on different host plants under laboratory conditions. *Acad. J. Entomol.* **2013**, *6*, 55–60.
37. Kedar, S.; Saini, R.; Ram, P. Suitability of different host plants for survival and development of solenopsis mealybug, *Phenacoccus solenopsis* Tinsley (Pseudococcidae: Hemiptera). *J. Entomol. Res.* **2013**, *37*, 47–50.

38. Çalışkan, A.F.; Kaydan, M.B.; Muştu, M.; Ulusoy, M.R. Demographic parameters and biological features of *Phenacoccus solenopsis* Tinsley (Hemiptera: Pseudococcidae) on four ornamental plants. *Phytoparasitica* **2016**, *44*, 75–82. [CrossRef]
39. Dogar, A.M.; Ayyaz, M.; Abbas, N.; Shad, S.A.; Naeem, A. Effect of host plants on life history traits of *Phenacoccus solenopsis* (Homoptera: Pseudococcidae). *Int. J. Trop. Insect Sci.* **2018**, *38*, 387–393. [CrossRef]
40. Younas, H.; Razaq, M.; Farooq, M.O.; Saeed, R. Host plants of *Phenacoccus solenopsis* (Tinsley) affect parasitism of *Aenasius bambawalei* (Hayat). *Phytoparasitica* **2022**, *50*, 669–681. [CrossRef]
41. Nabil, H.A. Ecological studies on cotton mealybug, *Phenacoccus solenopsis* Tinsley (Hemiptera: Sternorrhyncha: Coccoidea: Pseudococcidae) on eggplant at Sharkia Governorate, Egypt. *Egypt. Acad. J. Biol. Sci. A Entomol.* **2017**, *10*, 195–206. [CrossRef]
42. Ibrahim, S.S.; Abdelhalim Moharum, F.; Abd El-Ghany, N.M. The cotton mealybug *Phenacoccus solenopsis* Tinsley (Hemiptera: Pseudococcidae) as a new insect pest on tomato plants in Egypt. *J. Plant Prot. Res.* **2015**, *55*, 48–51. [CrossRef]
43. Shahid, M.R.; Arif, M.J.; Gogi, M.D.; Javed, N. Host-plant-preference and mortality analysis of *Phenacoccus solenopsis* in association with biochemical traits of different plant species. *Int. J. Agric. Biol.* **2017**, *19*, 211–218. [CrossRef]
44. Nagrare, V.S.; Naikwadi, B.; Deshmukh, V.; Kranthi, S. Biology and population growth parameters of the cotton mealybug, *Phenacoccus solenopsis* Tinsley (Hemiptera: Pseudococcidae), on five host plant species. *Anim. Biol.* **2018**, *68*, 333–352. [CrossRef]
45. Prasad, Y.G.; Prabhakar, M.; Sreedevi, G.; Ramachandra Rao, G.; Venkateswarlu, B. Effect of temperature on development, survival and reproduction of the mealybug, *Phenacoccus solenopsis* Tinsley (Hemiptera: Pseudococcidae) on cotton. *Crop Prot.* **2012**, *39*, 81–88. [CrossRef]
46. Huang, Y.-B.; Chi, H. Age-stage, two-sex life tables of *Bactrocera cucurbitae* (Coquillett) (Diptera: Tephritidae) with a discussion on the problem of applying female age-specific life tables to insect populations. *Insect Sci.* **2012**, *19*, 263–273. [CrossRef]
47. Bradley, E.; Tibshirani, R.J. An introduction to the bootstrap. *Monogr. Stat. Appl. Probab.* **1993**, *57*, 158.
48. Chi, H. TWOSEX-MSChart: A Computer Program for the Age-Stage, Two-Sex Life Table Analysis. 2023. Available online: http://140.120.197.173/Ecology/Download/TWOSEX-MSChart-setup.rar (accessed on 15 March 2024).
49. Akköprü, E.P.; Atlıhan, R.; Okut, H.; Chi, H. Demographic Assessment of Plant Cultivar Resistance to Insect Pests: A Case Study of the Dusky-Veined Walnut Aphid (Hemiptera: Callaphididae) on Five Walnut Cultivars. *J. Econ. Entomol.* **2015**, *108*, 378–387. [CrossRef] [PubMed]
50. Chi, H. TIMING-MSChart: A Computer Program for the Population Projection Based on Age-Stage, Two-Sex Life Table. 2023. Available online: http://140.120.197.173/Ecology/Download/TIMING-MSChart-exe.rar (accessed on 3 May 2024).
51. Goodman, D. Optimal life histories, optimal notation, and the value of reproductive value. *Am. Nat.* **1982**, *119*, 803–823. [CrossRef]
52. Chi, H.; Su, H.-Y. Age-Stage, Two-Sex Life Tables of *Aphidius gifuensis* (Ashmead) (Hymenoptera: Braconidae) and its host *Myzus persicae* (Sulzer) (Homoptera: Aphididae) with mathematical proof of the relationship between female fecundity and the net reproductive rate. *Environ. Entomol.* **2006**, *35*, 10–21. [CrossRef]
53. Fisher, R.A.S. *The Genetical Theory of Natural Selection*; Clarendon Press: Oxford, UK. [CrossRef]
54. Huang, Y.B.; Chi, H. The age-stage, two-sex life table with an offspring sex ratio dependent on female age. *J. Agric. For.* **2011**, *60*, 337–345.
55. Tuan, S.-J.; Lee, C.-C.; Chi, H. Population and damage projection of *Spodoptera litura* (F.) on peanuts (*Arachis hypogaea* L.) under different conditions using the age-stage, two-sex life table. *Pest. Manag. Sci.* **2014**, *70*, 805–813, Erratum in *Pest. Manag. Sci.* **2014**, *7*, 1936. [CrossRef]

plants

MDPI

Article

The Potential of Two Phytoseiid Mites as Predators of the Grape Erineum Mite, *Colomerus vitis*

Mahmoud M. Al-Azzazy [1,2] and Saleh S. Alhewairini [1,*]

1 Department of Plant Protection, College of Agriculture and Food, Qassim University, P.O. Box 6622, Buraidah 51452, Saudi Arabia; m.elazzazy@qu.edu.sa

2 Department of Agricultural Zoology and Nematology, College of Agriculture, Al-Azhar University, Cairo 11651, Egypt

* Correspondence: hoierieny@qu.edu.sa

Abstract: *Phytoseius plumifer* (Canestrini and Fanzago) and *Euseius scutalis* (Athias-Henriot) (Phytoseiidae) are generalist predatory mites important in controlling phytophagous mites on some agricultural crops. The biology of both species as potential biological control agents of the grape erineum mite, *Colomerus vitis* (Pagenstecher) (Eriophyidae) on grape leaf disks was studied in the laboratory at 33 ± 1 °C, 60%RH, 12:12 h L:D. The developmental time, survival, and reproductive parameters of *P. plumifer* and *E. scutalis* on *C. vitis*, date palm pollen as well as *C. vitis* plus date palm pollen were investigated. Both predators, *P. plumifer* and *E. scutalis*, thrived on the mixed diet of *C. vitis* and date palm pollen resulting in a shorter developmental time (6.16 and 6.69 days, respectively), higher oviposition rate (2.11 and 1.96 eggs/female/day, respectively), and higher intrinsic rate of increase (0.251 and 0.229 per female/day, respectively) than on any other diet. Date palm pollen was an adequate alternative food source for *P. plumifer* and *E. scutalis*. The results suggest that both predators have good potential to suppress *C. vitis* populations and that date palm pollen can support the population establishment of both predators in the absence or scarcity of the main prey in the environment. We discuss the relevance of our results for the biocontrol of *C. vitis*.

Keywords: biological control; *Phytoseius plumifer*; *Euseius scutalis*; mixed diet; alternative food

1. Introduction

Grapevine (*Vitis vinifera* L.) is an economically important fruit crop in the world and, globally, the third most valuable horticultural crop after potatoes and tomatoes [1]. Worldwide vineyard surface accounts for 7.327 million hectares, with a total grape production of 77.8 Mio.t that sustains dried grape and fresh grape markets and wine elaboration worldwide. However, most grapevine cultivars are susceptible to mite infestations, which considerably limit grape production. Moreover, grape mite infestations are becoming worse owing to the excessive use of pesticides and the destruction of biodiversity in recent years [2]. It is well known that some serious mite species occur on grape leaves. One of the most devastating pests globally infecting grapevines is grape erineum mite, *Colomerus vitis* (Pagenstecher), which may result in total crop damage in case of severe infestation [3]. The feeding of this mite causes a decrease in leaf area, photosynthetic rate, and chlorophyll contents and an increase in leaf fresh weight due to the hyperplasia and hypertrophy of mesophyll and epidermal cells and leaf deformities as well as reduction in the growth of green grapevine shoots and causes damage in nurseries and vineyards [4–7]. *Colomerus vitis*-infested grape leaves initially show white patches on the lower leaf surface. The young leaves then twist, become fragile, and experience the formation of a layer of white fluff plaque at the back, which appears blister-like on the upper surface. Analogous to rusts, yellow, fungi-like spots form in the subsequent period and, finally, become reddish brown. When *C. vitis* infestation is severe, the grapevines cannot renew and grow new buds, which inhibits the ability of leaves, and it affects grape production [8]. Furthermore, *C. vitis* infestation in grape leaves causes the spread of grapevine pinot gris virus (GPGV) [5,9], as well as grape berry necrosis virus (GINV) [9]. Therefore, grape plant health management is needed which enhances productivity, prevents crop loss, and contributes towards food security.

Currently, pest management practices are generally limited to chemical pesticide application; vineyards are subjected to very high levels of synthetic pesticides [10]. The use of synthetic pesticides to manage phytophagous mites results in the resistance of mites to major pesticides, the resurgence of secondary diseases and pests, the lethal effects of pesticides on natural enemies and other non-target organisms, and the presence of pesticide residues in crops, as well as negative impacts on biodiversity and on the environment and human health [11–13]. These negative impacts strongly limit the sustainability of farming systems and primarily necessitate the development of non-chemical methods of pest control. Additionally, due to concerns regarding the adverse effects of chemical pesticides, there is an increasing demand for pesticide-free fruits. Consequently, it is imperative to find effective biological control agents against *C. vitis* [3]. The utilization of predatory mites as a tool in phytophagous mites' management is important for sustainability and food security and a promising way to reduce the level of chemical pesticide use [14,15]. Phytoseiid mites are the most extensively studied and most applied biocontrol agents of phytophagous mites (particularly of the Eriophyoidea and Tetranychoidea), thrips and whiteflies in orchards, grapes, citrus crops, and greenhouses [16–20]. Most of all, phytoseiids are generalist predators and can survive and develop feeding on prey when they are available but are also capable of surviving on a broad range of other foods (plant exudates, pollen, nectar, fungi, etc.) when the prey are rare or absent [21]. *Phytoseius plumifer* (Canestrini and Fanzago) (Phytoseiidae) is an important generalist predator, and one of the most abundant natural enemies and an efficient predator of phytophagous mites on various crops in several countries [22,23]. It is found naturally on grapes in many countries around the world [19]. This predator is capable of preying on a range of prey-food types, including eriophyid mites, tetranychids, tarsonemids, and pollen as food [15,24–26]. Also, *Euseius scutalis* (Athias-Henriot) (Phytoseiidae) is a predator that can be considered in integrated pest management (IPM) programs against some pests. It appears to be adapted to several plants, such as fig, pomegranate, apple, okra, vines, avocado, citrus, etc. [27,28]. *E. scutalis* is a generalist predator capable of preying on a wide range of food items, including eriophyid mites, tetranychid mites, whiteflies, scale insects, and the eggs of some insects, in addition to pollen grains [29–33].

Habitually, generalist predacious mites can exploit different diets, including natural prey and pollen [34,35]. Furthermore, plant pollens can serve as a supply of nutrients and water complementary to a diet consisting of prey [36,37]. Many studies have shown that adding pollen to a crop can promote pest control by predacious mites [38]. The pollinivory of *P. plumifer* and *E. scutalis* offers the possibility to pre-establish populations in vineyards with the supplementation of pollen and allows them to sustain their populations in the grape crops when prey is scarce or absent and thus prevent severe declines in predatory mite populations during scarcities of primary prey. So far, no study has examined the potential of *P. plumifer* and *E. scutalis* as predators of the grape erineum mite, *C. vitis*. The main objective of the current laboratory study was to compare the potential of these predatory mite species as biocontrol agents of *C. vitis*. In particular, the developmental, survival, predation rate, and reproductive performance of the two phytoseiid mites was assessed, using *C. vitis*, date palm pollen, or a combination of both as food.

2. Results

2.1. Developmental Time and Survival of Immature Stages

Phytoseius plumifer and *Euseius scutalis* developed successfully on *C. vitis*, *date palm pollen*, and *C. vitis* plus date palm pollen. The mites developed the fastest when reared on *C. vitis* plus date palm pollen and the slowest on date palm pollen only (Table 1). That diet significantly affected the developmental time of *E. scutalis* but not that of *P. plumifer*. The predatory mite, *P. plumifer,* completed its development faster as compared to *E. scutalis* on all three diets. *P. plumifer* showed similar egg-to-adult developmental times of 6.16–6.37 days on the three diets, while *E. scutalis* required 6.69–8.44 days. Diet had no significant effect on the durations of the different developmental stages of *P. plumifer*. On the contrary, the predatory mite, *E. scutalis,* developed significantly faster on a mixed diet as compared to date palm pollen alone ($p < 0.005$) but not as compared to *C. vitis* alone ($p = 0.130$). The stages of this species developed faster on *C. vitis* alone as compared to date palm pollen alone ($p = 0.021$) but not as compared to a mixed diet ($p = 0.845$) (Table 1). On all tested diets, the survival of the immature stages was as high as 94.12 for *P. plumifer* and 91.80 for *E. scutalis* (Table 2). The interaction between diet and predator species was not significant, and there was not a noteworthy effect of predator species on survival from egg to adult. Tukey's HSD test indicated that juvenile survival was not significantly affected by diet.

Table 1. Average development duration in days of the immature stages of *Phytoseius plumifer* and *Euseius scutalis* feeding on three diets at 33 ± 1 °C, 60% ± 5% RH.

Predator Species	Diet	Sex	Egg	Larva	Protonymph	Deutonymph	Overall Developmental Time
Phytoseius plumifer	Date palm pollen.	Female	2.31 ± 0.22 a	1.05 ± 0.18 a	1.39 ± 0.14 a	1.62 ± 0.16 a	6.37 ± 0.65 a
		Male	2.25 ± 0.18 a	1.17 ± 0.12 a	1.31 ± 0.12 a	1.52 ± 0.14 a	6.25 ± 0.60 a
	mixed stages of *C. vitis*	Female	2.38 ± 0.18 a	1.05 ± 0.20 a	1.30 ± 0.18 a	1.56 ± 0.14 a	6.29 ± 0.47 a
		Male	2.35 ± 0.16 a	1.02 ± 0.10 a	1.24 ± 0.11 a	1.53 ± 0.12 a	6.14 ± 0.52 a
	mixed stages of *C. vitis* + date palm pollen	Female	2.30 ± 0.14 a	1.10 ± 0.04 a	1.29 ± 0.05 a	1.47 ± 0.05 a	6.16 ± 0.45 a
		Male	2.30 ± 0.13 a	1.08 ± 0.05 a	1.25 ± 0.04 a	1.43 ± 0.04 a	6.06 ± 0.52 a
Euseius scutalis	Date palm pollen	Female	2.36 ± 0.25 a	1.25 ± 0.20 a	2.26 ± 0.28 a	2.57 ± 0.18 a	8.44 ± 0.72 a
		Male	2.32 ± 0.22 a	1.22 ± 0.18 a	2.23 ± 0.20 a	2.84 ± 0.16 a	8.25 ± 0.62 a
	mixed stages of *C. vitis*	Female	2.32 ± 0.15 a	1.23 ± 0.17 a	1.64 ± 0.18 b	1.91 ± 0.12 b	7.10 ± 0.52 b
		Male	2.30 ± 0.12 a	1.22 ± 0.15 a	1.62 ± 0.16 b	1.85 ± 0.08 b	6.99 ± 0.66 b
	mixed stages of *C. vitis* + date palm pollen	Female	2.30 ± 0.18 a	1.21 ± 0.09 a	1.51 ± 0.09 b	1.67 ± 0.05 b	6.69 ± 0.64 b
		Male	2.28 ± 0.14 a	1.23 ± 0.07 a	1.50 ± 0.11 b	1.64 ± 0.07 b	6.65 ± 0.71 b

Different letters in each column denote significant difference within each species (ANOVA followed by Duncan's multiple range test: $p < 0.05$).

Table 2. Survival of immature stages of *Phytoseius plumifer* and *Euseius scutalis* feeding on three diets at 33 ± 1 °C, 60% ± 5% RH.

Predator Species	Diet	Stage Specific Survival (% ± SE)				Survival to Adulthood (% ± SE)
		Egg	Larva	Protonymph	Deutonymph	
Phytoseiu splumifer	Date palm pollen	93.25 ± 3.89	91.25 ± 4.12	91.08 ± 4.54	90.91 ± 4.19	90.14 ± 4.12
	C. vitis	94.54 ± 4.11	96.30 ± 3.96	95.16 ± 3.28	93.52 ± 3.85	92.23 ± 3.43
	C. vitis + pollen	97.35 ± 3.81	98.54 ± 3.35	97.10 ± 3.49	94.29 ± 3.18	94.12 ± 3.81
Euseius scutalis	Date palm pollen	92.24 ± 3.60	90.60 ± 4.87	89.42 ± 4.74	90.17 ± 3.61	90.37 ± 4.15
	C. vitis	93.97 ± 4.51	92.67 ± 4.45	92.25 ± 3.45	91.05 ± 3.53	91.80 ± 3.25
	C. vitis + Pollen	95.64 ± 3.78	94.30 ± 3.61	93.85 ± 3.56	92.14 ± 4.23	91.15 ± 2.89

No significative differences according to the Tukey HSD test.

2.2. Adult Longevity

Overall, no significant difference in the pre-oviposition period was observed between *C. vitis* and pollen alone, whereas the shortest pre-oviposition period was recorded on a mixed diet for *P. plumifer* and *E. scutalis*. (Table 3). Diet had a clear significant effect on the generation period and female longevity in both predator species. There were insignificant differences between the pollen and *C. vitis* and *C. vitis* diet treatments ($p = 0.0274$), while there was a significant difference between treatments of pollen diet alone, *C. vitis* and pollen, and *C. vitis* ($p = 0.0243$). The generation period and female adult longevity lasted

9.22 and 28.48 days, 9.08 and 33.13 days, and 8.37 and 33.43 days when *P. plumifer* fed on date palm pollen, *C. vitis*, and *C. vitis* plus pollen, respectively. The corresponding periods were 12.81 and 24.11 days, 10.40 and 28.30 days, and 9.14 and 28.96 days when *E. scutalis* fed on date palm pollen, *C. vitis*, and *C. vitis* plus pollen, respectively. For both predators, the oviposition period differed significantly among food types ($p = 0.064$). When the mixed diet was the food source, the total oviposition period was the longest, whereas it was the shortest when the predators fed on pollen alone (Table 3).

Table 3. Average development duration in days of *Phytoseius plumifer* and *Euseius scutalis* adults feeding on three diets at 33 ± 1 °C, 60% ± 5% RH.

Predator Species	Diet	Pre Oviposition	Oviposition	Post Oviposition	Longevity		Life Span	
					Female	Male	Female	Male
Phytoseius plumifer	Date palm pollen	2.85 ± 0.18 a	21.16 ± 0.1.24 a	4.47 ± 0.24 a	28.48 ± 1.16 a	26.89 ± 1.25 a	34.85 ± 1.35 a	33.14 ± 1.88 a
	mixed stages of *C. vitis*	2.79 ± 0.19 a	25.96 ± 0.79 b	4.38 ± 0.30 a	33.13 ± 0.96 b	31.53 ± 1.82 b	39.42 ± 1.46 b	37.42 ± 1.42 b
	mixed stages of *C. vitis* + date palm pollen	2.21 ± 0.14 b	26.87 ± 0.92 b	4.35 ± 0.38 a	33.43 ± 1.04 b	32.27 ± 0.1.14 b	39.59 ± 1.36 b	38.43 ± 1.69 b
Euseius scutalis	Date palm pollen	4.37 ± 0.31 a	16.58 ± 0.95 a	3.16 ± 0.34 a	24.11 ± 0.98 a	19.47 ± 2.34 a	32.55 ± 1.31 a	27.72 ± 2.51 a
	mixed stages of *C. vitis*	3.30 ± 0.24 a	21.12 ± 1.06 b	3.88 ± 0.46 a	28.30 ± 1.14 b	25.95 ± 0.98 b	35.40 ± 1.44 b	32.94 ± 1.70 b
	mixed stages of *C. vitis* + date palm pollen	2.45 ± 0.16 c	22.49 ± 0.83 b	4.02 ± 0.20 a	28.96 ± 1.26 c	27.43 ± 0.1.05 c	35.65 ± 1.33 c	34.08 ± 1.51 c

Different letter denotes significant difference within species (ANOVA followed by Duncan's multiple range test: $p < 0.05$).

2.3. Reproduction

Phytoseius plumifer and *E. scutalis* showed similar fecundity on either diet ($p = 0.8543$). In contrast, in both predatory mites, fecundity was significantly affected by diet ($p < 0.001$). On the grape erineum mite diet, total oviposition over the oviposition period averaged 53.94 ± 1.28 and 40.09 ± 1.22 eggs for *P. plumifer* and *E. scutalis*, respectively ($p > 0.878$). A combination of *C. vitis* and date palm pollen resulted in an average oviposition of 56.81 ± 1.30 and 44.16 ± 1.35 eggs over the oviposition period for the respective predators ($p > 0.985$). The addition of date palm pollen to the diet of *C. vitis* substantially increased the reproductive performance of both *P. plumifer* ($p = 0.004$) and *E. scutalis* ($p < 0.001$). In addition, *P. plumifer* and *E. scutalis* females fed on pollen alone exhibited a lower rate of fecundity than those feeding on grape erineum mite diet and the mixed diet. The post-oviposition periods of *P. plumifer* and *E. scutalis* did not differ between the three tested diets (Table 4).

Table 4. Fecundity of *Phytoseius plumifer* and *Euseius scutalis* feeding on three diets at 33 ± 1 °C, 60% ± 5% RH.

Diet	*Phytoseius plumifer*		*Euseius scutalis*	
	Total Fecundity ± SD	Daily Fecundity	Total Fecundity ± SD	Daily Fecundity
Date palm pollen	35.48 ± 1.12 Aa	1.67 ± 0.08	24.55 ± 1.07 Ab	1.48 ± 0.04
C. vitis	53.94 ± 1.28 Bb	2.07 ± 0.06	40.09 ± 1.22 Bc	1.89 ± 0.09
C. vitis + Pollen	56.81 ± 1.30 Bd	2.11 ± 0.09	44.16 ± 1.35 Be	1.96 ± 0.05

The capital letter denotes the significance within the same column, and the small letter denotes the significance within the same row at $p < 0.05$.

2.4. Predation of P. plumifer and E. scutalis

The larvae of both predators were inactive and did not feed during the experiment, and the feeding activity started immediately after the predators entered the protonymphal stages. Both *P. plumifer* and *E. scutalis* successfully suppressed the population of *C. vitis* on the small laboratory-rearing units, in the absence of date palm pollen. The mean daily predation rate was significantly affected by predator age and diet ($p < 0.001$).

For both predators, the adults consumed more prey compared with the nymph stages. There was a significant difference in the predation rate of adult *C. vitis* between the two species of predatory mites, *P. plumifer* and *E. scutalis*. The total number of *C. vitis* prey consumed by *P. plumifer* and *E. scutalis* immature and adult stages are shown in (Tables 5 and 6). In the absence of date palm pollen, immature females of *P. plumifer* significantly consumed a higher number of prey 106.86 ± 3.64 than *E. scutalis* 96.55 ± 2.17. The highest means for the daily predation of females were observed during the oviposition period, with the female of *P. plumifer* devouring an average of 2931.40 ± 16.62, while the female of *E. scutalis* consumed an average of 1994.52 ± 12.40. Thereafter, the daily consumption of predators fed on *C. vitis* decreased with age. At the end of the experiment, *P. plumifer* showed the highest predation rates on *C. vitis* in the absence of date palm pollen. The highest number of preys consumed during the life span was reported for *P. plumifer* females with 3525.63 ± 16.47 prey, while for *E. scutalis*, it was 2452.49 ± 10.53 prey (Table 5). So, it could be concluded that *P. plumifer* performance was better than *E. scutalis* against *C. vitis*. Providing date palm pollen with *C. vitis* resulted in a significant reduction in total prey consumption during *P. plumifer* and *E. scutalis* juvenile development from approx. 106.86 ± 3.64 and 96.55 ± 2.17 mites per predator female when mites were reared on *C. vitis* only to approx. 56.74 ± 1.47 and 46.60 ± 1.38 mites per predator female in the mixed diets for *P. plumifer* and *E. scutalis*, respectively. A similar trend was found for both predators during female longevity periods. Providing date palm pollen with *C. vitis* resulted in a significant reduction in the total prey consumption during *P. plumifer* and *E. scutalis* longevity from 3418.77 ± 14.50 and 2355.87 ± 15.85 mites per predator when mites were reared on *C. vitis* only to approx. 2052.00 ± 11.34 and 1247.65 ± 16.25 mites per predator in the mixed diets for *P. plumifer* and *E. scutalis*, respectively. Similarly, the highest means for the daily consumption rate of females were observed throughout the oviposition period, with the female of *P. plumifer* devouring an average of 1769.43 ± 10.42, while the female of *E. scutalis* devoured an average of 1037.83 ± 3.52. The highest number of preys consumed during the life span reported for *P. plumifer* females was 2108.74 ± 14.23 prey, while for *E. scutalis*, it was 36.56 ± 2.30 prey (Table 6).

Table 5. Predation rate by different stages of *Phytoseius plumifer* and *Euseius scutalis* feeding on grape erineum mite, *Colomerus vitis* at 33 ± 1 °C, 60% ± 5% RH.

Predatory Stage	Sex	*P. plumifer*		*E. scutalis*	
		No. of Attacked Mite Individuals			
		Total Average Mean ± SD	Daily Rate, Mean ± SD	Total Average, Mean ± SD	Daily Rate, Mean ± SD
Protonymph	Female	43.84 ± 1.68	33.72 ± 1.52	37.34 ± 2.43	22.76 ± 1.64
	Male	38.55 ± 2.12	31.08 ± 2.04	36.27 ± 2.50	22.38 ± 1.20
Deutonymph	Female	63.02 ± 2.45	40.39 ± 2.30	59.28 ± 3.16	31.03 ± 2.36
	Male	58.00 ± 1.89	37.90 ± 2.11	58.14 ± 2.70	31.42 ± 1.85
Pre-oviposition	Female	241.48 ± 3.15	86.55 ± 2.07	239.44 ± 3.65	72.55 ± 2.25
Oviposition	Female	2931.40 ± 16.62 a	112.91 ± 4.53 a	1994.52 ± 12.40 b	94.43 ± 4.16 b
Post-oviposition	Female	245.89 ± 2.30	56.13 ± 2.26	165.52 ± 3.35	42.66 ± 2.51
Longevity	Female	3418.77 ± 14.50 a	103.19 ± 3.08 a	2355.87 ± 15.85 b	83.24 ± 3.49 b
	Male	2845.36 ± 12.53	90.24 ± 3.60	1882.93 ± 14.14	72.56 ± 4.14
Life span	Female	3525.63 ± 16.47 a	89.43 ± 3.14 a	2452.49 ± 10.53 b	69.27 ± 3.30 b
	Male	2941.91 ± 12.90	78.61 ± 2.82	1977.34 ± 13.57	60.02 ± 3.14

Means followed by different letters in each row for total average and daily rate separately denote significant differences (ANOVA followed by Duncan's multiple range test: $p < 0.05$) (The comparation is made only with females).

Table 6. Predation rate by different stages of *Phytoseius plumifer* and *Euseius scutalis* feeding on a mixed diet of the grape erineum mite, *C. vitis* plus date palm pollen at 33 ± 1 °C, 60% ± 5% RH.

Predatory Stage	Sex	*P. plumifer*		*E. scutalis*	
		No. of Attacked Mite Individuals			
		Total Average, Mean ± SD	Daily Rate, Mean ± SD	Total Average, Mean ± SD	Daily Rate, Mean ± SD
Protonymph	Female	22.86 ± 1.21	17.72 ± 0.96	19.85 ± 0.71	13.14 ± 0.64
	Male	22.47 ± 1.06	17.97 ± 0.80	18.61 ± 0.75	12.40 ± 0.87
Deutonymph	Female	33.88 ± 1.12	23.04 ± 1.91	26.75 ± 1.18	16.01 ± 1.05
	Male	32.72 ± 1.36	22.88 ± 1.25	20.81 ± 0.98	12.68 ± 0.90
Pre-oviposition	Female	143.07 ± 2.11	64.73 ± 1.16	129.55 ± 1.23	52.87 ± 1.08
Oviposition	Female	1769.43 ± 10.42 a	65.85 ± 1.41 a	1037.83 ± 3.52 b	46.14 ± 3.16 b
Post-oviposition	Female	139.50 ± 2.09	32.06 ± 0.96	80.27 ± 1.46	19.96 ± 0.84
Longevity	Female	2052.00 ± 11.34 a	61.38 ± 2.59 a	1247.65 ± 16.25 b	43.08 ± 2.14 b
	Male	1439.65 ± 13.68	44.61 ± 1.83	1003.23 ± 10.72 b	36.56 ± 2.30 b
Life span	Female	2108.74 ± 14.23 a	53.26 ± 1.17 a	1294.25 ± 14.95	36.30 ± 2.48 b
	Male	1494.84 ± 10.89	38.89 ± 1.34	1042.65 ± 7.84 b	30.59 ± 2.27

Means followed by different letters in each row for total average and daily rate separately denote significant differences (ANOVA followed by Duncan's multiple range test: $p < 0.05$). (The comparation is made only with females).

2.5. Population Growth Parameters

Based on the above-mentioned findings, we calculated the life table parameters of *P. plumifer* and *E. scutalis* for each treatment. The highest life table parameter value was recorded when the predators were fed the mixed diet of *C. vitis* and date palm pollen. For both predators, the females reared on the mixed diet of *C. vitis* and date palm pollen showed the highest net reproductive rate (R_o), intrinsic rate of natural increase (r_m), and finite rate of increase (λ). On the other hand, *P. plumifer* and *E. scutalis* individuals reared on the mixed diet had the shortest mean generation time (Table 7).

Table 7. Population growth parameters of *Phytoseius plumifer* and *Euseius scutalis* feeding on three diets at 33 ± 1 °C, 60% ± 5% RH.

Parameters	*C. vitis*		*C. vitis* + Pollen		Date Palm Pollen	
	P. plumifer	*E. scutalis*	*P. plumifer*	*E. scutalis*	*P. plumifer*	*E. scutalis*
Net reproduction rate (R_o)	27.65	24.31	29.12	26.82	20.46	19.10
Mean generation time (T) (days)	18.57	19.65	17.21	18.72	22.36	25.48
Intrinsic rate of increase (r_m)	0.242	0.211	0.251	0.229	0.194	0.175
Finite rate of increase (λ)	1.246	1.212	1.377	1.285	1.194	1.186
Sex ratio	0.76	0.73	0.73	0.70	0.60	0.53
	(f = 23; m = 7)	(f = 22; m = 8)	(f = 22; m = 8)	(f = 21; m = 9)	(f = 18; m = 12)	(f = 16; m = 14)

2.6. Sex Ratio

The diet had a clear significant effect on the sex ratio in both predators. As shown in Table 7, sex ratio on all diets ranged from 53 to 76%. There were insignificant differences between the treatments of mixed diet (pollen plus *C. vitis*) and *C. vitis* diet alone. Also, there was a significant difference between treatments of pollen diet alone and other diets.

However, the maximum female-biased sex ratio was 76%, which was recorded for *P. plumifer* when fed on *C. vitis* only, while the minimum female-biased sex ratio was 53%, which was recorded for *E. scutalis* when fed on date palm pollen only.

3. Discussion

This study is the first documentation of the life history, predation capacity, fecundity, and life table parameters of *P. plumifer* and *E. scutalis* on the grape erineum mite as prey. It shows that the grape erineum mite is an acceptable prey and of high nutritional value for *P. plumifer* and *E. scutalis* resulting in a short developmental time and high fecundity rate. They were also able to develop and reproduce successfully when fed on fresh date palm pollen or on a mixed diet of grape erineum mite and date palm pollen. This is a significant step in the development of biocontrol strategies against the grape erineum mite. For both predators, larvae developed to the protonymphal stages without feeding. Non-feeding larvae behavior may be a mechanism for the avoidance of sibling cannibalism or reducing intraspecific competition. Similar findings have been observed in many phytoseiidae species [39].

The developmental time of the different life stages of *P. plumifer* on all three diets tested in the current study are considerably shorter than those stated by Moghadasi et al. 2006 for this predator when preying on spider mite, *Tetranychus urticae* Koch at 27 °C and 75–80% RH [40], or the eriophyid mite, *Rhyncaphytoptus ficifoliae* (Keifer) at 25 °C and 65% RH [25]. These researchers stated a mean total developmental time of *P. plumifer* females of 8.62 and 8.73 days on the respective prey species. The developmental time of female immatures of *P. plumifer* fed on fig spider mite *Eotetranychus hirsti* [41] at 35 °C and 60% RH are very close to the present results against *C. vitis*. On the other hand, the development of *P. plumifer* immature females was slightly longer in our study (6.37 days at 33 °C) than reported when fed on the eriophyid mite, *Aceria olivi* (5.67 d at 35 °C) [15]. In the present study, the developmental rate of *E. scutalis* was shorter (6.69 days) on a mixed date-palm-pollen–prey diet compared to date palm pollen alone, which is faster than on the other prey, including mites and insects such as *Oligonychus afrasiaticus*, *Eutetranychus orientalis*, *T. urticae*, *R. ficifoliae*, and *Insulaspis palidulla* (9.6, 8.19, 8.02, 7.00, and 6.75 days, respectively) [30,31,42], as well as pollens like sour orange pollen (*Citrus aurantium* L.), castor bean pollen (*Ricinus communis* L.), and alfalfa pollen (*Medicago sativa* L.) (7.90, 6.98, and 8.94 days, respectively) [43]. Muñoz-Cárdenas et al. (2014) noted a positive influence of a mixed diet of eggs from whiteflies and spider mites on the developmental time and fecundity of predatory mite *Balaustium leanderi* compared to either food alone [44]. A positive effect of a mixed diet on the development, predation capacity, reproduction, and life history parameters has been reported for other predacious mites as well [45].

The survival rate of immature stages was not significantly affected by diet, and it exceeded 90% for both predators on all three diets. The findings of Vervaet et al. (2022) support our results. They revealed that the survival rate of *Pronematusu biquitus* and *Homeopronematus anconai* during the immature stages exceeded 83% for both mites on three diets [45].

The pre-oviposition periods of *P. plumifer* and *E. scutalis* were close to those stated by Kasap and Şekeroğlu (2004) [28] and Hamedi et al. [46]. When *P. plumifer* and *E. scutalis* were fed on a mixed diet of *C. vitis* and date palm pollen, there was a significant increase in oviposition period, fecundity, and adult female longevity. Subsequently, predators' performance was strong. The oviposition period and female longevity of *P. plumifer* were parallel to the findings stated by Kouhjani-Gorji et al. (2012) [47], Louni et al. (2014) [25], Shakarami and Bazgir (2017) [41], and Al-Azzazy and Alhewairini (2020b) [15] for *P. plumifer* feeding on *T. urticae*, *R. ficifoliae*, *E. hirsti*, and *Tegolophus hassani*.

The oviposition duration and adult female longevity of *E. scutalis* (22.49 and 28.96) on a mixed diet were close to that reported against *Panonychus citri* (21.3, 28.6) [28] and (20.22, 29.57) when *E. scutalis* fed on Crawlers of *Bemisia tabaci* [48]. Although the addition of date palm pollen to the rearing units lowered the grape erineum mite predation by both predators, it substantially increased the fecundity of the predatory mites, and oviposition was always higher. The highest oviposition was obtained when *P. plumifer* was fed on a mixed diet (56.81 eggs/female). This value was higher when compared to mites that fed on *T. urticae* (49.10 eggs/female) [49], on *R. ficifoliae* (28.47 eggs/female) [25], on *E.hirsti* (35.71 eggs/female) [41], and on *Oxycenus niloticus* (50.80 eggs/female) [15]; furthermore, the results of Al-Azzazy and Alhewairini (2020b) for *P. plumifer* fed on *A. olivi* (57.46 eggs/female) [15] were higher than that obtained in this study. The maximum oviposition of *E. scutalis* was (44.16 eggs/female), which was higher than reported by Kasap and Şekeroğlu (2004) [28] (39.7 eggs/female) with feeding on *P. citri*. Also, *E. scutalis* has shown total fecundity of (17.13, 19.96, 23.16, 25.92, and 26.52 eggs/female) on golden shower tree pollen, caper bush pollen, *Tetranychus turkestani*, date palm pollen, and cattail pollen, respectively [35]. Moreover, the results of Bounfour and McMurtry (1987) for this predator fed on *Tetranycus pacificus* eggs (59.0 eggs/female at 35 °C) [27] were considerably higher than the value estimated in the current study. The high fecundity for both predators in the current study might be due to feeding on mixing prey with pollen. In *Amblydromalus limonicus*, Samaras et al. (2021) showed that mixing prey with pollen resulted in higher fecundity and r_m values, thus enhancing the medium-to long-term thrips-control potential [49].

In the current study, the addition of date palm pollen significantly lowered the predation rate of grape erineum mites by *P. plumifer* and *E. scutalis*, 59.81 and 52.77%, respectively. These results agree with those of Vervaet et al. (2022), who stated that *H. anconai* devoured fewer *Aculops lycopersici* adults in the presence of *Typha latifolia* L. pollen [45]. Similar findings have been obtained by Samaras et al. (2021), who showed the different effects of *Zea mays*, *Typha angustifolia* and *Pinus brutia* pollen on the predation rate of *A. limonicus* [49]. In the absence of date palm pollen, *P. plumifer* immature females consumed 106.86 prey of *C. vitis* in this study, while they devoured 108 individuals of *Aceria ficus* [50], 58.29 individuals of *R. ficifoliae* [25], and 127.46 individuals of *T. hassani*, 135.83 of *O. niloticus*, and 143.82 of *A. olivi* [15]. *Euseius scutalis* immature females devoured 96.62 prey of *C. vitis* in this study, while they consumed 40.78 individuals of *R. ficifoliae* and 65.30 of *A. ficus* [25].

The life table parameters of both predatory mites were clearly affected by diet. Several biological studies have confirmed that high-quality food sources result in higher values in life table parameters [51,52]. The rates of population growth were promising for *P. plumifer* fed on mixed stages of *C. vitis* and date palm pollen. This was proven by (r_m), which was 0.251. The reported intrinsic rate of increase for *P. plumifer* on *T. urticae* (0.200 at 26 °C) [31], *E. hirsti* (0.180 at 30 °C), *R. ficifoliae* (0.154 at 25 °C), and corn pollen (0.112 at 27 °C) [24] was lower than that obtained in this study when *P. plumifer* fed on mixed stages of *C. vitis* and date palm pollen. Kouhjani-Gorji et al. (2012) estimated (r_m) of 0.244 for *P. plumifer* fed on *T. urticae* at 35 °C [47]. Also Al-Azzazy and Alhewairini estimated (r_m) of 0.277, 0.288 and 0.298 for *P. plumifer* fed on *T. hassani*, *O. niloticus* and *A. olivi* at 35 °C, respectively [15]. This is somewhat higher than our estimate. The values of (R_o), (r_m), and (λ) of *P. plumifer* at 33 °C and 60% RH in the current study are higher than reported for *N. barkeri* against *C. vitis* at 35 °C and 50% RH [2]. *P. plumifer* performed better on *C. vitis* than *N. barkeri*, and this could be due to the moderate humidity level used in this study. In view of this, *P. plumifer* could be a useful biocontrol agent for *C. vitis*.

In the case of *E. scutalis* against *T. urticae*, *E. orientalis*, and *Oligonychus afrasiaticus* at 26 °C, the life table parameters (r_m, λ, R_o) values were 0.220, 0.175, and 0.161; 1.247, 1.192, and 1.175; and 26.73, 13.24, and 13.60, respectively [31]. On alfalfa pollen (*Medicago sativa* L.), 0.153, 1.150, and 18.51 [43] and on eriophyid mite, *A. ficus* and *R. ficifoliae*, at 28 °C, 0.218, 1.243, and 12.51 and 0.215, 1.240, and 12.02, respectively [30], were seen, while in the current study, they were 0.229, 1.254, and 26.82 at 33 °C. This indicates that *E. scutalis* performs well on *C. vitis* as a generalist predator. The rather high intrinsic rate of the natural increase (r_m) of *P. plumifer* and *E. scutalis* reached with a mixed diet could be highly favorable for mass production and augmentative release purposes. Therefore, date palm pollen can be considered an optimal supplementary food for *P. plumifer* and *E. scutalis*. In addition, any short-term negative impacts on predation rate due to preference of the pollen and floral nectar over the prey or predator satiation have been shown to be eventually overbalanced by an increase in the predation rate at the population level, i.e., long-term positive impacts of the mixed diet.

4. Materials and Methods

4.1. Stock Culture of Predators

The individuals of *Phytoseius plumifer* and *E. scutalis* were collected from unsprayed (for the previous 3 years) vineyards of Buraidah city (26.300366° N, 43.789661° E), Saudi Arabia, in the summer of 2021. The grape leaves containing the predators were cut and transferred to the laboratory. All predators were maintained separately on rearing units made of common bean (*Phaseolus vulgaris* L.) leaves which were placed underside facing up on daily moistened cotton in plastic trays (6 × 12), in an incubator at 33 ± 1 °C, 60% ± 5% RH, and with a 16:8 (L:D) h photoperiod. The edges were bordered with water-saturated tissue paper to provide the predators with water and prevent them from escaping. The grape leaves infested with *C. vitis* were used to feed the predatory mites five times a week, adding date palm pollen with a thin paintbrush as a supplementary food. Water was added to the cotton every day to keep the arena humid. Predator eggs were collected and transferred individually to the new rearing arenas to obtain cohorts of individuals of the same age.

Fifteen microscope slides were prepared with each species to confirm their identification. The identification of both the predators was confirmed according to Chant and McMurtry (2007) [53].

4.2. Stock Cultures of Prey

Grape erineum mites were collected from a vineyard grown at the Experimental Research Station of the College of Agriculture and Food (Qassim, Saudi Arabia). The specimens of *C. vitis* were transferred to the laboratory and reared on grape seedlings as a permanent source of prey, kept in a climate room at 30 ± 1 °C, 60% ± 5% RH, and with a 16:8 (L:D) h photoperiod. All the usual agriculture practices such as fertilization and irrigation were followed.

4.3. Experimental Set-Up

All experiments were performed in an incubator at 33 ± 1 °C, with a 12:12 h (L:D) photoperiod, at an average daily relative air humidity of 60 ± 5%. Freshly excised grape leaf disks (3 × 3 cm) were used as rearing arenas. The leaves were placed with the lower surface facing up on daily moistened cotton inside a Pyrex® Petri dish (5 cm in diameter, 2 cm high). The edges were bordered with water-saturated cotton to provide water and avoid mite escape.

4.4. Pollen Collection

Fresh date palm pollen (*Phoenix dactylifera* L.) was collected from trees planted in the Qassim region, Saudi Arabia, and oven-dried at 25 °C for one day and then stored at −18 °C. Before use in the trials, a small amount of date palm pollen was refrigerated at 4 °C for up to 2 weeks for early use in the experiments.

4.5. Effects of Diet on Life History Parameters of *P. plumifer* and *E. scutalis*

For each predator, 30 to 40 fresh eggs (less than 7 h old) from the stock culture were transferred individually to the grape leaf arena. The immature developmental time, survival, sex ratio, fecundity, and longevity of *P. plumifer* and *E. scutalis* were determined by feeding them with one of the following food sources: (1) date palm pollen, (2) mixed stages of *C. vitis*, and (3) mixed stages of *C. vitis* plus date palm pollen. The mixed stages of *C. vitis* were provided by a carefully examined small disk (0.5 cm in diameter) of severely infested grape leaves to record the total number of mites per disk before introducing it into the rearing arenas on the grape leaf disk for 24 h, after which the number of consumed preys was recorded. Date palm pollen grains were placed on the rearing arenas using a fine brush, three times per week. Observations were made twice a day to determine the developmental time (egg, larva, protonymph, and deutonymph) and the survival of immature stages. For each rearing arena, after the emergence of adults, a single male was put in the new female's rearing arena for mating. Males were then transferred into new leaf disks and individually reared until the end of their lifespan. Some cotton fibers were stuck on grape leaf disks to provide a suitable place for oviposition. The experimental units were monitored twice a day for any changes recorded until the death of the last female. This monitoring allowed us to determine the pre-oviposition, oviposition, and post-oviposition periods and the female and male longevity and fecundity. Once the adult stage was reached, the sex of the predators was determined. Whenever the quality of the rearing arena began to deteriorate, it was replaced with a fresh leaf disk. To test the sex ratio, daily, the eggs laid were placed individually into separate rearing units with a fine brush, and the hatched larvae were reared to adulthood to determine the sex ratio of the progeny.

4.6. Statistical Analysis

To assess immature development, pre-oviposition and post-oviposition periods, fecundity, adult longevity, predation, and life span of *P. plumifer* and *E. scutalis* and the effect of three food sources on these parameters, data were compared with analysis of variance (ANOVA), using SAS computer program version 9.2 (SAS, 2008). Means were separated by Duncan's multiple range test (DMRT) at $p < 0.05$. The means of survival percentages were separated using Tukey's honestly significant difference test (Tukey's HSD test). The life table parameters for both predators were constructed based on Birch (1948). The sex ratio for both predators was analyzed using a Chi-square test.

5. Conclusions

In conclusion, our results demonstrate that the grape erineum mite *C. vitis* is a suitable prey for both phytoseiids *P. plumifer and E. scutalis,* making them promising biocontrol agents of that pest. Furthermore, the addition of date palm pollen to the diet of *C. vitis* substantially increased the reproductive performance of both predators. Therefore, it can be concluded that the mixed diet of *C. vitis* and date palm pollen are good candidates for the mass rearing of *P. plumifer* and *E. scutalis* for use in augmentative biocontrol programs. Moreover, the addition of date palm pollen as an optimal supplementary food source can be an effective tool to boost the populations of *P. plumifer* and *E. scutalis* and enhance biocontrol even in the presence of a low grape erineum mite population.

Author Contributions: This manuscript was drafted by M.M.A.-A. and S.S.A. Laboratory work and statistical analysis were performed by M.M.A.-A. All authors have read and agreed to the published version of the manuscript.

Funding: This research was funded by Qassim University (QU-APC-2024-9/1).

Data Availability Statement: Data are contained within the article.

Acknowledgments: The researchers would like to thank the Deanship of Graduate Studies and Scientific Research at Qassim University for financial support (QU-APC-2024-9/1).

Conflicts of Interest: The authors declare no conflicts of interest.

References

1. Alston, J.M.; Sambucci, O. Grapes in the World Economy. In *The Grape Genome. Compendium of Plant Genomes;* Cantu, D., Walker, M., Eds.; Springer: Cham, Switzerland, 2019. [CrossRef]
2. Al-Azzazy, M.M. Biological performance of the predatory mite *Neoseiulus barkeri* Hughes (Phytoseiidae): A candidate for controlling of three mite species infesting grape trees. *Vitis* **2021**, *60*, 11–20. [CrossRef]
3. Carew, M.E.; Goodisman, M.A.; Hoffmann, A.A. Species status and population genetic structure of grapevine eriophyoid mites. *Entomol. Exp. Et Appl.* **2004**, *111*, 87–96. [CrossRef]

4. Javadi Khederi, S.; Khanjani, M.; Fayaz, B.A. Resistance of three grapevine cultivars to Grape Erineum Mite, *Colomerus vitis* (*Acari: Eriophyidae*), in field conditions. *Persian. J. Acarol.* **2014**, *3*, 63–75.
5. Javadi Khederi, S.; Khanjani, M.; Gholami, M.; de Lillo, E. Sources of resistance to the erineum strain of *Colomerus vitis* (*Acari: Eriophyidae*) in grapevine cultivars. *Syst. Appl. Acarol.* **2018**, *23*, 405–425. [CrossRef]
6. Javadi Khederi, S.; Khanjani, M.; Gholami, M.; de Lillo, E. Impact of the erineum strain of *Colomerus vitis* (*Acari: Eriophyidae*) on the development of plants of grapevine cultivars of Iran. *Exp. Appl. Acarol.* **2018**, *74*, 347–363. [CrossRef]
7. Bahirai, F.; Jafari, S.; Lotfollahi, P.; Shakarami, J. Eriophyoidea (*Acari: Trombidiformes*) of the Lorestan Province and first record of *Aceria querci* (Garnam, 1883) outside of the USA. *Persian. J. Acarol.* **2021**, *10*, 111–119.
8. Cooper, M.; Hobbs, M.; Strode, B.; Varela, L. Grape erineum mite: Postharvest sulfur use reduces subsequent leaf blistering. *Calif. Agric.* **2020**, *74*, 94–100. [CrossRef]
9. Malagnini, V.; de Lillo, E.; Saldarelli, P.; Beber, R.; Duso, C.; Raiola, A.; Zanotelli, L.; Valenzano, D.; Giampetruzzi, A.; Morelli, M.; et al. Transmission of grapevine Pinot gris virus by *Colomerus vitis* (*Acari: Eriophyidae*) to grapevine. *Arch. Virol.* **2016**, *161*, 2595–2599. [CrossRef]
10. Muneret, L.; Thiéry, D.; Joubard, B.; Rusch, A. Deployment of organic farming at a landscape scale maintains low pest infestation and high crop productivity levels in vineyards. *J. Appl. Ecol.* **2018**, *55*, 1516–1525. [CrossRef]
11. Geiger, F.; Bengtsson, J.; Berendse, F.; Weisser, W.W.; Emmerson, M.; Morales, M.B.; Ceryngier, P.; Liira, J.; Tscharntke, T.; Winqvist, C.; et al. Persistent negative effects of pesticides on biodiversity and biological control potential on European farmland. *Basic. Appl. Ecol.* **2010**, *11*, 97–105. [CrossRef]
12. Guedes, R.N.C.; Smagghe, G.; Stark, J.D.; Desneux, N. Pesticides induced stress in arthropod pests for optimized integrated pest management programs. *Annu. Rev. Entomol.* **2016**, *61*, 43–62. [CrossRef] [PubMed]
13. Rincón, R.A.; Rodríguez, D.; Coy-Barrera, E. Botanicals Against *Tetranychus urticae* Koch Under Laboratory Conditions: A Survey of Alternatives for Controlling Pest Mites. *Plants* **2019**, *8*, 272. [CrossRef] [PubMed]
14. Al-Azzazy, M.M.; Alhewairini, S.S. Effect of temperature and humidity on development, reproduction, and predation rate of *Amblyseius swirskii* (Phytoseiidae) fed on *Phyllocoptruta oleivora* (Eriophyidae) and *Eutetranychus orientalis* (Tetranychidae). *Int. J. Acarol.* **2020**, *46*, 304–312. [CrossRef]
15. Al-Azzazy, M.M.; Alhewairini, S.S. A life table analysis to evalu ate biological control of four mite species associated with olive trees using the predatory mite *Phytoseius plumifer* (*Acari: Phytoseiidae*) in Saudi Arabia. *Pak. J. Agric. Sci.* **2020**, *57*, 299–305. [CrossRef]
16. Abou Jawdah, Y.; Ezzeddine, N.; Fardoun, A.; Kharroubi, S.; Sobh, H.; Atamian, H.S.; Skinner, M.; Parker, B. Biological Control of Three Major Cucumber and Pepper Pests: Whiteflies, Thrips, and Spider Mites, in High Plastic Tunnels Using Two Local Phytoseiid Mites. *Plants* **2024**, *13*, 889. [CrossRef] [PubMed]
17. Gerson, U.; Smiley, R.L.; Ochoa, R. *Mites. (Acari) for Pest Control*; Blackwell Publishing: Oxford, UK, 2003; p. 537. [CrossRef]
18. McMurtry, J.A.; Moraes, G.J.D.; Sourassou, N.F. Revision of the lifestyles of phytoseiid mites (Acari: Phytoseiidae) and implications for biological control strategies. *Syst. Appl. Acarol.* **2013**, *18*, 297–320. [CrossRef]
19. Tixier, S.; Baldassar, A.; Duso, C.; Kreiter, S. Phytoseiidae in European grape (*Vitis vinifera* L.): Bio-ecological aspects and keys to species (*Acari: Mesostigmata*). *Zootaxa* **2013**, *3721*, 101–142. [CrossRef] [PubMed]
20. Knapp, M.; Houten, Y.; Baal, E.; Groot, T. Use of predatory mites in commercial biocontrol: Current status and future prospects. *Acarologia* **2018**, *58*, 72–82. [CrossRef]
21. Eini, N.; Jafari, S.; Fathipour, Y.; Prager, S. Experienced generation-dependent functional and numerical responses of *Neoseiulus californicus* (*Acari: Phytoseiidae*) long-term reared on thorn apple pollen. *Acarologia.* **2023**, *63*, 539–552. [CrossRef]
22. Nadimi, A.; Kamali, K.; Arbabi, M.; Abdoli, F. Selectivity of three miticides to spider mite predator, *Phytoseius plumifer* (*Acari: Phytoseiidae*) under laboratory conditions. *Agric. Sci. China* **2009**, *8*, 326–331. [CrossRef]
23. Jafari, S. Phytoseiid Mites of the Lorestan Province and Determining the Predation Efficiency of *Neoseiulus barkeri* (Phytoseiidae). Ph.D. Thesis, Faculty of Agriculture, Tarbiat Modares University, Tehran, Iran, 2010; p. 192.
24. Khodayari, S.; Fathipour, Y.; Kamali, K. Life history parameters of *Phytoseius plumifer* (*Acari: Phytoseiidae*) fed on corn pollen. *Acarologia* **2013**, *53*, 185–189. [CrossRef]
25. Louni, M.; Jafari, S.; Shakarami, J. Life table parameters of *Phytoseius plumifer* (Phytoseiidae) fed on *Rhyncaphytoptus ficifoliae* (Diptilomiopidae) under laboratory conditions. *Syst. Appl. Acarol.* **2014**, *19*, 275–282. [CrossRef]
26. Khodayari, S.; Fathipour, Y.; Sedaratian, A. Prey stage preference, switching and mutual interference of *Phytoseius plumifer* (Acari: Phytoseiidae) on *Tetranychus urticae* (*Acari: Tetranychidae*). *Syst. Appl. Acarol.* **2016**, *21*, 347–355. [CrossRef]
27. Bounfour, M.; McMurtry, J. Biology and ecology of *Euseius scutalis* (Athias-Henriot) (*Acarina: Phytoseiidae*). *Hilgardia* **1987**, *55*, 1–23. [CrossRef]
28. Kasap, İ.; Şekeroğlu, E. Life history of *Euseius scutalis* feeding on citrus red mite *Panonychus citri* at various temperatures. *BiolControl* **2004**, *49*, 645–654. [CrossRef]
29. Nomikou, M.; Janssen, A.; Schraag, R.; Sabelis, M.W. Phytoseiid predators as potential biological control agents for *Bemisia tabaci*. *Exp. Appl. Acarol.* **2001**, *25*, 271–291. [CrossRef] [PubMed]
30. Momen, F.M.; Abdel-Khalek, A. Influence of diet on biology and life- table parameters of the predacious mite *Euseius scutalis* (A. H.) (*Acari: Phytoseiidae*). *Arch. Phytopathol.* **2008**, *41*, 418–430. [CrossRef]

31. Al-Shammery, K.A. Different biological aspects of the predaceous mite *Euseius scutalis* (*Acari: Gamasida: Phytoseiidae*) and the effects due to feeding in three tetranychid mite species in Hail, Saudi Arabia. *Asian J. biol. Sci.* **2010**, *8*, 77–81. [CrossRef]
32. Maoz, Y.; Gal, S.; Argov, Y.; Domeratzky, S.; Melamed, E.; Gan-Mor, S.; Palevsky, E. Efficacy of indigenous predatory mites (Acari: Phytoseiidae) against the citrus rust mite *Phyllocoptruta oleivora* (*Acari: Eriophyidae*): Augmentation and conservation biological control in Israeli citrus orchards. *Exp. Appl. Acarol.* **2014**, *63*, 295–312. [CrossRef]
33. Stathakis, T.I.; Kapaxidi, E.V.; Papadoulis, G.T.; Papanikolaou, N.E. Predation by *Euseius scutalis* (*Acari: Phytoseiidae*) on *Tetranychus urticae* and *Eutetranychus orientalis* (*Acari: Tetranychidae*): Effect of prey density and developmental stage. *Syst. Appl. Acarol.* **2021**, *26*, 1940–1951. [CrossRef]
34. Xin, T.; Zhang, Z. Suitability of pollen as an alternative food source for different developmental stages of *Amblyseius herbicolus* (Chant) (*Acari: Phytoseiidae*) to facilitate predation on whitefly eggs. *Acarologia* **2021**, *61*, 790–801. [CrossRef]
35. Shishehbor, P.; Rahmani Piyani, A.; Riahi, E. Effects of different pollen diets in comparison to a natural prey, *Tetranychus turkestani* (Acari: Tetranychidae), on development, survival, and reproduction of *Euseius scutalis* (*Acari: Phytoseiidae*). *Syst. Appl. Acarol.* **2022**, *27*, 2111–2122. [CrossRef]
36. Coll, M.; Guershon, M. Omnivory in terrestrial arthropods: Mixing plant and prey diets. *Annu. Rev. Entomol.* **2002**, *47*, 267–297. [CrossRef] [PubMed]
37. Ghasemzadeh, S.; Leman, A.; Messelink, G.J. Biological control of *Echinothrips americanus* by phytoseiid predatory mites and the effect of pollen as supplemental food. *Exp. Appl. Acarol.* **2017**, *73*, 209–221. [CrossRef] [PubMed]
38. Duarte, M.V.A.; Venzon, M.; Bittencourt, M.C.D.; Rodriguez-Cruz, F.A.; Pallini, A.; Janssen, A. Alternative food promotes broad mite control on chilli pepper plants. *BioControl* **2015**, *60*, 817–825. [CrossRef]
39. Metwally, A.M.; Abou-Awad, B.A.; Al-Azzazy, M.M. Life table and prey consumption of the predatory mite Neoseiulus cydnodactylon Shehata and Zaher (Acari: Phytoseiidae) with three mite species as prey. *J. Plant. Dis. Prot.* **2005**, *112*, 276–286. Available online: https://www.jstor.org/stable/45154911 (accessed on 1 January 2023).
40. Moghadasi, M.; Hajizadeh, J.; Saboori, A.; Nowzari, J. Biology of *Phytoseius plumifer* (Canestrini & Fanzago) (*Acari: Phytoseiidae*) feeding on *Tetranychus urticae* Koch (Acari: Tetranychidae). In *Conference: 14th National and 2nd International Conference of Biology*; At Tarbiat Modares University: Tehran, Iran, 2006.
41. Shakarami, J.; Bazgir, F. Effect of temperature on life table parameters of *Phytoseius plumifer* (Phytoseiidae) fed on *Eotetranychus hirsti* (Tetranychidae). *Syst. Appl. Acarol.* **2017**, *22*, 410–422. [CrossRef]
42. Abou-Elella, G.M.; Saber, S.A.; El-Sawi, S.A. Biological aspects and life tables of the predacious mites, *Typhlodromips swirskii* (Athias-Henriot) and *Euseius scutalis* (Athias-Henriot) feeding on two scale insect species and plant pollen. *Arch. Phytopathol.* **2013**, *46*, 1717–1725. [CrossRef]
43. Al-Shammery, K.A. Plant pollen as an alternative food source for rearing *Euseius scutalis* (*Acari: Phytoseiidae*) in Hail, Saudi Arabia. *J. Entomol.* **2011**, *8*, 365–374. [CrossRef]
44. Muñoz-Cárdenas, K.; Fuentes, L.S.; Cantor, R.F.; Rodríguez, C.D.; Janssen, A.; Sabelis, M.W. Generalist red velvet mite predator (*Balaustium* sp.) performs better on a mixed diet. *Exp. Appl. Acarol.* **2014**, *62*, 19–32. [CrossRef]
45. Vervaet, L.; Parapurath, G.; De Vis, R.; Leeuwen, T.V.; De Clercq, B. Potential of two omnivorous iolinid mites as predators of the tomato russet mite, *Aculops lycopersici*. *J. Pest. Sci.* **2022**, *95*, 1671–1680. [CrossRef]
46. Hamedi, N.; Fathipour, Y.; Saber, M. Sublethal effects of abamectin on the biological performance of the predatory mite, *Phytoseius plumifer* (*Acari: Phytoseiidae*). *Exp. Appl. Acarol.* **2011**, *53*, 29–40. [CrossRef] [PubMed]
47. Kouhjani-Gorji, M.; Fathipour, Y.; Kamali, K. Life table parameters of *Phytoseius plumifer* (Phytoseiidae) fed on two-spotted spider mite at different constant temperatures. *Int. J. Acarol.* **2012**, *38*, 377–385. [CrossRef]
48. Saleh, E.B.Y.; Mostafa, M.A.; Desuky, W.M.H.; El-Kawas, H.M.G. Thermal Units of the Predatory Mite, *Euseius scutalis* (Athisa-Henriot) (*Acari: Phytoseiidae*) Fed on Crawlers of *Bemisia tabaci* (Genn.) (*Hemiptera: Aleyrodidae*). *Egypt. J. Biol. Pest Control* **2015**, *25*, 663–667.
49. Samaras, K.; Pappas, M.L.; Pekas, A.; Wäckers, F.; Broufas, G.B. Benefits of a balanced diet? Mixing prey with pollen is advantageous for the phytoseiid predator *Amblydromalus limonicus*. *Biol. Control* **2021**, *155*, 104531. [CrossRef]
50. Rasmy, H.; Elbanhawy, E.M. The Phytoseiidae mite *Phytoseius plumifer* as a predator of the Eriophyid mite *Aceria ficus* (Acarina). *Entomophaga* **1974**, *19*, 427–430. [CrossRef]
51. Bouras, S.L.; Papadoulis, G.T. Influence of selected fruit tree pollen on life history of *Euseius stipulates* (*Acari: Phytoseiidae*). *Exp. Appl. Acarol.* **2005**, *36*, 1–14. [CrossRef]
52. Jiale, L.V.; Yang, K.; Wang, E.; Xuenong, X.U. Prey diet quality affects predation, oviposition and conversion rate of the predatory mite *Neoseiulus barkeri* (*Acari: Phytoseiidae*). *Syst. Appl. Acarol.* **2016**, *21*, 279–287. [CrossRef]
53. Chant, D.A.; McMurtry, J.A. *Illustrated Keys and Diagnoses for the Genera and Subgenera of the Phytoseiidae of the World*; Indira Publishing House: West Bloomfield, MI, USA, 2007.

plants

MDPI

Article

Role of Endophytic Entomopathogenic Fungi in Mediating Host Selection, Biology, Behavior, and Management of Tarnished Plant Bug, *Lygus lineolaris* (Hemiptera: Miridae)

Justin George *, James P. Glover, Omaththage P. Perera and Gadi V. P. Reddy

Southern Insect Management Research Unit, Agricultural Research Service, United States Department of Agriculture, Stoneville, MS 38776, USA; james.glover@usda.gov (J.P.G.); op.perera@usda.gov (O.P.P.); gadi.reddy@usda.gov (G.V.P.R.)

* Correspondence: justin.george@usda.gov

Abstract: Non-insecticidal control strategies using entomopathogens, nematodes, and endophytes provide sustainable and safer alternatives for managing crop pests. This study investigated the potential of different fungal endophytes, specifically *Beauveria bassiana* strains, in colonizing cotton plants and their efficacy against tarnished plant bug, *Lygus lineolaris*. The effect of endophytes on plant growth parameters and cotton yield were measured during different plant growth stages. The entomopathogenicity of these fungi was studied in diet cup bioassays using *L. lineolaris* adults. The behavior of adult males and females toward endophytic cotton squares was analyzed using olfactometer assays. The experiments showed that the fungal endophytes colonized the plant structures of cotton plants, which resulted in an increase in the number of cotton squares, plant height, and weight compared to control plants. *B. bassiana* strains/isolates such as GHA, NI-8, and JG-1 caused significant mortality in *Lygus* adults compared to controls. Also, male and female *Lygus* adults exhibited repellence behavior towards endophytic cotton squares containing JG-1 isolate of *B. bassiana* and to other *B. bassiana* strains such as NI-8, GHA, and SPE-120. No differences were observed in the survival and development of *L. lineolaris* second-instar nymphs on endophytic cotton, and no yield differences were observed in the field experiments.

Keywords: fungal endophytes; entomopathogens; *Beauveria bassiana*; plant volatiles; host plant selection; cotton; host plant resistance

1. Introduction

Endophytes are plant-associated microorganisms that colonize tissues and reproductive structures of plants for part of their life cycle without causing damage to their host [1,2]. These naturally occurring microorganisms are described as bacteria, fungi, archea, and protists. Among these groups, fungal and bacterial endophytes are the most studied taxa [3]. Fungal endophytes colonize above-ground plant tissues such as stems, leaves, flowers, and seeds [1,4], whereas mycorrhizal fungi colonize the plant rhizosphere. These fungal endophytes can also act as entomopathogens and play an essential role as biocontrol agents against a myriad of insect pests and pose little or no harm to non-target insects and other beneficial organisms [5,6]. Studies have reported that these fungal endophytes can improve the plant response to biotic and abiotic stress through induction of systemic resistance or the production of insecticidal, antifungal, or antiviral compounds, in addition to its effectiveness against insect and mite pests [7]. Fungal endophytes have also attracted the attention of the scientific community in recent years due to their potential beneficial effects on vegetation, including in the facilitation of plant growth in metal-polluted environments. Studies have reported that both fungal and bacterial endophytes can co-exist in the same leaf or stem tissues of the host plant [8,9]. Fungal endophytes have been reported in a variety of field crops, including vegetables, fruits, maize, cotton, coffee, jute, and cocoa [6].

Endophytes provide numerous benefits to their host plant. The application of facultative endophytes in promoting plant growth and pest management can be improved by artificial inoculation of endophytes into plants by seed treatment, surface spray, and root inoculation. These fungal endophytes may act as multifaceted tools in plant growth and protection from pests and diseases and facilitate improved nutrient transfer and increased yield [10–12]. Endophytes act as yield promoters, soil nutrient distributers, abiotic stress, and drought tolerance enhancers in plants, as well as an indirect defense for insect pests, nematodes, and diseases [6,13]. Some are known to serve as pathogenic agents in plants by infecting lepidopterous larvae, aphids, thrips, and other insect pests, which are of great concern in agriculture [14]. Endophytes can also act as plant-defending mutualists and provide protection against herbivores [15,16]. These endophytic entomopathogenic fungi, such as *B. bassiana*, *Clonostachys rosea*, *Metarhizium anisopliae*, and *Lecanicillium lecanii*, have been reported to control pests and reduce their damage in crops [17]. Colonization of corn and sorghum by *B. bassiana* resulted in reduced tunneling by *Ostrinia nubilalis* Hubner and *Sesamia calamistis* Hampson [18,19]. Studies have shown that cotton aphid, *Aphis gossypii* (Glover), populations are significantly reduced in cotton colonized by endophytic strains, *B. bassiana* Vuillemin and *Phialemonium inflatum* (Burnside) strain TAMU-490 [20,21]. Evidence has also shown that the development of the cotton bollworm *Helicoverpa zea* (Boddie) is significantly reduced when exposed to *B. bassiana* [21,22]. Because of the benefits endophytes could offer to agricultural crops, it is important to explore endophytes as an alternative control strategy in disease and pest management programs [6].

The effect of fungal endophytes on the performance and feeding behavior of insect herbivores, including phloem feeders such as aphids and whiteflies [20,23], dipterans [24], beetles [25,26], caterpillars [21,27–29], grasshoppers [30], and hymenopteran gall wasps [31], has been reported. Fewer studies have explored the behavioral responses of hemipteran pests towards fungal endophytic plants, as they directly feed on fruiting structures and cause economic damage to the crop. Polyphagous pests such as plant bugs (Miridae) and stink bugs (Pentatomidae) are a major problem in row crops, and their feeding results in significant yield loss. Sword et al. [32] reported that *Lygus hesperus* and *Nezara viridula* showed strong negative responses to flower buds and cotton bolls and were deterred prior to contact with endophyte-colonized plants.

The tarnished plant bug, *Lygus lineolaris* (Palisot de Beauvois) (Hemiptera: Miridae), is considered the most important economic pest of cotton in the mid-southern United States, feeding on more than 700 plant species [33]. It is the most widely established *Lygus* species found on agricultural crops in North America and was reported to infest 6.9 million acres of cotton in 2021. In this study, we investigated the endophytic and entomopathogenic activity of five different fungal endophytes and an arbuscular mycorrhizal fungus on the olfaction behavior, mortality, host plant resistance, and the pest biology of tarnished plant bug, *L. lineolaris*, a major cotton pest in the Mississippi delta. Greenhouse and field experiments explored the effect of these fungal endophytes on cotton growth-promoting factors, plant bug damage, and cotton yield.

2. Results

2.1. Plant Growth Measurements of Fungal Endophytic Plants

Cotton plant growth parameters such as percentage germination, number of true leaves, cotton squares, plant height and dry root mass, stem and whole plant were measured to understand the effect of fungal endophytes on plant growth and physiology. No significant effects were observed in the percentage germination of different endophyte-coated seeds ($F_{6,277} = 1.44$, $p = 0.19$, $n = 40$) (Table 1). No differences were observed in the number of true leaves produced 14 days after germination ($F_{6,212} = 1.21$, $p = 0.31$, $n = 30$). The mean number of true leaves was in the range of 3.2–3.5 leaves/plant (Table 1). A significant difference was observed in the number of squares produced on the plants. The plants containing fungal endophyte SPE-120 produced a higher number of squares (4.2 $\pm$ 0.20) than the control plants (2.4 $\pm$ 0.24) and those containing other fungal endophytes studied

($F_{6,224} = 6.64$, $p < 0.001$, $n = 32$) (Table 1). TAMU-490 plants were the shortest in length compared to all the other plants studied ($F_{6,224} = 11.92$, $p < 0.001$, $n = 32$). No significant differences were observed in the dry weight of plant roots from the different endophytic plants ($F_{6,69} = 1.66$, $p = 0.14$, $n = 10$), whereas a significant effect was observed in the dry weight of stem and leaves ($F_{6,69} = 12.10$, $p < 0.001$, $n = 10$) (Table 1). The arbuscular mycorrhizal fungal (AMF) plants had the highest stem dry weight (30.5 ± 1.98) compared to control plants (17.3 ± 1.11). A similar trend was observed for the whole plant dry weight, and the AMF plants had the highest dry weight compared to control plants and other endophytic colonized plants ($F_{6,69} = 11.12$, $p < 0.001$, $n = 10$) (Table 1). Sporulation was observed on the various plant parts cultured such as leaf, stem, root, and squares (Figure 1), and the fungal spores were verified using Q-PCR.

Table 1. Cotton plant growth parameter measurements following the application of fungal endophytes or arbuscular mycorrhizal fungus as seed application. Plant growth parameters measured included percentage germination, number of true leaves, squares, plant height, and dry weight of roots, stems, and whole plant. Data were analyzed by ANOVA followed by Tukey's HSD for mean comparisons. Treatments that have no letters in common within the row were significantly different ($\alpha = 0.05$).

Growth Parameters	Fungal Endophytes							F-Value	*p*-Value
	TAMU 490	GHA	NI-8	JG-1	SPE-120	AMF	Control		
% germination	80	77.5	95	75	70	82.5	77.5	1.44	0.19
True leaf count	3.6 ± 0.13	3.5 ± 0.11	3.6 ± 0.12	3.3 ± 0.15	3.5 ± 0.15	3.2 ± 0.15	3.2 ± 0.14	1.21	0.31
Square count	3.2 ± 0.18 ^{bc}	3.0 ± 0.23 ^{bc}	3.9 ± 0.20 ^{ab}	2.8 ± 0.19 ^{c}	4.2 ± 0.20 ^{a}	3.2 ± 0.19 ^{abc}	2.4 ± 0.24 ^{c}	6.64	<0.0001
Plant height (cm)	20.8 ± 0.86 ^{b}	26.9 ± 0.52 ^{a}	27.0 ± 0.66 ^{a}	25.2 ± 0.48 ^{a}	25.4 ± 0.31 ^{a}	27.1 ± 0.53 ^{a}	25.5 ± 0.53 ^{a}	11.92	<0.0001
Root dry weight (gm)	6.8 ± 0.50	6.8 ± 0.71	7.5 ± 0.47	6.1 ± 0.43	7.0 ± 0.55	6.8 ± 0.51	5.4 ± 0.46	1.66	0.14
Stem dry weight (gm)	23.6 ± 1.00 ^{b}	21.6 ± 1.70 ^{bc}	26.6 ± 1.48 ^{ab}	23.4 ± 0.77 ^{b}	17.3 ± 1.08 ^{c}	30.5 ± 1.98 ^{a}	17.3 ± 1.11 ^{c}	12.10	<0.0001
Whole plant dry weight (gm)	30.4 ± 1.02 ^{bc}	28.4 ± 1.83 ^{bcd}	34.1 ± 1.71 ^{ab}	29.5 ± 0.73 ^{bc}	24.3 ± 1.33 ^{cd}	37.3 ± 2.31 ^{a}	22.7 ± 1.23 ^{d}	11.12	<0.0001

Figure 1. (**A**) Plant sample cultures collected from endophytic plants to test the presence of fungal endophytes in roots, stems, leaves, and squares of cotton plants; (**B**,**C**) fungal endophyte growing from leaves of endophytic plants; (**D**) fungus growing from squares of endophytic cotton plants.

2.2. *Lygus lineolaris* Mortality Following Direct Exposure to Fungal Spores

Lygus lineolaris adults exposed to different concentrations of endophytic fungal spores exhibited differential mortality rates. The commercial GHA strain of *B. bassiana* (BotaniGard 22 WP) showed significantly higher mortality of *L. lineolaris* adults at concentrations of 10^8 and 10^9 than the control ($F_{3,11} = 25.12$, $p < 0.001$, $n = 3$) (Figure 2A). No significant mortality effects of GHA strain of *B. bassiana* were observed at 10^7 concentrations. The fungal endophyte JG-1 containing *B. bassiana* had a significant impact on the mortality of *L. lineolaris* adults at a concentration of 10^8 and 10^9 compared to controls ($F_{3,11} = 7.93$, $p < 0.01$, $n = 3$) (Figure 2B). Different concentrations of TAMU-490 and SPE-120 did not affect the mortality of *L. lineolaris* adults ($F_{3,11} = 1.16$, $p = 0.38$, $n = 3$) ($F_{3,11} = 0.95$, $p = 0.45$, $n = 3$, respectively) (Figure 2C,E). NI-8 showed significantly higher mortality of *L. lineolaris* at 10^9 concentrations compared to the control and other concentrations tested ($F_{3,11} = 20.18$, $p < 0.001$, $n = 3$, respectively) (Figure 2D). Sporulation of *L. lineolaris* adult cadavers was observed for many of the fungal endophytes used in the study.

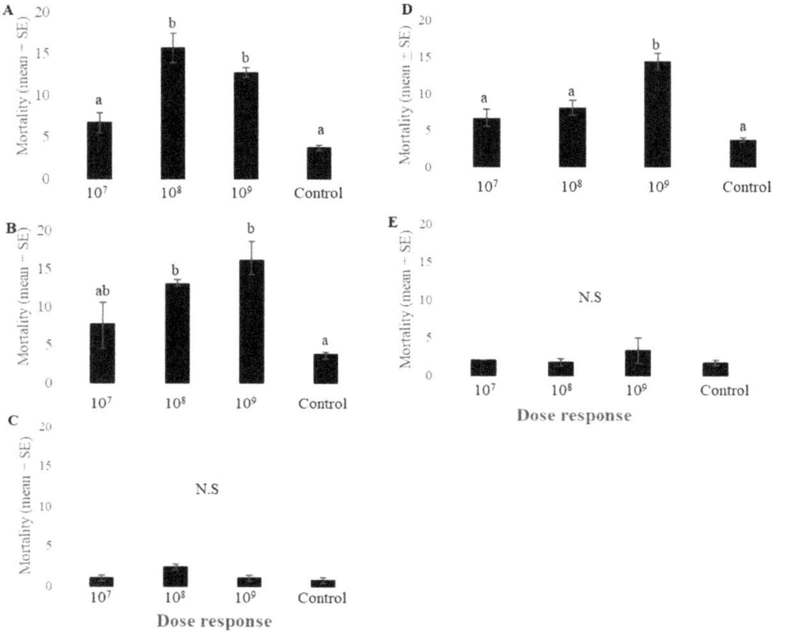

Figure 2. Mortality of *Lygus lineolaris* adults (mean ± SE) following exposure to different concentrations of endophytic fungal spores. (**A**) GHA, (**B**) JG-1, (**C**) TAMU-490, (**D**) NI-8 and (**E**) SPE-120 compared to water controls. Data were analyzed by ANOVA followed by Tukey's HSD for mean comparisons ($n = 30$). Treatments that have no letters in common were significantly different ($\alpha = 0.05$).

2.3. Detection and Validation of Endophytic Fungal Species in Treated Plants

Alignment of ITS1/2 nucleotide sequences from PDA cultures and stock fungal cultures indicated the presence of the fungal strains used for inoculation in stems and roots of cotton plants (Supplemental Figure S1a,b). ITS nucleotide sequences of different *B. bassiana* isolates contained nucleotide polymorphisms, insertions, or deletions specific to each isolate. More than 90% of the sequence reads generated from stock fungal isolates of *B. bassiana* (GHA, JG1, NI8, and SPE) and *P. inflatum* (TAMU) assembled to produce an ITS sequence specific to each fungal isolate. Nucleotide sequence reads generated from fungal cultures from cotton square tissues had 65–80% sequence reads mapping to the ITS sequences of corresponding stock isolates, indicating that the fungal inoculums had successfully established in the cotton plants. Nucleotide sequences of fungal cultures from square tissue that did not map to the stock isolates may indicate that either fungal endophytes other than the isolates used to inoculate seeds were also present in the square tissues or sequence reads had mismatches or low-quality nucleotide calls that did not meet the 96% similarity cut off. Evolutionary relationships among the ITS sequences of the fungal strains are shown in the Supplemental Figure S2. The branch representing *P. inflatum* was supported in 100% of the bootstrap replicates, while the bootstrap support for grouping different isolates of

B. bassiana ranged from 0 (NI8) to 86 (SPE). This low support is due to the low number of differences between ITS nucleotide sequences of *B. bassiana* isolates.

2.4. Survival and Development of Lygus lineolaris Nymphs on Fungal Endophytic Plants

The development of *L. lineolaris* nymphs on cotton pinhead squares was overall low for all the endophytic plants studied, including the control. No significant differences were observed in the development and adult emergence of *L. lineolaris* on these endophytic plants ($F_{6,69} = 0.83$, $p = 0.55$, $n = 10$) (Figure 3). A higher mortality of nymphs was observed on all the plants used in the experiments.

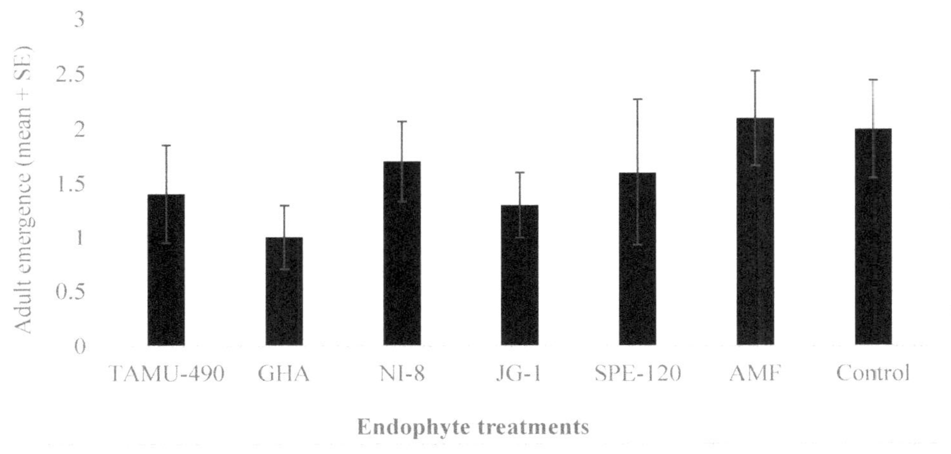

Figure 3. *Lygus lineolaris* adult emergence on endophyte-treated cotton plants (mean ± SE). Second instars were introduced to the plants, and adult emergence was recorded after 8 days. Data were analyzed by ANOVA using JMP software (v. 10, SAS Inc., Cary, NC, USA). No significant differences were observed in the adult emergence of *Lygus lineolaris* on different endophytic cotton plants.

2.5. Olfactometer Studies Using Lygus lineolaris Adults Towards Endophytic Cotton Squares

Y-tube olfactometer studies investigated the behavioral responses of *L. lineolaris* male and female adults towards endophytic cotton squares. Male *L. lineolaris* adults showed significant repellence behavior towards cotton squares from fungal endophytes such as SPE-120, JG-1, and GHA compared to their corresponding controls (($\chi^2 = 5.0$; df = 1; $p = 0.025$), ($\chi^2 = 7.2$; df = 1; $p = 0.007$), ($\chi^2 = 5.0$; df = 1; $p = 0.025$), respectively) (Figure 4A). No significant differences were observed in the behavior of male *L. lineolaris* towards cotton squares of TAMU-490, NI-8 and AMF compared to their controls. No differences were observed in the behavior of male *L. lineolaris* towards double controls that were performed to avoid any bias in their movement towards light or position preferences (Figure 4A).

Female *L. lineolaris* adults showed a significantly higher repellence behavior towards cotton squares from fungal endophytes such as JG-1 and NI-8 compared to their corresponding controls (($\chi^2 = 9.8$; df = 1; $p = 0.001$), ($\chi^2 = 5.0$; df = 1; $p = 0.025$), respectively) (Figure 4B). No significant differences were observed in the behavior of female *L. lineolaris* adults towards cotton squares of AMF, SPE-120, TAMU-490 and GHA compared to their controls. No differences were observed in the behavior of female *L. lineolaris* towards double controls that were performed to avoid any bias in their movement towards light or position preferences (Figure 4B).

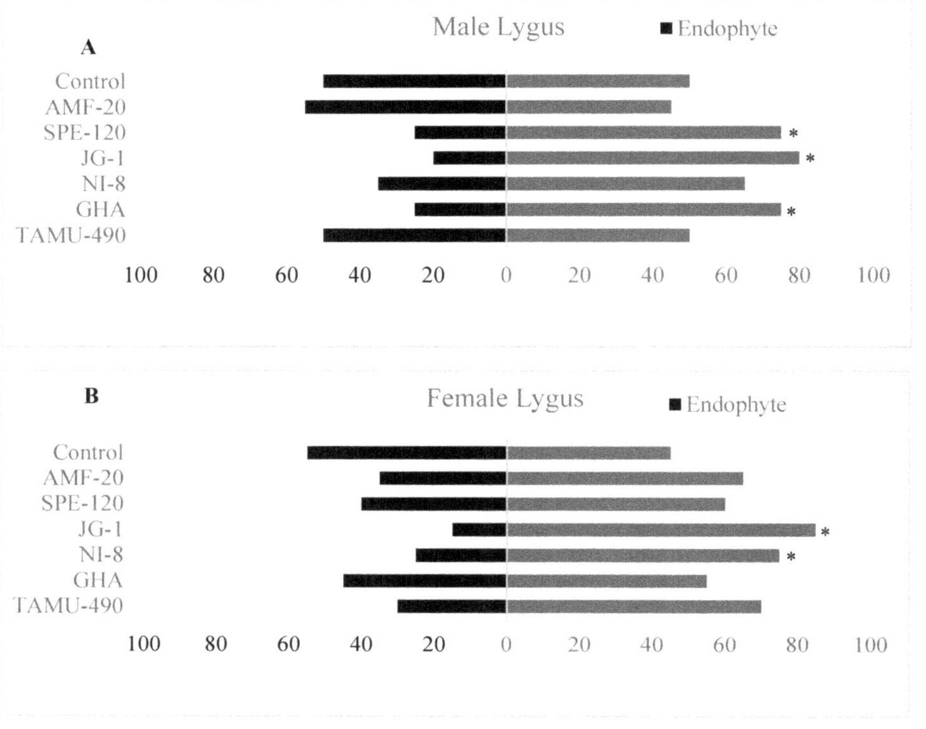

Figure 4. Behavioral response of *Lygus lineolaris* (**A**) males and (**B**) females towards endophytic cotton squares in olfactometer assays. * indicates a significant difference in their behavioral choice towards endophytic squares compared with control squares ($\alpha = 0.05$).

2.6. Performance of Fungal Endophytes against Tarnished Plant Bugs and Bollworms under Field Conditions

Field sampling of tarnished plant bug nymphs and adults on plant terminals showed a significantly fewer number of first–third-instar nymphs on SPE-120-treated plants compared to TAMU-490 and control plants. No early instar nymphs were observed on GHA, NI-8, and JG-1 endophyte-treated plants ($p < 0.05$) (Figure 5A). No significant differences were observed in the number of late instar nymphs (4th–5th) or adults observed on different endophyte-treated plants and control ($p > 0.05$) (Figure 5B,C). Also, no differences were observed in the number of different instar cotton bollworm larvae observed on the treated and control plants (Figure 5D). Some treatment plants did not receive any stages of tarnished plant bugs or cotton bollworm larvae.

In the experiment where second-instar *L. lineolaris* nymphs were introduced in the top five nodes of cotton plants under field conditions, no differences were observed in the development of third-instar or fourth-instar plant bug nymphs ($p > 0.05$) (Figure 6A,B). A combination of seed treatment alone, foliar spray alone, or in combination had no effect on their development. Endophytic seed treatment of GHA combined with endophyte foliar spray and SPE-120 seed treatment combined with foliar spray had the lowest number of fifth-instar *L. lineolaris* nymphs than other treatments post threshold and foliar endophyte spray ($p < 0.05$) (Figure 6C). No differences were observed in the development and adult emergence of *L. lineolaris* on the different endophytic treatments ($p > 0.05$) (Figure 6D). Cotton was harvested from individual treatment plots. No significant differences were observed in the yield of cotton produced under different endophyte seed treated with spray alone or combination treatments ($p > 0.05$) (Figure 7). Even though the yield was numerically higher in SPE-120 seed + spray treatments than controls, the yield did not show a statistical difference.

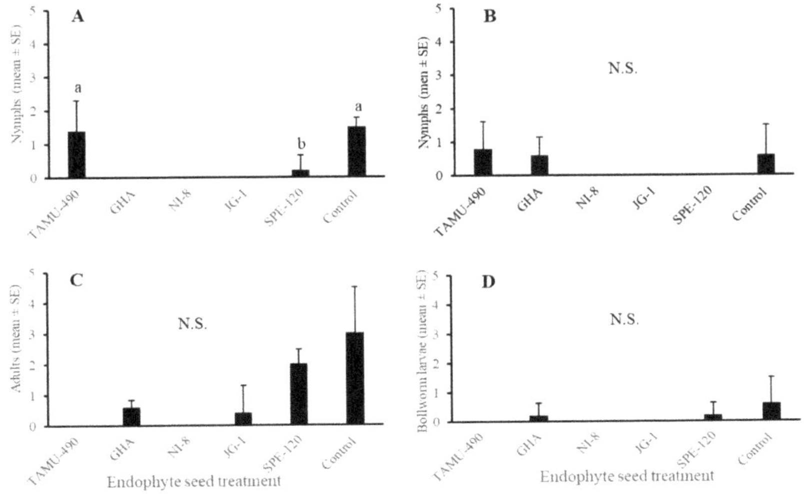

Figure 5. Mean (± SEM) *Lygus lineolaris* and *Helicoverpa zea* larval counts 3 d post-treatment. (**A**) first–third-instar *Lygus lineolaris*; (**B**) fourth- and fifth-instar *Lygus lineolaris*; (**C**) Adult tarnished plant bug; and (**D**) pooled *Helicoverpa zea* larvae. Tarnished plant bugs were recorded as small or large based on the absence or presence of wing pads. Bars with different letters indicate significant differences ($\alpha = 0.05$) between treatments by model contrast analysis based on a generalized linear mixed-effect model.

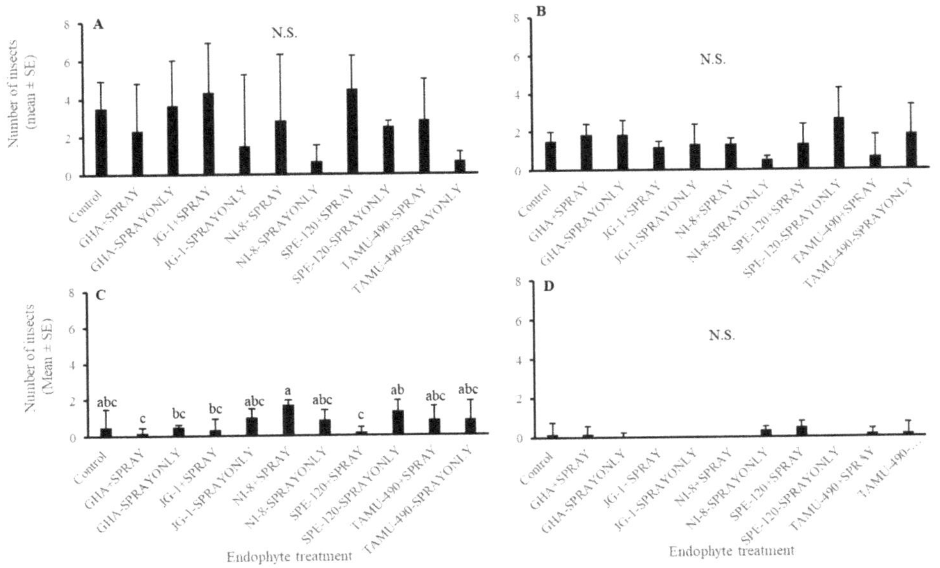

Figure 6. Mean (±SEM) number of second-instar *Lygus lineolaris* caged on the top five nodes of cotton plants in the second week of bloom and allowed to develop for one week. (**A**) Third-instar *Lygus lineolaris* nymphs; (**B**) Fourth-instar *Lygus lineolaris* nymphs; (**C**) Fifth-instar *Lygus lineolaris* nymphs; and (**D**) adult *Lygus lineolaris*. Bars with different letters indicate significant differences ($\alpha = 0.05$) between treatments by model contrast analysis based on a generalized linear mixed-effect model.

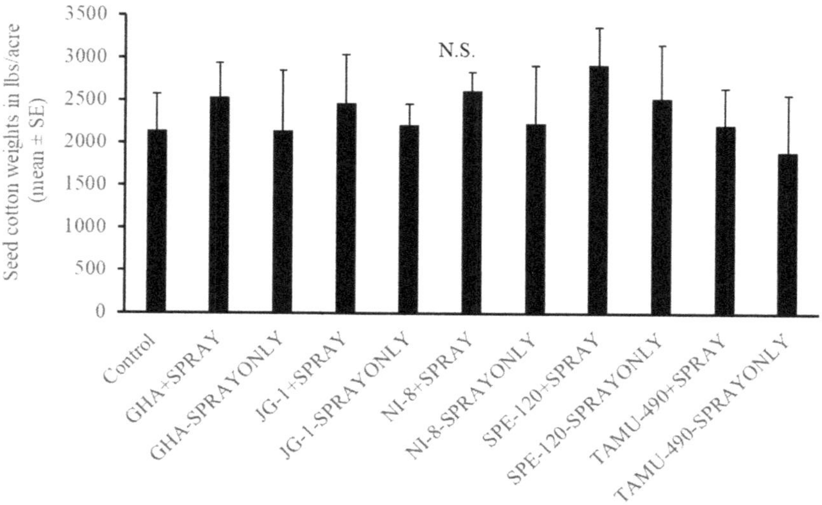

Figure 7. Mean (±SEM) seed cotton weights for seed coated and topical application of cotton endophyte-treated plants. Control plants were sprayed with water. Treatments consisted of seeds treated with spores of endophyte alone and in combination with a foliar endophyte spray. No significant differences ($\alpha = 0.05$) observed between treatments by model contrast analysis based on a generalized linear mixed-effect model.

3. Discussion

Ascomycetous insect-pathogenic fungi such as *Beauveria* and *Metarhizium* can act as endophytes and symbionts in addition to their pathogenicity against many arthropod species. These endophytic entomopathogenic fungi can influence plant growth and the infestation rate of different pests and diseases. Also, the endophytes may help in mitigating biotic and abiotic stress associated with plant growth and herbivore infestation. In this study, we tested the endophytic activity of different known and unknown strains of *B. bassiana* to investigate their effect on cotton growth characteristics, yield, pest infestation rates, behavioral responses of *L. lineolaris* adults to endophytic cotton squares, and yield parameters of different endophytic seed treatments and spray applications. Our PCR analysis showed that the different commercial strains (NI-8, BotaniGard 22 WP, SPE-120) and isolated strains (JG-1) of *B. bassiana* were able to successfully colonize as fungal endophytes on different plant tissues such as leaf, stem, roots, and fruiting squares of cotton plant (Figure 1 and Figure S1a,b). Prior studies have reported the epiphytic and endophytic growth of *B. bassiana* strain GHA on tomato plants [34].

Previous studies have shown that fungal endophytes such as *B. bassiana* and *P. inflatum* affect the host plant selection behavior of *Lygus hesperus* (western tarnished plant bug) and *Nezara viridula* (Southern stink bug), which are two major pests in cotton and soybean [32]. Our research investigated how the fungal endophytes and arbuscular mycorrhizal fungus affect plant growth and yield, as well as the development of *L. lineolaris* nymphs under controlled greenhouse and field conditions. We observed a significant difference in the number of cotton squares produced by *B. bassiana* endophyte strains such as SPE-120 and NI-8 compared to the controls. Simultaneously, no differences were observed in the number of squares by TAMU-490 and GHA *B. bassiana* strain compared to the controls. The stem dry weight and whole plant weight were significantly higher for arbuscular mycorrhizal fungus AMF than the control, TAMU-490, and some of the *B. bassiana* endophytes such as SPE-120, JG-1, and GHA. Previous studies have also reported that the hyphal network formation of AMFs contributes towards plant growth [35] and increased availability and translocation of nutrients [36]. No significant differences were observed in the root weight of AMF treatments compared to controls or other endophytes studied. It was observed that the AMFs had a lot more fibrous fine roots compared to other treatments.

Mortality assays using different concentrations of commercial, isolated *B. bassiana* strains, and *P. inflatum* showed high mortality of *L. lineolaris* adults within 5–7 days. Previous studies have reported the entomopathogenic activity of GHA (BotaniGard 22 WP) and NI-8 against different pest species, including *L. lineolaris* [37–40]. As previously reported, the *B. bassiana* strain GHA was effective against *L. lineolaris* at 10^8 and 10^9 concentrations, and NI-8 caused *L. lineolaris* mortality at 10^9 concentrations. However, the entomopathogenic and endophytic activity of newly isolated *B. bassiana* strain JG-1 has not been tested or previously reported against *L. lineolaris*. JG-1 showed significantly higher mortality of *L. lineolaris* adults at 10^8 and 10^9 concentrations compared to other *B. bassiana* strains such as GHA and NI-8. SPE-120 and TAMU-490 showed no entomopathogenicity towards *L. lineolaris* adults. SPE-120 is labeled as a soil and plant enhancer containing *B. bassiana*, though its pathogenicity is not reported against any insect species. TAMU-490 has been reported to influence the host plant selection behavior of *L. hesperus* and *Nezara viridula*, though their pathogenicity is not reported against *L. lineolaris*. We observed no difference in the development of *L. lineolaris* nymphs to adults under greenhouse conditions (Figure 3). However, significantly fewer fifth-instar nymphs were observed on some of the *B. bassiana* containing endophytic plants under field conditions.

Endophytes can mediate herbivore–plant interactions by altering the plant volatiles and by producing alkaloid-based defensive compounds in the plant tissues [41,42] or through altering the nutritional quality of plants [43,44]. These alkaloids, flavonoids, and phenolic compounds act as a defense against pathogen infections and are reported to have antibiotic, antiparasitic, and antioxidant activities [45]. The selection of host plants by herbivores may include pre-alighting cues such as plant volatiles, post-alighting cues such as leaf surface chemistry, and mechanoreceptor cues such as trichomes and leaf texture. Sword et al. [32] reported strong negative responses by *L. lineolaris* and *N. viridula* against flower buds and squares of endophytic cotton plants containing *P. inflatum* and *B. bassiana*. We also observed similar negative responses exhibited by *L. lineolaris* males towards *B. bassiana* strains such as SPE-120, GHA, and the new strain JG-1. However, we did not observe any negative response of *L. lineolaris* males or females towards TAMU-490. Also, it was observed that *L. lineolaris* female adults showed a negative response towards *B. bassiana* strains such as JG-1 and NI-8. *B. bassiana* strain JG-1 elicited negative responses from both male and female *L. lineolaris*. Adult emergence was significantly low in JG-1 treatment. The mortality of *Lygus* adults was also higher for JG-1 at the 10^9 concentration. *B. bassiana* strain JG-1 exhibited the characteristics of an endophytic entomopathogenic fungi by eliciting negative responses from *L. lineolaris*, and by causing high mortality of *L. lineolaris* adults in the mortality assays.

Endophytes have been reported to improve drought tolerance and reduce attacks by pests, thereby increasing overall crop yield. Two patent claims have been reported on the use of fungal endophytes that show resistance to drought, cold, salt, fungi, bacteria, and pests in cotton (*Dothideomycetes* spp.) and show improved tolerance to drought and pests [32,46]. Greenhouse studies showed a higher number of squares produced by endophytic cotton plants containing *B. bassiana* strains, such as SPE-120 and NI-8, compared to the control. In our field studies, we did not observe any significant differences in cotton yield between the different fungal endophytes studied. However, the yield was noticeably higher in some of the *B. bassiana* treated strains such as NI-8, GHA, and SPE-120, but was not statistically significant (Figure 7).

Laboratory and greenhouse assays clearly showed the establishment of fungal endophytes on cotton plants and their presence in different plant tissues, including fruiting sites. The presence of endophytes in cotton squares resulted in negative responses from *L. lineolaris* adults (Figure 4). Also, it caused developmental delays in *L. lineolaris* nymphs (Figure 6C). Very few early-instar *L. lineolaris* nymphs were observed on NI-8-, GHA-, JG-1-, and SPE-120-treated endophytic cotton plants compared to untreated cotton under field conditions (Figure 5A). However, these did not translate to differences in cotton yield. There could be multiple biotic and abiotic factors that could influence the yield parameters under field conditions. Gehring et al. [47] reported that mycorrhizal fungal associations result in increased stress tolerance, and the adaptive capabilities of the plant increased substantially [48]. Changes in plant volatile profile following endophyte colonization may play an important role in repelling herbivores and reducing plant damage. Sword et al. [32] reported differences in the host selection behavior of *L. hesperus* towards endophytic cotton squares containing *P. inflatum* and *B. bassiana*. However, later studies using solid-phase micro-extraction (SPME) showed no significant differences in the leaf volatile emission profile of these endophytic plants [49]. Multiple factors associated with colonization and persistence of endophytes in plants and their co-existence with native plant microbiomes may influence the establishment and performance of these endophytes. Further research is required to understand these interactions between fungal endophytes and cotton plants, and how they affect host plant resistance and plant growth characteristics.

4. Materials and Methods

4.1. Fungal Endophytes

Fungal endophytes were selected based on previous research studies and new strains identified in our research. The fungal endophyte *Phialemonium inflatum* (=*Paecilomyces inflatus*) (strain TAMU-490) was originally isolated from surface-sterilized cotton leaves as part of a survey of naturally occurring foliar fungal endophytes in College Station, TX, USA. A culture of TAMU-490 was received from Sword's lab (Texas A&M, College Station, TX, USA). SPE-120 is a soil and plant enhancer containing *B. bassiana* that forms a symbiotic relationship with plants (Jabb Incorporated in Raleigh, NC, USA). SPE-120 is applied as an inoculant that becomes a symbiotic endophyte in the plant and improves plant health, quality and yield in soybeans, corn, potato, and wheat. JG-1 is a *B. bassiana* strain isolated locally from corn plants in Stoneville, MS, USA. Another endophyte used for comparison was the GHA strain of *B. bassiana* (BotaniGard 22 WP), a commercial mycoinsecticide approved by EPA for use against many insect pests. NI-8, an isolate of *B. bassiana* naturally infecting tarnished plant bug in the Mississippi Delta, was also tested to study its efficacy as an endophyte in cotton against tarnished plant bugs. Endophytes TAMU-490, NI-8, GS-1, and JG-1 were previously isolated from multiyear field surveys of naturally occurring fungal endophytes in cotton and corn [50,51]. An endomycorrhizal fungi containing *Rhizophagous intraradices* (300 propagules/gm) (Wallace organic Wonder, Greene, RI, USA) was also used as a comparison in some of the studies to understand its effects on *L. lineolaris* and cotton yield. All the fungal endophytes were cultured, and spores were used to inoculate untreated cotton seeds.

4.2. Spore Culturing and Seed Preparation

Spore culturing: Spore powder production was accomplished by utilizing a biphasic culture system, as described by Portilla et al. [52] and Glover et al. [53]. Harvested technical powder was analyzed for conidia concentration (spores per milliliter) and germination. Fresh potato dextrose agar plates were utilized to extract fungal plugs from 3-day old hyphal growth, which were then suspended in 1000 mL of potato dextrose broth. This mixture was agitated for seven days at 27 °C in an incubator shaker (Excella E25, New Brunswick Scientific Co., Inc., Edison, NJ, USA) to create the inoculum for bioassays. Plastic bags (560 mm × 385 mm × 225 mm) containing 1000 g of white rice and 600 mL of water were autoclaved and inoculated with 150 mL aliquots of the prepared inoculum. The bags were kept in an environmental chamber (27 °C, 50% RH, dark 0/24 h L/D photoperiod) for seven to ten days to allow for full colonization, with the bags being flipped every 24 h. The colonized rice was dried in paper bags (30 cm × 17 cm × 43 cm Barrel, Kraft, Chicago, IL, USA) for seven to ten days until the moisture content was sufficiently low ($aw \leq 0.3$). Conidia were manually separated from the dried rice using graded sieves (Grainger, Sieve SS Frame 8, SS Mesh # 15, 30, and 100). The harvested spores were plated on PDA, incubated for 24 h to assess germination, and counted (spores mm^2). The produced spores were then stored at −80 °C at the Southern Insect Management Research Unit (SIMRU) in Stoneville, MS, USA.

Seed preparation: Seeds were surface-sterilized by immersion in 70% ethanol for 3 min with constant stirring using a magnetic stir bar, immediately followed by 3 min in 2% sodium hypochlorite (NaOCl), followed by three washes in sterile water, based on Lopez et al. [21]. Wash water was plated on PDA media to confirm surface sterilization efficiency. The seeds were then coated with one concentration of the four fungi and two commercial products, in addition to sterile water used as the control. Spore concentrations for each fungus were zero (water control), and $\geq n \times 10^7$ spores/mL based on concentrations used in previous studies of endophytic entomopathogens [21,54]. A hemocytometer 0.1 mm deep (Hausser-Bright Line, Horsham, PA, USA) was used to calculate the conidia concentrations of the resulting stock solutions from diluting 1.0 g of harvested spores containing 6.1×10^{11} spores suspended in 50 mL of 0.04% Tween-80 (P8074, Sigma-Aldrich, St. Louis, MO, USA) and methyl cellulose (M7027, Sigma-Aldrich, St. Louis, MO, USA) 2% *v*/*v*. Aliquots of 2 mL suspensions ($n \times 10^7$) provided a concentration of $\geq 550 \pm 25$ viable spores per mm^2 with 500 ± 15 viable spores (95% viability) for all treatments. Fungal-endophyte-coated seeds were dried under a forced air fan on aluminum foil for 12 h and planted within 24 h of the experimental coating application.

Bollgard II™ (DP1646, Delta and Pine Land Company™, Scott, MS, USA) seeds were planted in a potting mix containing 1:2 proportion of potting mix/topsoil. Four sets of ten plants each were planted for each treatment for a total of 40 plants/treatment. The plants were kept under greenhouse conditions under a 16:8 light/dark cycle. The plants were watered regularly and fertilized biweekly. Samples were collected from stems, leaves, squares, and roots 45 days after planting to check for the presence of fungal endophytes by culturing them on an agar medium. Following sporulation, samples of the fungal spores and hyphae were collected and analyzed using PCR.

Plant growth measurements were performed during different growth stages of cotton, including seed germination (5 days after planting), number of true leaves (14 days after germination), plant height (35 days after germination), number of squares (35 days after germination), and dry weights of roots, stem, and whole plant. Roots were cut 7.5 cm above the bottom tip of the taproot. After cutting the roots, they were thoroughly washed using tap water to remove all soil and placed into labeled brown paper bags. The remaining plant stem and leaves were transferred to another brown paper bag. The paper bags were put into an oven at 70 °C for 20 h. After all of the roots and stems were dried, they were weighed on a scale, and dry weight was recorded. Data were analyzed by ANOVA followed by Tukey's HSD for mean comparisons using JMP statistical software (v. 10, SAS Inc, Cary, NC, USA). Treatments that have no letters in common within the row were significantly different ($\alpha = 0.05$).

4.3. Detection and Validation of Endophytic Fungal Species in Treated Plants

Pieces of cotton square tissue from plants inoculated with *B. bassiana* strains JG1, GHA, SPE and, *P. inflatum* were surface-sterilized with 0.5% sodium hypochlorite solution for 2 min and washed with sterile distilled water three times, followed by three washes with 75% ethanol under a laminar flow sterile hood. Small pieces of surface-sterilized square tissue were cut with a sterile razor blade. One set of square tissue from each plant was placed in 115 mm Petri dishes containing PDA-agar medium to facilitate the growth of endophytes and another set was frozen at −80 °C in sterile 1.5 mL centrifuge tubes. Genomic DNA was extracted from the fungal cultures resulting from plant tissues using MasterPure tissue and cell lysis reagent kit (Epicentre Technologies, Madison, WI, USA) following the manufacturer's protocol as outlined in the work of Perera et al. [55]. Briefly, two sterile stainless steel ball bearings (3 mm) were placed in each tube containing plant or fungal hyphae and homogenized for 2 min using a bead beater (Biospec Products, Bartlesville, OK, USA). Proteinase K was added to the homogenate to a final concentration of 2 ng/µL and was incubated at 65 °C for two hours, followed by digestion with RNAse at 37 °C for 30 min and precipitation of proteins with MPC solution by incubating on ice for 30 min. The tubes were centrifuged at 16,000× *g* for 15 min at 4 °C, and the supernatant containing DNA was transferred to new tubes. An equal volume of 100% isopropanol was added to the supernatant, mixed well, and the DNA was precipitated by centrifuging at 16,000× *g* for 15 min. DNA pellets were rinsed three times with 70% ethanol, air dried, and resuspended in 35 µL of Tris-HCl, pH 7.5. Genomic DNA was also extracted using the above protocol from the parent fungal stocks used to inoculate cotton plants.

Internal transcribed spacers 1 (ITS1). 5.8S rRNA and ITS2 of the ribosomal RNA gene were PCR-amplified from each DNA sample using a forward primer designed for 18S rRNA (4046_18SF: 5′-CGCTACTACCGATTGAATGGCTC-3′) and a reverse primer for 28S rRNA (4053_28SR: 5′-TCCTCCGCTTATTGATATGC-3′). Standard Taq polymerase and 1× buffer (New England Biolabs, Ipswich, MA, USA) was used on a PTC-200 thermal cycler with a thermal cycling profile containing 30 s initial denaturation at 95 °C, followed by 35 cycles of 10 s denaturation at 95 °C, 15 s annealing at 52 °C, 60 s extension at 72 °C and a final extension of 5 min at 72 °C. Amplicons were resolved in a 1.0% agarose gel to verify amplification.

Ribosomal RNA amplicons from the samples were cleaned by binding to AmPure XP paramagnetic beads (Beckman Coulter, Indianapolis, IN, USA) at a DNA/beads ratio of 1:1.8. Nucleotide sequences of the purified amplicons were obtained by direct sequencing with Flongle flowcell from Oxford Nanopore sequencing technology (Oxford Nanopore Technologies, New York, NY, USA) using the library construction reagent set SQK-LSK114 and the native DNA barcode kit SQK-NDB-96. Nucleotide sequences were analyzed using CLC Genome WorkBench v22.02 (Qiagen, Redwood City, CA, USA).

Nucleotide sequences generated from fungal stocks and published ITS sequences from *B. bassiana* (AB576868, MG548313, and LC768985), and *P. inflatum* (MH857776) were used as references to map the ITS1/2 sequence reads produced from fungal cultures obtained from cotton square tissues. Mismatch cost of 2, linear gap cost of 3 for insertions and deletions, length fraction of 0.90, and similarity fraction of 0.96 were used in the mapping of Oxford Nanopore sequences to the reference sequences. Phylogenetic analysis was performed using Molecular Evolutionary Genetic Analysis (MEGA) 11 [56] based on a multiple-sequence alignment of the ITS1/2 regions. Gene phylogenetic relationships were estimated using the Maximum Likelihood (ML) method [57,58] applying the Kimura-2-Parameter model of sequence evolution [59], that implemented a discrete Gamma shape parameter and node support inferred from 10,000 bootstrap pseudo replications [60] of the aligned nucleotide sequences. ITS1/2 nucleotide sequences of *B. bassiana* and *P. inflatum* obtained from the NCBI database were also used as references. *Cephalotheca sulfurea* ITS sequence (AB278194) was used as the outgroup.

4.4. *Lygus lineolaris* Mortality Following Direct Exposure to Fungal Spores

Four-day-old, tarnished plant bug adults were exposed to serial dilutions of spores from six endophytes at concentrations of ($n \times 10^9$, $n \times 10^8$, $n \times 10^7$). Mortality and sporulation were monitored daily over a 10-d period. The bioassays were carried out in an environmental chamber (Percival Scientific, Perry, IA, USA) set at 28 °C, with a photoperiod of 16:8 h (L:D) and 65% RH. Each treatment (endophyte concentration × fungal endophyte) was replicated three times, involving a total of 270 adults. Spores harvested from four candidate endophytes (TAMU-490, N-I8, GS-1, JG-1) and two commercial products SPE-120 (SBb-2.5 inoculant) and GHA (BotaniGard 22 WP) were diluted with deionized water to achieve the specified concentrations ($n \times 10^9$, $n \times 10^8$, $n \times 10^7$) and estimated based on dilution of $n \times 10^9$ concentration. The number of spores applied was adjusted for viability across all concentrations tested. Aliquots of 5 mL suspensions ($n \times 10^9$) provided a concentration of ≥450 ± 75 viable spores per mm^2 with 398 ± 25 viable spores (95% viability) for all endophytes and commercial products tested.

Spore concentrations per milliliter were determined using a hemocytometer (0.1 mm deep, Hausser-Bright Line, Horsham, PA, USA) by diluting 1.0 g of harvested spores, which contained 3.1×10^{11} spores in 50 mL of 0.04% Tween-80 (Sigma-Aldrich P8074). Conidia suspensions were sprayed onto disposable glass microscope coverslips (Fisherbrand™, Thermo Fisher Scientific, Waltham, MA, USA) using a handheld sprayer, covering an area of 38.5 mm^2. Spore concentration (spores per mm^2) was determined by counting the deposited spores. This process was repeated three times at a final concentration of ($n \times 10^9$), with specific dilutions for each endophyte: 5.1, 1.2, 8.1, 1.3, 6.1, and 9.0 for TAMU-490, NI-8, GS-1, JG-1, SPE-120, and GHA, respectively. Tarnished plant bugs were monitored daily for ten days under a light microscope. Bugs were deemed dead if no movement was observed when prodded with a natural hair paint brush and alive if movement was detected. Bugs remained in their original diet cups, with mycosis checked daily by noting the first appearance of external hyphal growth. Data were analyzed using ANOVA and Tukey's HSD for mean comparisons with JMP statistical software (v. 10, SAS Inc., Cary, NC, USA), considering treatments significantly different if they had no common letters within the row ($\alpha = 0.05$)

4.5. Survival and Development of Lygus lineolaris Nymphs on Fungal Endophytic Plants

Greenhouse experiments were performed on individual cotton plants. Cotton terminals were isolated with insect enclosure bags constructed from organza ($22 \times 22 \times 8$ mm ~240 μm mesh, JoAnn's Fabrics, Hudson, OH, USA) that enclosed the plant terminals affixed with pipe cleaners (top five nodes) containing many squaring sites. The second week of bloom (≈65 d) was chosen to reflect not only a physiologically vulnerable growth stage of cotton but also a location that is commonly associated with economic damage from tarnished plant bug. Nymphal development of *L. lineolaris* was measured by introducing second-instar nymphs to different fungal endophytic plants. Ten second-instar *L. lineolaris* nymphs were introduced to the terminal of each plant with multiple developing squares in a mesh bag, and a total of ten plants were used for each treatment. Adult emergence was monitored starting 5 days after introducing the nymphs, and the total number of adults that emerged after 10 days is reported. Data were analyzed by ANOVA followed by Tukey's HSD for mean comparisons using JMP statistical software (v. 10, SAS Inc., Cary, NC, USA).

4.6. Olfactometer Studies Using Lygus lineolaris Adults Towards Endophytic Cotton Squares

Olfactometer studies were performed to study the behavior of male and female *L. lineolaris* towards endophytic cotton squares. Twenty male and twenty female *L. lineolaris* were used in each olfactometer experiment, and a total of 40 adults were studied for each endophytic fungus. Charcoal-purified air was pushed through the arms of the olfactometer, and the airflow rate was set to 250 mL/min and checked using an ADM flow meter. The adults were released at the main arm of the olfactometer, and their response to control cotton squares or endophytic cotton squares was recorded after 10 min. If no responses were observed by 10 min, the insect was discarded, and a new insect was used. Responses were recorded from 20 male and 20 female adults. After every 5 insect responses, the position of the olfactometer was switched to avoid preferences. A clean olfactometer was used after every 10 behavioral responses were completed. Experiments were performed under dark conditions, as preliminary studies have shown that *L. lineolaris* responds better to odorants under dark conditions. Behavior preferences of adults in the Y-tube olfactometer were statistically analyzed by the χ^2 test of independence ($\alpha = 0.05$) in JMP (v. 10, SAS Inc., Cary, NC, USA) that gave 95% confidence intervals.

4.7. Field Planting, Yield Measurements, and Performance of Fungal Endophytes against Tarnished Plant Bugs and Bollworms

The field experiment was conducted in the summer of 2023 at the Southern Insect Management Research Unit (SIMRU) research farm near Stoneville, MS, USA. The experiment was laid out in a randomized complete block design with three replications. Each plot consisted of 4 rows (101.6 cm in width) approximately 14 m in length. Bollgard II™ (DP1646, Delta and Pine Land Company™, Scott, MS, USA) cotton seed was planted in early June on 91 m rows and 96 cm row centers at a field site of ≈2.1 ha, resulting in a plant stand of ≈106,250 plants/ha (42,500 plants/acre). Ad hoc applications of herbicides and a plant growth regulator (mepiquat chloride, Loveland Products, Inc., Morgantown, KY, USA) were applied equally to all plots in the study. Transform insecticide (Sulfoxaflor, Corteva Agriscience, Indianapolis, IN, USA), Diamond insecticide (Novaluron, Makhteshim Agan of North America, Inc., Raleigh, NC, USA), and Orthene insecticide (Acephate, AMVAC, Newport Beach, CA, USA) at labeled rates was used ca. every 10 d to maintain the plots pest-free until the first week of bloom. Insecticide applications were discontinued 14 days prior to the first and weekly insect sampling. Insect sampling occurred weekly throughout the six weeks of bloom, and treatment decisions were based on tarnished plant bug densities estimated using a black drop cloth. Mechanical harvest occurred on 11 October 2023 (119 d emergence to harvest).

Treatments included an untreated control (control), endophyte-coated seed only, endophyte spray only, and a seed-coated and sprayed combination. The plots were arranged in a randomized complete block design with four replications. Prior to bloom, all *L. lineolaris* samples were taken with a standard 38 cm sweep net (25 sweeps per plot) once per week. After the first bloom, all plots were sampled with a 0.76 m black drop cloth, with two drops per plot taken weekly. Thresholds for current management practices were based on Mississippi State University Extension Services recommendations of three per 0.76 row m when utilizing a black drop cloth [61]. The efficacy of endophytes was evaluated by taking two drops, using a black drop cloth 6 row feet in length, from the center of two rows of each plot. Total *L. lineolaris* (nymphs and adults) were reported from each plot. *L. lineolaris* were recorded as small (first–third instar) and large (fourth–fifth instar) based on the presence or absence of wing pads and fully developed adults. Cotton bollworm larval stages were pooled. Control plants were sprayed with water. Treatments consisted of treated seeds (coated with spores) of endophyte alone and in combination with a foliar endophyte spray. The plots were sprayed once plot averages reached the threshold for *L. lineolaris*. Data were subject to statistical analysis using one-way ANOVA in SAS software (version 9.4, SAS Institute Inc., Cary, NC, USA). All data were analyzed as randomized complete block designs. Each treatment consisted of 4 to 6 replicates. Data were analyzed using the General Linear Model, and means were compared using the Tukey–Kramer HSD test for least squared means ($\alpha = 0.05$).

5. Conclusions

Endophytic entomopathogenic fungi can influence plant growth and the infestation rate of different pests and diseases. Our greenhouse and field experiments showed that these fungal endophytes can colonize different plant parts, impact the plant growth of cotton, and affect the biology and behavior of *L. lineolaris* adults and nymphs. All the different *B. bassiana* strains tested had some effect on the development or mortality of *L. lineolaris*. The new *B. bassiana* strain JG-1 significantly affected the olfactory response of male and female *L. lineolaris* adults and caused significant adult mortality in the bioassays. Volatile collection studies and GC-MS analysis may provide more details of the induced plant volatiles that cause negative responses from *L. lineolaris* adults. Also, further experiments may help to optimize JG-1 as an entomopathogenic endophytic fungi for practical application against different agricultural pests.

Supplementary Materials: The following supporting information can be downloaded at: https://www.mdpi.com/article/10.3390/plants13152012/s1.

Author Contributions: Conceptualization and experiment design, J.G., J.P.G. and G.V.P.R.; validation, J.G., J.P.G. and O.P.P., data analysis, J.G., J.P.G. and O.P.P.; provided materials, J.G., J.P.G. and O.P.P.; writing—original draft, J.G. and J.P.G.; writing—review, and editing, J.G., J.P.G., O.P.P. and G.V.P.R. All authors have read and agreed to the published version of the manuscript.

Funding: This work was supported by USDA-ARS Research Project# 6066-22000-091-00D Ecologically Sustainable Approaches to Insect Resistance Management in Bt Cotton.

Data Availability Statement: Data is contained within the article or Supplementary Material.

Acknowledgments: Special thanks to Shundalyn Moore and Russ Godbold (USDA-ARS, Stoneville, MS) for technical assistance in laboratory and field studies. Also, we would like to thank Maribel Portilla for supplying *B. bassiana* NI-8 spores and Gregory Sword (Texas A&M) for providing *P. inflatum* spores for the experiments. USDA is an equal opportunity provider and employer. The findings and conclusions in this publication are those of the author(s) and should not be construed to represent any official USDA or U.S. Government determination or policy. Mention of trade names or commercial products in this publication is solely for the purpose of providing specific information and does not imply recommendation or endorsement by the U.S. Department of Agriculture.

Conflicts of Interest: The authors declare no competing interests.

References

1. Hardoim, P.R.; Van Overbeek, L.S.; Berg, G.; Pirttilä, A.M.; Company, S.; Campisano, A.; Döring, M.; Sessitsch, A. The hidden world within plants: Ecological and evolutionary considerations for defining functioning of microbial endophytes. *Microbiol. Mol. Biol. Rev.* **2015**, *79*, 293–320. [CrossRef] [PubMed]
2. Puri, A.; Padda, K.P.; Chanway, C.P. Evidence of nitrogen fixation and growth promotion in canola (*Brassica napus* L.) by an endophytic diazotroph *Paenibacillus polymyxa* P2b-2R. *Biol. Fertil. Soils.* **2016**, *52*, 119–125. [CrossRef]
3. Kumar, K.K.; Dara, S.K. Fungal and bacterial endophytes as microbial control agents for plant-parasitic nematodes. *Inter. J. Environ. Res. Public Health* **2021**, *18*, 4269. [CrossRef]
4. Suryanarayanan, T.S. Endophyte research: Going beyond isolation and metabolite documentation. *Fungal Ecol.* **2013**, *6*, 561–568. [CrossRef]
5. Smith, R.J.; Pekrul, S.; Grula, E.A. Requirement for sequential enzymatic activities for penetration of the integument of the corn earworm (*Heliothis zea*). *J. Invert. Pathol.* **1981**, *38*, 335–344. [CrossRef]
6. Bamisile, B.S.; Dash, C.K.; Akutse, K.S.; Keppanan, R.; Afolabi, O.G.; Hussain, M.; Qasim, M.; Wang, L. Prospects of endophytic fungal entomopathogens as biocontrol and plant growth promoting agents: An insight on how artificial inoculation methods affect endophytic colonization of host plants. *Microbiol. Res.* **2018**, *217*, 34–50. [CrossRef] [PubMed]
7. Mantzoukas, S.; Eliopoulos, P.A. Endophytic entomopathogenic fungi: A valuable biological control tool against plant pests. *Appl. Sci.* **2020**, *10*, 360. [CrossRef]
8. Posada, F.; Aime, M.C.; Peterson, S.W.; Rehner, S.A.; Vega, F.E. Inoculation of coffee plants with the fungal entomopathogen *Beauveria bassiana* (*Ascomycota: Hypocreales*). *Mycol. Res.* **2007**, *111*, 748–757. [CrossRef]
9. Fürnkranz, M.; Lukesch, B.; Müller, H.; Huss, H.; Grube, M.; Berg, G. Microbial diversity inside pumpkins: Microhabitat-specific communities display a high antagonistic potential against phytopathogens. *Microb. Ecol.* **2012**, *63*, 418–428. [CrossRef]
10. Jaber, L.R.; Enkerli, J. Effect of seed treatment duration on growth and colonization of *Vicia faba* by endophytic *Beauveria bassiana* and *Metarhizium brunneum*. *Biol. Control* **2016**, *103*, 187–195. [CrossRef]
11. Jaber, L.R.; Enkerli, J. Fungal entomopathogens as endophytes: Can they promote plant growth? *Biocontrol Sci. Technol.* **2017**, *27*, 28–41. [CrossRef]
12. Lugtenberg, B.J.; Caradus, J.R.; Johnson, L.J. Fungal endophytes for sustainable crop production. *FEMS Microbiol. Ecol.* **2016**, *92*, 194. [CrossRef] [PubMed]
13. Schardl, C.L.; Leuchtmann, A.; Spiering, M.J. Symbioses of grasses with seedborne fungal endophytes. *Annu. Rev. Plant Biol.* **2004**, *55*, 315–340. [CrossRef] [PubMed]
14. Akutse, K.S.; Fiaboe, K.K.; Van den Berg, J.; Ekesi, S.; Maniania, N.K. Effects of endophyte colonization of *Vicia faba* (Fabaceae) plants on the life–history of leafminer parasitoids *Phaedrotoma scabriventris* (Hymenoptera: Braconidae) and *Diglyphus isaea* (Hymenoptera: Eulophidae). *PLoS ONE* **2014**, *9*, e109965. [CrossRef] [PubMed]
15. Saikkonen, K.; Gundel, P.E.; Helander, M. Chemical ecology mediated by fungal endophytes in grasses. *J. Chem. Ecol.* **2013**, *39*, 962–968. [CrossRef] [PubMed]
16. White, J.F., Jr.; Belanger, F.A.; Meyer, W.I.; Sullivan, R.F.; Bischoff, J.F.; Lewis, E.A. Minireview article: Clavicipitalean fungal epibionts and endophytes-development of symbiotic interactions with plants. *Symbiosis* **2002**, *33*, 3.
17. Hu, S.; Bidochka, M.J. Root colonization by endophytic insect-pathogenic fungi. *J. Appl. Microbiol.* **2021**, *130*, 570–581. [CrossRef] [PubMed]
18. Arnold, A.E.; Lewis, L.C. Ecology and evolution of fungal endophytes, and their roles against insects. In *Insect-Fungal Associations: Ecology and Evolution*; Oxford University Press: New York, NY, USA, 2005; Volume 3, pp. 74–96.
19. Reddy, N.P.; Khan, A.P.; Devi, U.K.; Sharma, H.C.; Reineke, A. Treatment of millet crop plant (*Sorghum bicolor*) with the entomopathogenic fungus (*Beauveria bassiana*) to combat infestation by the stem borer, *Chilo partellus* Swinhoe (Lepidoptera: Pyralidae). *J. Asia-Pacific Entomol.* **2009**, *12*, 221–226. [CrossRef]

20. Lopez, C.D.; Zhu-Salzman, K.; Ek-Ramos, M.J.; Sword, G.A. The entomopathogenic fungal endophytes *Purpureocillium lilacinum* (formerly *Paecilomyces lilacinus*) and *Beauveria bassiana* negatively affect cotton aphid reproduction under both greenhouse and field conditions. *PLoS ONE* **2014**, *9*, e103891.
21. Lopez, D.C.; Sword, G.A. The endophytic fungal entomopathogens *Beauveria bassiana* and *Purpureocillium lilacinum* enhance the growth of cultivated cotton (*Gossypium hirsutum*) and negatively affect survival of the cotton bollworm (*Helicoverpa zea*). *Biol. Control* **2015**, *89*, 53–60. [CrossRef]
22. Leckie, B.M.; Ownley, B.H.; Pereira, R.M.; Klingeman, W.E.; Jones, C.J.; Gwinn, K.D. Mycelia and spent fermentation broth of *Beauveria bassiana* incorporated into synthetic diets affect mortality, growth, and development of larval *Helicoverpa zea* (Lepidoptera: Noctuidae). *Biocontrol Sci. Technol.* **2008**, *18*, 697–710. [CrossRef]
23. Menjivar, R.D.; Cabrera, J.A.; Kranz, J.; Sikora, R.A. Induction of metabolite organic compounds by mutualistic endophytic fungi to reduce the greenhouse whitefly *Trialeurodes vaporariorum* (Westwood) infection on tomato. *Plant Soil* **2012**, *352*, 233–241. [CrossRef]
24. Akutse, K.S.; Maniania, N.K.; Fiaboe, K.K.; Van den Berg, J.; Ekesi, S.J. Endophytic colonization of *Vicia faba* and *Phaseolus vulgaris* (Fabaceae) by fungal pathogens and their effects on the life-history parameters of *Liriomyza huidobrensis* (Diptera: Agromyzidae). *Fungal Ecol.* **2013**, *6*, 293–301. [CrossRef]
25. Newcombe, G.; Shipunov, A.; Eigenbrode, S.D.; Raghavendra, A.K.; Ding, H.; Anderson, C.L.; Menjivar, R.; Crawford, M.; Schwarzländer, M. Endophytes influence protection and growth of an invasive plant. *Commun. Integr. Biol.* **2009**, *2*, 29–31. [CrossRef] [PubMed]
26. Biswas, C.; Dey, P.; Satpathy, S.; Satya, P.; Mahapatra, B.S. Endophytic colonization of white jute (*Corchorus capsularis*) plants by different *Beauveria bassiana* strains for managing stem weevil (*Apion corchori*). *Phytoparasitica* **2013**, *41*, 17–21. [CrossRef]
27. Thakur, A.; Kaur, S.; Kaur, A.; Singh, V. Enhanced resistance to *Spodoptera litura* in endophyte infected cauliflower plants. *Environ. Entomol.* **2013**, *42*, 240–246. [CrossRef] [PubMed]
28. Zhou, W.; Starr, J.L.; Krumm, J.L.; Sword, G.A. The fungal endophyte *Chaetomium globosum* negatively affects both above-and belowground herbivores in cotton. *FEMS Microbiol. Ecol.* **2016**, *92*, fiw158. [CrossRef]
29. Zhou, W.; Wheeler, T.A.; Starr, J.L.; Valencia, C.U.; Sword, G.A. A fungal endophyte defensive symbiosis affects plant-nematode interactions in cotton. *Plant Soil* **2018**, *422*, 251–266. [CrossRef]
30. Gurulingappa, P.; Sword, G.A.; Murdoch, G.; McGee, P.A. Colonization of crop plants by fungal entomopathogens and their effects on two insect pests when in planta. *Biol. Control* **2010**, *55*, 34–41. [CrossRef]
31. Quesada-Moraga, E.; Munoz-Ledesma, F.J.; Santiago-Alvarez, C. Systemic protection of *Papaver somniferum* L. against *Iraella luteipes* (Hymenoptera: Cynipidae) by an endophytic strain of *Beauveria bassiana* (*Ascomycota: Hypocreales*). *Environ. Entomol.* **2009**, *38*, 723–730. [CrossRef]
32. Sword, G.A.; Tessnow, A.; Ek-Ramos, M.J. Endophytic fungi alter sucking bug responses to cotton reproductive structures. *Insect Sci.* **2017**, *24*, 1003–1014. [CrossRef] [PubMed]
33. George, J.; Glover, J.P.; Gore, J.; Crow, W.D.; Reddy, G.V.P. Biology, ecology, and pest management of the tarnished plant bug, *Lygus lineolaris* (Palisot de Beauvois) in southern row crops. *Insects* **2021**, *12*, 807. [CrossRef] [PubMed]
34. Nishi, O.; Sushida, H.; Higashi, Y.; Iida, Y. Epiphytic and endophytic colonisation of tomato plants by the entomopathogenic fungus *Beauveria bassiana* strain GHA. *Mycology* **2021**, *12*, 39–47. [CrossRef] [PubMed]
35. Bowles, T.M.; Barrios-Masias, F.H.; Carlisle, E.A.; Cavagnaro, T.R.; Jackson, L.E. Effects of arbuscular mycorrhizae on tomato yield, nutrient uptake, water relations, and soil carbon dynamics under deficit irrigation in field conditions. *Sci. Total Environ.* **2016**, *566*, 1223–1234. [CrossRef] [PubMed]
36. Rouphael, Y.; Franken, P.; Schneider, C.; Schwarz, D.; Giovannetti, M.; Agnolucci, M.; De Pascale, S.; Bonini, P.; Colla, G. Arbuscular mycorrhizal fungi act as biostimulants in horticultural crops. *Sci. Hortic.* **2015**, *196*, 91–108. [CrossRef]
37. Portilla, M.; Snodgrass, G.; Luttrell, R. Effects of morning and night applications of Beauveria. bassiana strains NI8 and GHA against the tarnished plant bug in cotton. In Proceedings of the Beltwide Cotton Conference, New Orleans, LA, USA, 5–8 January 2014; pp. 729–734.
38. Gad, A.A.; Nada, M.S. Effect of entomopathogenic fungi *Beauveria bassiana* on the cellular immunity and biochemistry of green bug *Nezara viridula* L. *J. Biopest.* **2020**, *13*, 135–144. [CrossRef]
39. Lopes, R.B.; Laumann, R.A.; Blassioli-Moraes, M.C.; Borges, M.; Faria, M. The fungistatic and fungicidal effects of volatiles from metathoracic glands of soybean-attacking stink bugs (Heteroptera: Pentatomidae) on the entomopathogen *Beauveria bassiana*. *J. Invertebr. Pathol.* **2015**, *132*, 77–85. [CrossRef]
40. Sosa-Gómez, D.R.; Moscardi, F. Laboratory and field studies on the infection of stink bugs, *Nezara viridula*, *Piezodorus guildinii*, and *Euschistus heros* (Hemiptera: Pentatomidae) with *Metarhizium anisopliae* and *Beauveria bassiana* in Brazil. *J. Invertebr. Pathol.* **1998**, *71*, 115–120. [CrossRef] [PubMed]
41. Clay, K.; Holah, J. Fungal endophyte symbiosis and plant diversity in successional fields. *Science* **1999**, *285*, 1742–1744. [CrossRef]
42. Faeth, S.H. Are endophytic fungi defensive plant mutualists? *Oikos* **2002**, *98*, 25–36. [CrossRef]
43. Bernays, E.A. Plant sterols and host-plant affiliations of herbivores. *Insect-Plant Interact. (1992)* **2017**, *22*, 53–66.
44. Jallow, M.F.; Dugassa-Gobena, D.; Vidal, S. Influence of an endophytic fungus on host plant selection by a polyphagous moth via volatile spectrum changes. *Arthropod Plant Interact.* **2008**, *2*, 53–62. [CrossRef]
45. Strobel, G.A. Endophytes as sources of bioactive products. *Microbes Infect.* **2003**, *5*, 535–544. [CrossRef] [PubMed]

46. Chitnis, V.R.; Suryanarayanan, T.S.; Nataraja, K.N.; Prasad, S.R.; Oelmüller, R.; Shaanker, R.U. Fungal endophyte-mediated crop improvement: The way ahead. *Front. Plant Sci.* **2020**, *11*, 561007. [CrossRef] [PubMed]
47. Gehring, C.A.; Sthultz, C.M.; Flores-Rentería, L.; Whipple, A.V.; Whitham, T.G. Tree genetics defines fungal partner communities that may confer drought tolerance. *Proc. Natl. Acad. Sci. USA* **2017**, *114*, 11169–11174. [CrossRef]
48. Lau, J.A.; Lennon, J.T.; Heath, K.D. Trees harness the power of microbes to survive climate change. *Proc. Natl. Acad. Sci. USA* **2017**, *114*, 11009–11011. [CrossRef] [PubMed]
49. Gale, C.C.; Suh, C.P.; Perez, J.; Lesne, P.; Wilson, C.; Kramer, Z.; Madamba, C.; Sword, G.A. Sampling volatile organic compounds from individual cotton leaves to test effects of fungal endophyte treatments. *Southwest. Entomol.* **2021**, *46*, 299–304. [CrossRef]
50. Ek-Ramos, M.J.; Zhou, W.; Valencia, C.U.; Antwi, J.B.; Kalns, L.L.; Morgan, G.D.; Kerns, D.L.; Sword, G.A. Spatial and temporal variation in fungal endophyte communities isolated from cultivated cotton (*Gossypium hirsutum*). *PLoS ONE* **2013**, *8*, e66049. [CrossRef]
51. McGuire, M.R.; Leland, J.E.; Dara, S.; Park, Y.H.; Ulloa, M. Effect of different isolates of *Beauveria bassiana* on field populations of *Lygus hesperus*. *Biol. Control* **2006**, *38*, 390–396. [CrossRef]
52. Portilla, M.; Jones, W.; Perera, O.; Seiter, N.; Greene, J.; Luttrell, R. Estimation of median lethal concentration of three isolates of *Beauveria bassiana* for control of *Megacopta cribraria* (Heteroptera: Plataspidae) bioassayed on solid *Lygus* spp. diet. *Insects* **2016**, *7*, 31. [CrossRef]
53. Glover, J.P.; Nufer, M.I.; Perera, O.P.; Portilla, M.; George, J. Entomopathogenicity of Ascomycete fungus *Cordyceps militaris* on the cotton bollworm, *Helicoverpa zea* (Boddie) (Lepidoptera: Noctuidae). *J. Fungi.* **2023**, *9*, 614. [CrossRef]
54. Gurulingappa, P.; McGee, P.A.; Sword, G. Endophytic *Lecanicillium lecanii* and *Beauveria bassiana* reduce the survival and fecundity of *Aphis gossypii* following contact with conidia and secondary metabolites. *Crop. Prot.* **2011**, *30*, 349–353. [CrossRef]
55. Perera, O.P.; Little, N.S.; Abdelgaffar, H.; Jurat-Fuentes, J.L.; Reddy, G.V. Genetic knockouts indicate that the ABCC2 protein in the bollworm *Helicoverpa zea* is not a major receptor for the Cry1Ac insecticidal protein. *Genes* **2021**, *12*, 1522. [CrossRef]
56. Tamura, K.; Stecher, G.; Kumar, S. MEGA11: Molecular evolutionary genetics analysis version 11. *Mol. Biol. Evol.* **2021**, *38*, 3022–3027. [CrossRef]
57. Nei, M.; Kumar, S. *Molecular Evolution and Phylogenetics*; Oxford University Press: New York, NY, USA, 2000.
58. Beerli, P.; Felsenstein, J. Maximum likelihood estimation of a migration matrix and effective population sizes in n subpopulations by using a coalescent approach. *Proc. Natl. Acad. Sci. USA* **2001**, *98*, 4563–4568. [CrossRef]
59. Kimura, M. A simple method for estimating evolutionary rates of base substitutions through comparative studies of nucleotide sequences. *J. Mol. Evol.* **1980**, *16*, 111–120. [CrossRef] [PubMed]
60. Felsenstein, J. Confidence limits on phylogenies: An approach using the bootstrap. *Evolution* **1985**, *39*, 783–791. [CrossRef] [PubMed]
61. Catchot, A.L.; Crow, W.D.; Dodds, D.; Gore, J.; Musser, F.R.; Irby, T.; Cook, D.R.; Layton, M.B.; Larson, E. *Insect Control Guide for Agronomic Crops*; Insect Control Guide Committee; Mississippi State University Extension Service Publication: Starkville, MS, USA, 2020; Volume 104, p. 2471.

plants

MDPI

Review

Weed Role for Pollinator in the Agroecosystem: Plant–Insect Interactions and Agronomic Strategies for Biodiversity Conservation

Stefano Benvenuti

Department of Agriculture, Food and Environment, University of Pisa, Via del Borghetto, 80, 56124 Pisa, Italy; stefano.benvenuti@unipi.it

Abstract: The growing interest in safeguarding agroecosystem biodiversity has led to interest in studying ecological interactions among the various organisms present within the agroecosystem. Indeed, mutualisms between weeds and pollinators are of crucial importance as they influence the respective survival dynamics. In this review, the mutualistic role of flower visitors and the possible (often predominant) abiotic alternatives to insect pollination (self- and wind-pollination) are investigated. Mutualistic relations are discussed in terms of reward (pollen and/or nectar) and attractiveness (color, shape, scent, nectar quality and quantity), analyzing whether and to what extent typical weeds are linked to pollinators by rigid (specialization) or flexible (generalization) mutualistic relations. The entomofauna involved is composed mainly of solitary and social bees, bumblebees, Diptera, and Lepidoptera. While some of these pollinators are polylectic, others are oligolectic, depending on the shape of their mouthparts, which can be suited to explore the flower corollas as function of their depths. Consequently, the persistence dynamics of weed species show more successful survival in plants that are basically (occasional insect pollination) or totally (self and/or wind pollination) unspecialized in mutualistic relations. However, even weed species with typical abiotic pollination are at times visited during periods such as late summer, in which plants with more abundant rewards are insufficiently present or completely absent. Many typically insect-pollinated weeds can represent a valid indicator of the ecological sustainability of crop management techniques, as their survival dynamics are closely dependent on the biodiversity of the surrounding entomofauna. In particular, the presence of plant communities of species pollinated above all by butterflies (e.g., several Caryophyllaceae) gives evidence to the ecological compatibility of the previous agronomic management, in the sense that butterflies require certain weed species for oviposition and subsequent larva rearing and, therefore, provide further evidence of plant biodiversity in the environment.

Keywords: biological conservation; functional biodiversity; weed management; wildflowers; sustainability

1. Introduction

Weeds are predominantly self-pollinated [1]; insect-pollinated weeds are also frequently found in agricultural ecosystems [2]. Self-pollination is of crucial importance for rapid seed formation; in accordance with the time-limitation hypothesis [3], weeds require a certain degree of allogamy to maintain a genetic base capable of adapting to the dynamics of agronomic disturbance. Cross-pollination of angiosperms evolved in ancient natural ecosystems, from entomophily to anemophily [4], probably to reduce dependence on biotic factors whose presence is affected by erratic climatic conditions [5]. This widely accepted hypothesis is supported by the evidence of rudimental and inefficient that nectaries are often present even in typically wind-pollinated species [6]. However, despite this evolutionary trend, many agroecosystem plants base their survival dynamics on insect pollination. This mutualistic component of a part of the agroecosystem weed communities assumes a crucial ecological importance in terms of pollinator biodiversity even in this highly anthropized environment. The insects involved are often defined as "flower visitors" rather than pollinators, as their ecological role has not yet been fully clarified. It is frequently unclear whether the insect activity on the flowers allowing a contact pollen-gynoecium is accidental or mutualistic. However, the high frequency of flower visits constitutes a valid parameter for the almost certainty of this plant–insect mutualism [7]. Thus, for many insects and pollinated weeds, basically wildflowers, an evolution strategy towards flowers attractiveness can be discerned [8]. However, there are numerous cases of predominantly self- or wind-pollinated weeds that are visited by insects virtually capable of bringing about mutualistic or accidental gamy [9]. This ambophily is regarded as a transitional state intermediate between biotic and/or abiotic pollen movement [10]. This dual strategy allows the risk that erratic biotic and/or abiotic conditions may reduce the gene flow essential to evolve biotypes suitable for surviving in the agroecosystem. Such a strategy allows a shift from predominant self-pollination to predominant cross-pollination, as observed in numerous species [11]. A good example is offered by *Solanum ptycanthum* (Solanaceae), whose pollination strategy depends on the extent of disturbance in its growth environment, with predominantly insect-pollinated biotypes in natural ecosystems and predominantly self-pollinated biotypes in agricultural ecosystems [12]. It is not clear whether in this and other similar cases the greater attractiveness of the wild biotype is due to greater or lesser development and functioning of the nectaries. But it is worth noting that the presence of nectaries is not strictly necessary to induce insect flower visits, since pollen also constitutes a food source for a vast range of insects. Scanty or absent nectar production, thus, does not rule out the possibility that a species may be insect-pollinated, as it is pollen grain size that makes a species suited to insect pollination [13]. But it has been found that self-compatibility and self-pollination are associated with reduced pollen limitation, presumably because the capacity for self-fertilization decreases reliance on cross-pollination by pollinators [14].

Recent years have seen increasing interest in gene flow mechanisms (biotic and/or abiotic) between the various weed species, partly for agronomic reasons, such as herbicide resistance [15], and due to environmental concerns, as in the case of pollen transfer between genetically modified crops and potentially hybridizable weeds [16]. In addition, the plant and flower visitor interaction arising from insect pollination has aroused concern on account of increased awareness of the concept of safeguarding biodiversity in the agroecosystem [17] and in other anthropized ecosystems [18]. Attention is focusing on a possible cause–effect relationship between the rarefaction and/or disappearance of some species and their dependence on entomofauna, whose ecological role is often overlooked.

The purpose of this study was to survey the state of the art of flora–fauna interaction in weed communities in terms of insect pollination, investigate the ecological importance of these mechanisms in survival dynamics, and determine agronomic strategies that can be adopted to safeguard biodiversity in the agricultural environment.

2. Pollinator Biodiversity and Reward

Most flower visitors are social and solitary bees, bumblebees, Diptera, and Lepidoptera, as shown in Table 1. Each insect species feeds on pollen and/or nectar of given plant species as a function of its respective mouthparts [19], which, in many cases, have evolved in a manner that enables the insect to reach and feed on solid (pollen) or liquid (nectar) food. These food resources are produced in specific structures that are highly diversified among the various plant species [20].

Table 1. Flower visitor observations on several weed species of the agroecosystem.

Weed Species	Observed Flower Visitors	Source
Abutilon theophrasti Medic.	Solitary bees, Tachinidae	Benvenuti, personal observ.
Agrostemma githago L.	Lepidoptera, solitary bees	[21]
Anagallis arvensis L.	Solitary bees, Bombylidae, Coleoptera	Benvenuti, personal observ.
Anthemis cotula L.	Solitary bees, Lepidoptera	Benvenuti, personal observ.
Arum italicum Mill.	Diptera	[22]
Asclepias syriaca L.	*Bombus*, *Xylocopa*, Sphecidae	[23]
Aster squamatus (Spreng.) Hier.	Bees	Benvenuti, personal observ.
Blackstonia perfoliata (L.) Huds.	Bombylidae	Benvenuti, personal observ.
Borago officinalis L.	Bees, solitary bees	Benvenuti, personal observ.
Calystegia sepium (L.) R.Br.	Solitary bees, Coleptera	Benvenuti, personal observ.
Centaurea cyanus L.	Bees, solitary bees. Lepidoptera	[24]
Centaurium erytrea Rafn	Syrphidae	[25]
Chrysanthemum segetum L.	Bees, solitary bees; Bomyilidae; Lepidoptera	[26]
Chrysanthemum coronarium L.	Bees, solitary bees; Bomyilidae; Lepidoptera	Benvenuti, personal observ.
Cirsium arvense (L.) Scop.	Solitary bees, Lepidoptera	[27]
Consolida regalis Gray.	*Bombus*, Lepidoptera	[26]
Convolvulus arvensis L.	Solitary bees, Coleptera	Benvenuti, personal observ.
Conyza canadensis L.	Bees	Benvenuti, personal observ.
Cuscuta campestris Yunk.	Bees, wasps	Benvenuti, personal observ.
Cychorium inthibus L.	Diptera, solitary bees	Benvenuti, personal observ.
Datura stramonium L.	*Bombus* spp.	Benvenuti, personal observ.
Dianthus carthusianorum L.	Lepidoptera	[26]
Dipsacus fullonum L.	Lepidoptera, solitary bees	Benvenuti, personal observ.
Echium vulgare L.	*Bombus* spp.	[26]
Euphorbia esula L.	Solitary bees	[28]
Euphorbia helioscopia L.	Bees	Benvenuti, personal observ.
Geranium molle L.	Diptera	Benvenuti, personal observ.
Lamium purpureum L.	Bees, solitary bees, Bombylidae	Benvenuti, personal observ.
Lavatera punctata All.		[26]
Legousia speculum-veneris (L.) *Chaix*	Bees, Syrphidae, Bombylidae	[24]
Linaria vulgaris Mill.	*Bombus*	[29]
Lycnis flos-cuculi L.	Bees, solitary bees, Lepidoptera	[26]
Matricaria chamomilla L.	Bees, solitary bees, Bombylidae, Lepidoptera	Benvenuti, personal observ.
Nigella damascena L.	Bees	[26]
Oenotera biennis L.	Lepidoptera	[30]
Ornithogalum spp.	Solitary bees, Coleoptera	Benvenuti, personal observ.

Table 1. *Cont.*

Weed Species	Observed Flower Visitors	Source
Papaver rhoeas L.	Bees, solitary bees, *Xilocopa*	Benvenuti, personal observ.
Polygonum lapathifolium L.	Bees, solitary bees	Benvenuti, personal observ.
Portulaca oleracea L.	Bees, Bombylidae	Benvenuti, personal observ.
Senecio vulgaris L.	Bees	Benvenuti, personal observ.
Silene alba (Mill.) Krause	Nocturnal and diurnal Lepidoptera	[31]
Silene dioica (L.) Clairv.	Bombus, bees, Lepidoptera, Syrphidae	[32]
Silene noctiflora L.	Nocturnal Lepidoptera	[33]
Sinapis arvensis L.	Bees, solitary bees; Bomyilidae; Lepidoptera	Benvenuti, personal observ.
Stellaria media (L.) Vill.	Bees, solitary bees	Benvenuti, personal observ.
Veronica persica Poir.	Solitary bees, ants	Benvenuti, personal observ.
Viola spp.	Solitary bees	[34]
Xanthium strumarium L.	Bees	Benvenuti, personal observ.

Among flower visitors, honeybees (*Apis mellifera*, Figure 1) and solitary bees (Figure 2) are predominant. Within the Mediterranean environment, solitary bees are represented above all by Andrenidae, Anthophoridae, Apidae, Melittidae, Colletidae, Halictidae, and Megachilidae [35]. The reward consists of nectar and/or pollen, with the latter being transported in different body places and after being packed in special pollen baskets situated on the insects' legs. But it is rare for bees to collect both pollen and nectar simultaneously, as energy economy prompts bees to visit species with a predominance of one or the other reward [36]. Natural ecosystems have an abundance of wildflowers, which have typically evolved nectar production as a reward. On the contrary, the conventional agroecosystem has a predominance of species poor of nectar, so that insects are rewarded with pollen. For example, *Papaver rhoeas*, in spite of their flower appearance, has no nectaries, and pollen is the only food source for pollinators, the latter being indispensable for seed set as this species is completely self-incompatible [37].

However, although species with brightly colored flowers are the most frequently visited, even many common weeds with less gaudy flowers also constitute a useful food source, especially during periods when the natural environment offers fewer species in flower [38]. Large numbers of species flower in spring [39], and they do compete with one another in producing nectar [40], while in the subsequent months, late-flowering species with less noticeable flowers may be of interest to the pollinator, even though the reward is less advantageous. Above all, in late summer, flower visitors may be observed on species that previously exerted poor visual attraction, as occurs in many Asteraceae, such as *Senecio vulgaris*, *Sonchus* spp., *Aster squamatus*, *Conyza canadensis*, and even *Xanthium strumarium*. The latter species, which is typically wind- and/or self-pollinated, constitutes a curious case in that the ecological role of flower visitors on pollination can be considered negligible, as the position of its male flowers, separate from the female flowers, suggests that pollen movement towards the gynaeceum may be purely accidental. The presence of flower visitors (in particular bees and bumblebees) has also been noted almost exclusively in late summer on other species with small and/or poorly attractive flowers, such as *Polygonum laphatipholium*, *P. aviculare*, *Cuscuta campestris*, *Portulaca oleracea*, *Stellaria media*, and *Abutilon theophrasti*. Similar phenomena have been observed on species with gaudy but usually self-pollinated flowers, such as *Convolvulus arvensis* and *Calystegia saepium* [41] and *Datura stramonium* [42].

Figure 1. Honeybees observed on the flowers of common weeds of the agroecosystem: 1 = *Euphorbia helioscopia*, 2 = *Polygonum laphatifolium*, 3 = *Linaria vulgaris*, 4 = *Nigella damascena*, 5 = *Lamium purpureum*, 6 = *Picris echioides*, 7 = *Galinsoga parviflora*, 8 = *Malva sylvestris*, 9 = *Cichorium inthybus*, 10 = *Papaver rhoeas*, 11 = *Cirsium arvense*, 12 = *Crepis vesicaria*.

Figure 2. Bumblebees (2, 3, 8, 11), solitary bees (1, 6, 7, 9, 12), and wasps (4, 5, 10) observed on common weeds of the agroecosystem: 1 = *Borago officinalis*, 2 = *Datura stramonium*, 3 = *Verbascum sinuatum*, 4 = *Orobanche crenata*, 5 = *Cuscuta campestris*, 6 = *Echium vulgare*, 7 = *Anagallis arvensis*, 8 = *Scabiosa columbaria*, 9 = *Malva sylvestris*, 10 = *Daucus carota*, 11 = *Lavatera punctata*, 12 = *Chrysanthemum coronarium*.

A less important, but still underestimated, role is played by Diptera (Figure 3) (overall Syrphidae, Tachinidae, Sarcophagidae, and Bombyliidae) [43]. In Bombyliidae, on the other hand, the mouthparts appear to have evolved to allow utilization of nectar by means of a long proboscis that can penetrate inside small flowers [44]. Visits by Bombyliidae have been noted, above all, on flowers that would be difficult for other insects to reach due to their small and elongated floral calyx, as in the case of Gentianaceae (*Centaurium erythraea* and *Blackstonia perfoliata*), Campanulaceae (*Legousia speculum veneris* and *Jasione montana*), Lamiaceae (*Lamium purpureum* and *L. amplexicaule*), Primulaceae (*Anagallis arvensis* and *A. foemina*), and other species.

Figure 3. Diptera Empididae (1, 2), Tachinidae (3, 12), Syrphidae (4, 5, 6, 7, 8, 10), and Bombyiliidae (9, 11) observed on common weeds of the agroecosystem: 1 = *Geranium molle*, 2 = *Linum perenne*, 3 = *Inula viscosa*, 4 = *Silene alba*, 5 = *Papaver hybridum*, 6 = *Anthemis arvensis*, 7 = *Crysanthemum segetum*, 8 = *Scandix pecten-veneris*, 9 = *Cruysanthemum coronarium*, 10 = *Cephalaria transsylvanica*, 11 = *Matricaria chamomilla*, 12 = *Tordylium apulum*.

Some insects, above all, Lepidoptera, have a long proboscis that enables them to visit flowers with an elongated calyx even when the nectaries are hidden at the base of the calyx (Figure 4). Such insects include Lycaenidae, Pieridae and Sphingidae, Papilionidae, Nymphalidae, and Satyrids (Figure 5). Visits are frequent on flower species whose flowers have a particular shape, such as Dipsaceae (*Knautia arvensis* and *Dipsacus fullonum*) and Caryophyllaceae (*Agrostemma githago*, *Lychnis flos-cuculi*, *Silene* spp., etc.). However, pollen transport by butterflies is decidedly less efficient in comparison to bees [45]. Species belonging to the order of Coleoptera (Figure 6) are even less efficient, as such species, similarly to those of the order of Thysanoptera, often lack pollen transport specialization and appear to act more as pollen predators rather than potential pollinators [46]. Similarly poor efficiency is seen in ants, even though they are flower visitors of many species [47], and this appears to be due to the low pollen germination after ant contact [48]. Their lack of hair and their very limited plant–plant movement on account of their inability to fly, at least in most species, suggests that these Hymenoptera are likely to be only occasional pollinators. Indeed, it was observed that ants may negatively affect plant fitness by reduced intensity of pollinator visits and that ants are repelled from flowers of many plant species (overall in tropical environments), although this repellence is clearly not ubiquitous [49].

Figure 4. Butterfly (*Gonepteryx rhamni*) observed on *Dianthus cartusianorum* flower during nectar suction through their long-proboscid, evolved to be able to reach the nectaries of elongated floral calyxes.

Figure 5. Lepidoptera observed on common wildflowers of the agroecosystem: 1 = *Crysanthemum segetum*, 2 = *Dianthus carthusianorum*, 3 = *Crepis vesicaria*, 4 = *Anthemis arvensis*, 5 = *Cephalaria transsylvanica*, 6 = *Senecio erraticus*, 7 = *Cirsium arvense*, 8 = *Dipsacus fullonum*, 9 = *Ranunculus arvensis*, 10 = *Orlaya grandiflora*, 11 = *Sinapis arvensis*, 12 = *Lavatera punctata*.

Similar scarce pollination-efficiency is evidenced by some nectar-robbing species such as *Bombus occidentalis*, which visits *Linaria vulgaris* where it collects nectar by poking holes into the corolla without penetrating inside it [50], and are likely to be of equally negligible ecological impact, since they are cheaters rather than mutualists [51].

Figure 6. Coleoptera observed on common weeds of the agroecosystem: 1 = *Ornithogalum umbellatum*, 2 = *Ammi majus*, 3 = *Mentha suaveolens*, 4 = *Crysanthemum segetum*, 5 = *Campanula rapunculus*, 6 = *Lavatera punctata*, 7 = *Plantago lanceolata*, 8 = *Dianthus cartusianorum*, 9 = *Daucus carota*, 10 = *Malva sylvestris*, 11 = *Anagallis arvensi*, 12 = *Ornithogalum umbellatum*.

3. Generalization or Specialization?

It is widely believed that common weeds owe their time and space persistence to their lack of specialization [52], except for some species that are increasingly rare in conventional agroecosystems [53]. The evolutionary trend from generalization to specialization noted in natural ecosystems [54] does not appear to be suited to the requirements of weeds in an agricultural environment, where "plasticity" (despecialization) seems to be more important [55]. Indeed, it implies a lower risk of pollinator lack because of the high degree of disturbance dynamics, as typically occurs in the agroecosystem. Therefore, the specialization of some plant species towards certain pollinators could explain why they are increasingly uncommon [56]. Pesticide application, which is typical of conventional agricultural systems, can interfere with the efficacious but fragile mechanisms that involve rigid flora–fauna mutualisms.

Indeed, "conventional" agroecosystems are characterized by the dominance of self- and wind-pollination, while, on the contrary, insect pollination is more frequent in natural ecosystems as a function of the greater abundance and biodiversity of pollinators typically available in undisturbed environments. However, even predominantly insect-pollinated species have different degrees of specialization depending on the possible pollinators [57], as floral symmetry plays an important role in plant–pollinator systems [58]. Zygomorphic flowers, such as *Consolida regalis*, *Echium vulgaris*, *Lamium amplexicaule*, and *Stachys arvensis*, are visited mainly by specialized long-tongued bees (Melittidae, Megachilidae, Anthrophoridae, and Apidae) as a consequence of the particular position of the nectaries [59]. Actinomorphic flowers, on the other hand, are visited by a higher number of visitor species. For example, almost all of the Asteraceae species (e.g., *Centaurea cyanus*, *Chrysanthemum myconis*, and *Cirsium arvense*) showed a higher degree of unspecialized visitors, such as short-tongued bees (Colletidae, Andrenidae, and Halictidae) and flies (personal observation). Another example is that of *Raphanus raphanistrum* (Brassicaceae), which exhibits traits typical of generalized pollination, including radially symmetric flowers, exposed reproductive organs, and an upright flower [60].

A further type of specialization, with visits limited to a restricted botanical group, may represent a characteristic of the pollinators themselves, as observed in Italy for *Heriades truncorum* (Megachilidae), which is almost always seen on Asteraceae weeds [61]. But it cannot necessarily be assumed that the insects most frequently observed on flowers are the most efficient pollinators, since pollen transport is strongly dependent on the shape and hair of the insect but also on the speed of visits, with rapid speed proving to be less efficient. For example, the typical rapid visits by long-tongued bees may result in reduced pollen transport [62]. Furthermore, efficiency of pollination is also influenced by the electrostatic forces of pollen, which can assure adhesion to the pollinator even if the insect may lack hair [63].

Overall, however, specialization is indisputably linked to flowering phenology [64], which may or may not be compatible with the biological cycle of the pollinator. Thus, with cool season flowering species, potential pollinators are represented by insects that are capable of maintaining a certain degree of activity even at low temperatures, as in the case of early Amaryllidaceae pollinated by cold-tolerant Andrenidae [65]. The role of temperature as a limiting factor is confirmed by observations on various species of Campanulaceae, which show decreasing frequency of visits with increasing altitude [66]. In addition, the literature indicates that pollinators tend to favor peak or earlier flowering, whereas predispersal seed predators tend to favor off-peak or later flowering [67].

An interesting example of plant–pollinator mutualistic specialization is found in *Silene noctiflora*, a gynomonoecious annual whose individuals produce both hermaphroditic and pistillate flowers. It flowers only during the night and is pollinated exclusively by nocturnal moths [68]. A similar system, albeit less exclusive, is seen in *Silene alba* [69], which tends to open its flowers at the end of the day, thereby allowing pollination both by diurnal (bees, flies, and wasps) and nocturnal visitors (the latter being mostly Sphyngid and noctuid moths). This mixed system is typical of numerous other species belonging to the same family of Caryophyllaceae, as in the genera *Agrostemma*, *Saponaria*, *Dianthus*, and *Vaccaria* [21].

Flora–fauna specialization does not depend only on the shape and manner of opening of the flowers but also on nectar composition, in terms of sugars and amino acids, as well as the nectar secretion rate, which is measurable in the field with various techniques [70]. For example, butterflies are attracted by the flowers richest in amino acids, as their diet is based exclusively on nectar and must therefore allow sufficient protein synthesis. Solitary and social bees are more attracted by an elevated sugar content, as they also feed on pollen and therefore do not need an additional protein supply (Gardener and Gilman, 2002). It has also been hypothesized that specialist nectarivores can assimilate sucrose, whereas some opportunistic nectar feeders digest only the simple exoses [71]. A crucial role is also played by amino acid typology. The predominance of phenylalanine and/or gamma-aminobutyric acid tends to attract long-tongued bees and flies (overall Syrphidae), whereas asparagine and tryptophane are rather repellent to these insect species [72]. Investigations aiming to obtain experimental evidence of an ecological function of nectar composition have been conducted on a vast range of species [73], with results suggesting that some amino acids elicit different responses in insect receptors. However, amino acid concentration in nectar is not exclusively a function of the genotype. It can be influenced by agronomic management, as in the case of nitrogen fertilization, which has been shown to increase nectar amino acid concentration in *Agrostemma githago* [74]. Natural factors such as arbuscular mycorrhizal fungi can, likewise, increase flower visitor numbers (overall visits by Diptera and Hymenoptera) in *Centaurea cyanus* [75]. Finally, the ecological significance of the toxic nectar secreted, for example, by *Heliotropium europaeum*, *Cuscuta* spp., *Solanum nigrum*, and *Euphorbia* spp. is still poorly understood. It has been hypothesized that bees are more resistant to alkaloids than adult Lepidoptera, and that alkaloids in nectar encourage pollination by specialist rather "flower-inconstant" pollinators [76].

Pollinator attraction is linked to the mechanisms involved in recognition of appropriate flowers. Recognition is crucial in that it avoids confusion in pollen transfer, which must take place within the same species as far as possible [77]. Flower shape and color both play an important role in facilitating recognition. Color is perceived differently by the insect as compared to the human eye, and light reflectance at wavelengths invisible to humans (roughly 300–400 nm) is well perceived [78]. Many Brassicaceae reflect ultraviolet color in order to attract the attention of pollinators [79]. Some flower colors appear to be correlated with certain categories of pollinators, although this cannot always be generalized due to poor convergence of data from different environments [80]. Specialist bumblebees have been noted to show a preference for purple, and this example would appear to confirm the so-called "pollination syndrome" theory [81]. In some cases, recognition is facilitated by color patterns: thus, the "search images" system possessed by insects [82] can be aided by characteristic black spots at the base of the petal, functioning as a "nectar guide" [83], as observed, for example, in *Papaver rhoeas*. But bright and gaudy colors are not always an indispensable condition for attracting flower visitors. Some Euforbiaceae have pale green flowers that do not stand out within the surrounding vegetation, yet they are frequently visited, as in the case of *Euphorbia esula* [84]. Moreover, a further and often decisive mechanism for identification and recognition of flowers consists of production and emission of fragrant molecules composed essentially of terpenoids and benzenoids [85].

4. Weed Pollination Strategies

The main characteristic of arable weeds is their ability to persist despite the vast range of agronomic disturbances. One of the crucial strategies adopted by weeds is rapid seed set. Rapidity is maximized with self-pollination, as autogamous seed set is not dependent on the occurrence of any (biotic or abiotic) event [86]. Weeds are regarded as pioneer flora of early stages of secondary successions [87], and their frequent annual cycle is often correlated with self-pollination [88] according to the "time-limitation" hypothesis of crucial importance in the typically highly disturbed agroecosystem. For example, *Amaranthus retroflexus*, characterized by flowers devoid of attractiveness and by almost total self-pollination [89], represents the ideotype of weeds as it already produces mature seeds just a few weeks after emergence [90]. Species whose corolla is a marked attractant (in terms of size and/or color) may also be mainly self-pollinated, as in *Convolvulus arvensis* [41], *Stellaria media* [91], *Portulaca oleracea* [92], and *Anagallis arvensis* [93]. However, it is likely that this predominance of self-fertilization is found in biotypes present in agricultural environments, while outcrossing is more widespread in biotypes present in less disturbed environments. This means that the frequent disturbances of the agroecosystem probably favored coevolution towards predominantly self-pollinated biotypes. This is the case of *Datura stramonium*, which is pollinated in its original environments by hawkmoths (*Manduca quinquemaculata*), Halictids, and honeybees (*Apis mellifera*) [94], whereas the biotypes present in cultivated fields are almost totally autogamous [95]. This coevolution with agronomic disturbances has led to a decrease in nectary function and a corresponding elongation of the androecium and gynoecium to allow contact between anthers and stigma (herkogamy), thereby favoring self-pollination [96]. Despite this, some exceptions that evolved towards self-incompatibility are observed, as in the case of *Papaver rhoeas* [97] and *Ranunculus repens* [98]. But, in general, the absolute self-incompatibility in the agroecosystem plant communities is unusual. Indeed, a variable frequency of distribution between self- and cross-pollination is more frequently found, as in the cases of *Anthemis cotula* [99], *Sinapis arvensis*, [100], *Raphanus raphanistrum* [101], and other insect-pollinated species typically widespread in the agroecosystem [102]. In the case of *Raphanus raphanistrum*, flower color (typically white or yellow) plays an ecological role in favoring or discouraging cross-pollination despite partial self-compatibility. Thus, one important visitor is a lepidopteran (*Pieris rapae*), which mainly visits yellow flowers, resulting in a predominance of cross-pollination in these populations [103], whereas self-pollination is predominant in white flower varieties. This dual typology of biotypes could represent an example of optimal trade-off between the advantages of one or the other gene flow mechanism, highlighting plastic and evolutionary changes in floral traits. Such a trade-off is not exclusive to this weed species but is widespread in many other species as well, and can probably be interpreted as a strategy for maintaining populations with diversified biological characteristics [104].

Overall, pollen self-incompatibility is one of the various strategies adopted by plant species to avoid the pollination of different flowers growing on the same plant, especially in the case of individuals bearing numerous blooms [105]. Such pollination, which would be pointless in terms of gene flow, is known as geitonogamy [106]. Other geitonogamy avoidance strategies include spatial and/or temporal separation of pollen and stigma. In general, geitonogamy avoidance is advantageous in favoring adaptability to dynamic environmental conditions [107]. A trade-off between geitonogamy and xenogamy (cross-pollination) is observed in *Daucus carota*, an andromonoecious and protandrous species [108]. In this case, separation of the male and female phases is complete at the level of the flower and umbel, but the two phases overlap at the level of the full plant, creating conditions for geitonogamy even if insect visits between the umbels of adjacent plants lead to xenogamy. Another curious characteristic of this species is the possible ecological role of the dark central floret of the inflorescences, for which a possible "fly catcher" role has been suggested, although its function has not yet been fully clarified [108]. Additional examples of a balance between self-pollination and insect pollination are observed in *Cynoglossum officinale*, *Echium vulgare* [109], and some species belonging to the botanical genus *Delphinum* [110], in which the extremely variable number of flowers results in diversified probability of geitonogamy. Plants with a greater number of flowers tend to favor self-pollination as there is an increased probability that the pollinators will sequentially visit (geitonogamy) flowers of the same plant. A further diversification of breeding frequency is found in *Echium vulgare*, as the protandrous flowers produce more nectar and receive higher rates of visitation during their male than during their female phase [111], although nectar production in this species is also closely dependent on environmental conditions [112].

Among geitonogamy avoidance strategies, a particularly drastic mechanism is displayed by dioecious species, in which the separation of individuals into different sexes makes self-fertilization impossible. An example is seen in *Silene dioica*, which is visited by bumblebees, hoverflies (Syrphidae), butterflies (mainly Pieridae), and honeybees [113]. But since the invasiveness of this species in the agroecosystem is negligible, as compared to other monoecious species belonging to the same genus, it can be concluded that such a strategy is unsuccessful [114].

5. Insect-Pollinated Weeds as Indicators of Biodiversity and Agroecosystem Health

A clear-cut distinction between entomopollinated and nonentomopollinated species cannot easily be drawn, as the true ecological role of insect visits has not yet been clarified for each individual species. This uncertainty is aggravated by the above-described differences among biotypes present in the wild (more dependent) versus those in the agroecosystem (less dependent). However, it is generally agreed that while the reward may consist of pollen (Figure 7), both quantitative and qualitative (sugar concentration) nectar production is linked to entomofauna through mutualistic specialization. Since nectar production requires considerable energy requirement, in short-lived weeds, characterized by annual cycle, nectar is less abundant than in perennial species [115], and if it is not collected, it is reabsorbed by the plant for its own metabolism [116]. It can, therefore, be stated that species with a well-developed nectary rely mainly on insect visits for their survival, with the visits being crucial for seed set. Thus, weed species characterized by this feature (essentially, wildflowers) face a greater risk within the agroecosystem, because the high level of disturbance of the agricultural environment tends to restrict the availability of their pollinators. Pesticide toxicity and its residues play a crucial role in this regard by severely affecting the chance for survival of entomofauna and, consequently, of insect-pollinated flora.

Figure 7. Honeybee observed during a visit to an inflorescence of *Daucus carota*: note the balls of pollen accumulated on the hind legs, typically yellow in this species.

The frequency of wildflower species in the agroecosystem plant communities can, therefore, represent a valid indicator of their ecological sustainability. It is now recognized that the floristic diversity of the agroecosystem provides an assessment of the agroecological impact [117]. This is particularly true about entomopollinated species, as their presence presupposes a level of biodiversity extended to the animal kingdom, and in this context, it should not be forgotten that evolution towards self-pollination occurred precisely in situations of a lack of pollinators [118].

The mutualistically more specialized species, such as wildflowers, constitute the most reliable ecological assessment parameters since these are the species whose presence is most severely threatened by disruption of the balance of the agroecosystem. Although the decreased presence or disappearance of some species may be due to other agronomic causes (herbicides, heightened aggression by more competitive weeds, crop seed selection, etc. [119]), it is highly probable that the declining numbers of pollinators have been a contributing factor in the increasing rarity of some species. As has been widely noted, in the past few decades, the decline of biodiversity has involved many previously widespread plant and animal organisms [120]. Throughout Europe, only plant species whose persistence dynamics do not rely on biotic action for pollination are only occasionally cited as rare weeds [121]. On the contrary, lists of declining species include numerous wildflowers, which are threatened by their dependence on flower visitors for seed set [122]. It has also been shown that the frequency of flower visitors on wildflowers is closely related to the quantity of viable seeds produced [123]. Weed species that are only scantily present have difficulty in attracting insects, as it has been noted both in natural ecosystems [124] and in the agroecosystem [125] that insects prefer to visit more numerous species. In arable fields, a large number of insect-pollinated wildflowers have now become rare or are in decline, such as *Agrostemma githago*, *Centaurea cyanus*, *Papaver argemone*, *Ranunculus ar-*

vensis, [126], *Chrysanthemum sagetum*, *Matricaria recutita*, *Legousia hybrida*, *Silene alba*, *Viola arvensis* [127], *Consolida regalis*, *Silene noctiflora*, *Lamium amplexicaule*, [128], *Myosotis arvensis*, *Viola tricolor* [129], *Legousia speculum veneris*, *Anchusa arvensis*, [130], *Nigella arvensis*, and *Ornithogalum umbellatum* [131].

As stated above, the risk of decline is greatest when mutualistic interaction is specialized. Thus, many of the abovementioned species belong to the Cariophyllaceae, a botanical family often characterized by rigid mutualistic interactions set in motion by butterflies [132]. This type of mutualism is highly fragile, because Lepidoptera require a twofold plant-related availability: food source (visitable flora endowed with nectaries), and suitable conditions for reproduction (appropriate flora for oviposition and feeding of larval forms). Each butterfly species is dependent on restricted plant groups, often belonging to a single botanical family, a single genus, or even a single species (Table 2). Generally, reference is made to a hierarchy of preferences since some species may be preferred to others within a given botanical grouping, as in the case of *Papilio machaon*, which oviposits exclusively on Apiaceae (Figure 8). If certain host plants have poor invasiveness within the various ecosystems, this inevitably leads to very scanty presence not only of the respective plant species but also of the correlated butterfly species [133].

Table 2. Some examples of weed host selected by several butterfly species.

Weed-Host	Butterfly Species	Source
Amaranthus spp., *Chenopodium* spp.	*Pholisora catullus*	[134]
Apiaceae	*Papilio machaon*	[135]
Aristolochia spp.	*Pachliopta aristolochiae*	[136]
Carex spp.	*Lopinga achine*	[137]
Asteraceae, *Malvaceae*, others	*Vanessa cardui*	[134]
Avena fatua L.	*Cercyonis pegala*	[134]
Bidens pilosa L.	*Nathalis iole*	[134]
Brassica nigra (L.) W.D.J.Koch	*Pontia bekeri*	[134]
Brassicaceae	*Anthocaris sara*	[134]
Brassicaceae	*Euchloe ausoides*	[134]
Brassicaceae	*Pieris rapae*	[134]
Brassicaceae	*Pontia protodice*	[134]
Cardamine spp., *Nasturtium* spp.	*Anthocaris cardamines*	[138]
Caryophyllaceae	*Euphyia picata*	[139]
Centaurea spp.; *Plantago* spp.	*Melitaea didyma*	[139]
Chenpopodium album L.	*Brephidium exilis*	[134]
Cirsium spp., *Centaurea solstitialis*	*Phycioides mylitta*	[134]
Cirsium spp., *Centaurea* spp.	*Tymelicus sylvestris*	[139]
Echium vulgare L.	*Ethmia terminella*	[139]
Galium spp.	*Ryparia purpurata*	[139]
Galium spp.	*Macroglossa stellatarum*	[139]
Graminaceae	*Brintesia circe*	[139]
Graminaceae	*Lerodea eufala*	[134]

Table 2. *Cont.*

Weed-Host	Butterfly Species	Source
Hypericum spp.	*Deilephila elpenor*	[139]
Lamiaceae	*Perizoma alchemillata*	[139]
Lytrum spp., *Epilobium* spp.	*Hyles vespertilio*	[139]
Malva sylvestris, Alcea rosea	*Heliopetes ericetorum*	[134]
Malvaceae, Chenopodium album L.	*Pyrgus communis*	[134]
Malvaceae, Papilionaceae	*Strymon melinus*	[134]
Malvaceae, Urticaceae	*Vanessa annabella,*	[134]
Melilotus officinalis Lam., *M. alba* Medic.	*Colias eurytheme*	[134]
Papillionaceae	*Cyaniris semiargus*	[139]
Plantaginaceae	*Euphydryas chalcedona*	[140]
Plantago lanceolata L.	*Mellicta hatalia*	[141]
Plantago major L. and *P. lanceolata* L.	*Euphydryas editha*	[134]
Polygonaceae	*Lycaena phlaeas*	[139]
Polygonum persicaria L.	*Lycaena helloides*	[134]
Rumex crispus L.	*Lycaeana xanthoides*	[134]
Rumex spp.	*Lycaena cupreus*	[134]
Scrophulariaceae	*Junonia coenia*	[134]
Scrophulariaceae, Verbenaceae	*Junonia coenia*	[140]
Silene spp.; *Lycnis* spp.	*Hadena rivularis*	[139]
Sonchus oleraceus L.	*Helicoverpa armigera*	[142]
Urtica spp.	*Vanessa atlanta*	[139]
Veronica spp.	*Stenoptilia pterodactyla*	[139]
Viola spp.	*Argynnis paphia*	[139]

While specificity between pollinator and host weed is may be variable, it is quite typical of the different families of Lepidoptera, being found preferentially or obligatorily linked to restricted botanical groupings. For example, with Satytiridae (e.g., *Brintesia circe*), there is only scanty specialization as the pollinator/host-plant relation is observed in many ubiquitarian species of Graminaceae, but requirements are more stringent for *Macroglossa stellatarum* (Sphyngidae), which needs one of the various species of the genus *Galium* (Rubiaceae). The risk of butterfly/plant-host coextinction cannot be disregarded and has already been reported in some parts of the world [143]. It is also interesting to note that numerous Lepidoptera, known as myrmecophilous butterflies, have mutualistic relations with ants, as ants defend butterfly eggs and the subsequent caterpillars against predator attack [144]. It has been observed that oviposition of myrmecophilous butterflies takes place preferentially on plants most frequently visited by ants [145], showing that the presence of butterflies testifies to an even more extensive level of biodiversity [146]. In contrast, bumblebee reproduction, while similarly limited by scanty availability of undisturbed environments, is less specialized as it generally takes place in soil [147]. Lack of specificity is also noted in most solitary bees, with some species nesting in soil while others also nest in plant residue cavities [148].

Figure 8. Caterpillar of the *Papilio machaon* butterfly specialized to lay eggs on plants of the Apiaceae family (in this case, the toxic *Conium maculatum*).

A special form of ecological interaction is found in pollinators that require the presence of other pollinators for their reproduction. For example, in many species of Diptera Bombylidae, which have a very thin and elongated mouthpiece allowing them to collect nectar from small flowers such as *Anagallis arvensis*, *Legousia speculum veneris*, *Centaurium erytrea*, and *Lamium purpureum* (Benvenuti, personal observation), parasitic oviposition takes place in the nest of several species of solitary bees [149], where the larvae feed on resources intended to be food for the bees. Indeed, Diptera Bombylidae exhibit mutualistic behavior towards insect-pollinated plant species but have parasitic behavior towards other pollinators.

Often, rare weeds are pollinated above all by Lepidoptera, as in the case of many Caryophyllaceae [150]. Consequently, they represent the most valid indicator of the biodiversity of the agroecosystem since their presence testifies to a complex level of flora–fauna interactions. But exceptions to this rule are found for some butterflies, such as *Pieris rapae*, which often choose the same species both for pollinating and for oviposition as well as for subsequent rearing of larvae. This exemplifies mutualism and parasitism simultaneously [151], which may result in a sort of conflict of interest [152] for the pollinator itself. However, apart from these exceptions, butterfly presence and diversity depend on the landscape context in that their survival dynamics are linked to availability of the required host species in the environment [153]. Analogously, the widespread presence of Syrphid (Diptera) has also been considered a good indicator of plant biodiversity [154], on account of the abundance of different environments suitable for their reproduction. Some species of spider can also be considered as a further parameter for assessment of ecosystem integrity, as they hide on flowers to prey on pollinators. Such a phenomenon has been observed on *Aslepias syriaca* [155], *Leucanthemum vulgare* [156], and many other species. Thus, spiders represent the tip of the ecological pyramid of this flower–pollinator–predator food chain, and this ecosystem appears be particularly vulnerable [157].

Confirmation of the reliability of insect-pollinated flora as an indicator of ecological sustainability of the agronomic cultural practices adopted comes from the observation that organic agricultural systems lead to an increase in insect-pollinated species [158]. Future ecological assessments could be based on monitoring the spider species known to be the preferential predators of pollinating insects.

6. Long-Term Plant–Animal Biodiversity Sustainability

Weed management is crucial for biodiversity sustainability [159], as also highlighted by organic agricultural systems [160]. Perhaps the starting point is diversification of land management, since species richness, genetic variability, and extinction probability are closely linked to landscape traits such as habitat diversity, structural heterogeneity, patch dynamics, and perturbations [161]. In other words, arable weed diversity increases with landscape complexity [162]. Landscape planning is crucial for biodiversity [163]. It is important to keep in mind that distances between nesting environments and food sources must not be excessive, since beyond a certain distance, the trip energy will no longer be advantageous due to excessive energy consumption for flight between the nest and flora to be visited. Other, more suitable environments will therefore be sought [164]. This implies that the geometry of the agroecosystem plays an important role in ensuring that insect-pollinated weeds achieve sufficient seed set.

Woods represent a fundamental reserve of environments suitable for pollinator survival, especially if they are established in a mosaic pattern [165] within an agricultural setting. A similar positive effect is produced by cattle grazing [166] because this kind of land use presupposes both forage resources and nesting resources [167]. With regard to the potential for nesting, at least as far as numerous species of solitary bees and bumblebees, it is important to plan uncropped areas near the crops in such a way as to allow undisturbed aboveground nesting [168]. In addition, the introduction of long-term crops such as Lucerne (*Medicago sativa*) can guarantee prolonged periods (3–4 years) free from soil disturbance. Honeybees are severely damaged by microencapsuled pesticides, whose microgranules adhere to the insect's hair and are thereby transferred into the hives, where their toxic effects can kill the larvae [169]. Therefore, the use of such pesticides should be strictly avoided.

The likelihood of pollinator survival can be increased by the presence of field margins, hedges [170], and other buffer zones [171] or set-aside fields [172]. Such areas not only offer a suitable environment for soil-nesting pollinators, but also for Lepidoptera that require certain weeds on which to oviposit [173]. Wildflowers linked to mutualistic relations with the pollinators represent the ideotype of field margins as they not only provide a suitable ecological niche for an elevated number of pollinators, but they also ensure positive benefits for the agricultural landscape. Thus, it has been shown that the introduction of complex mixed wildflower strips leads to an increase in butterflies, which are drawn by the presence of host plants for oviposition and nectar as a food source [174]. The use of native wildflowers achieves the best ecological response in safeguarding pollinator biodiversity, above all, with regard to specialized pollinators (i.e., short- or long-tongued bees) linked to the wildflowers involved [175]. The introduction of exotic pollinators [176] tends to impair the plant–pollinator balance of the agroecosystem overall following increasingly evident climate changes [177]. This occurs because the balance of weed–pollinator competition is disrupted by altered mutualistic plant–animal interactions [178]. Even though it is not always easy to distinguish the cause from the effect, i.e., to determine whether the imbalance is triggered by a lack of wildflowers or pollinators [179], limiting or preventing the introduction of non-native animal or plant organisms into such environments represents a strategy of paramount importance. It may, therefore, help to avert the tendency to "biological globalization" and the ensuing genetic erosion or loss of native insect-pollinated plants and/or the relative flower visitors.

7. Conclusions

Every weed species is characterized by a particular survival strategy and ruderal species (early flowering, abundant and prolonged seed production) derived from an evolutionary direction capable of giving them ideal weed traits [180] to persist in the agroecosystem. In this context, the rigid plant–pollinator mutualisms are undoubtedly a disadvantage since they presuppose the presence of a consistent pollinator quantity and biodiversity. Despite this, in many cases, there is a transition towards weed communities that display an increasing presence of species whose survival strategies depend on pollinators with varying degrees of specialization. The growing need to maintain and/or restore agroecosystem biodiversity has focused attention on the insect-pollinated weeds that are among the first to decline or even disappear in the agricultural landscape. Paradoxically, it is now widely believed that even weeds perform an "ecosystem service" dedicated to the survival of pollinators essential for the productivity of the various insect-pollinated crops [181]. Within pollinators, butterflies are particularly subjected to rarefaction since their survival does not depend exclusively on the presence of pollinated plants but also on the further availability of host plants essential for their oviposition. Thus, it is important to identify valid ecological indicators to monitor the health of the agroecosystem. The presence of crab spiders has been proposed as a valid indicator of the level of the agroecosystem biodiversity. Indeed, these arthropods feed on the pollinators that wait camouflaged on the flowers, thus highlighting the biodiversity of both pollinated plants and pollinators [182] (Figure 9). However, further studies are required to determine more precisely, for each species, the pollinator requirement for seed set, especially for specialized weeds. This improved knowledge would not only aid research based on biological indicators, but it would also help to optimize the biodiversity restoration programs of degraded agroecosystems.

Figure 9. Observed crab spider with their evident success in catching a solitary bee on *Onobrychis viciifolia* inflorescence.

Funding: This research received no external funding.

Conflicts of Interest: The author declares no conflicts of interest.

References

1. Sutherland, S. What makes a weed: Life story traits of native and exotic plants in the USA. *Oecologia* **2004**, *141*, 24–39. [CrossRef]
2. Bretagnolle, V.; Gaba, S. Weeds for bees? A review. *Agron. Sustain. Dev.* **2015**, *35*, 891–909. [CrossRef]

3. Aarssen, L.W. Why are most selfing annuals? A new hypothesis for the fitness benefit of selfing. *Oikos* **2000**, *89*, 606–612. [CrossRef]
4. Mitchell, R.J.; Irwin, R.E.; Flanagan, R.J.; Karron, J.D. Ecology and evolution of plant–pollinator interactions. *Ann. Bot.* **2009**, *103*, 1355–1363. [CrossRef]
5. Bawa, K.S. Pollination, seed dispersal and diversification of angiosperms. *Trends Ecol. Evol.* **1995**, *10*, 311–312. [CrossRef]
6. Culley, T.M.; Weller, S.G.; Sakai, A.K. The evolution of wind pollination in angiosperms. *Trends Ecol. Evol.* **2002**, *7*, 361–369. [CrossRef]
7. Ne'eman, G.; Jürgens, A.; Newstro-Lloyd, L.; Potts, S.G.; Dafni, A. A framework for comparing pollinator performance: Effectiveness and efficiency. *Biol. Rev.* **2010**, *85*, 435–451. [CrossRef]
8. Bronstain, J.L.; Alarcón, R.; Geber, M. The evolution of plant-insect mutualisms. *New Phytol.* **2006**, *172*, 412–425. [CrossRef]
9. Stephens, R.E.; Gallagher, R.V.; Dun, L.; Cornwell, W.; Sauquet, H. Insect pollination for most of angiosperm evolutionary history. *New Phytol.* **2003**, *240*, 880–891. [CrossRef]
10. Abrahamczyk, S.; Struck, J.H.; Weigend, M. The best of two worlds: Ecology and evolution of ambophilous plants. *Biol. Rev.* **2023**, *98*, 391–420. [CrossRef]
11. Holsinger, K.E. Mass-action models of plant mating systems: The evolutionary stability of mixed mating systems. *Am. Nat.* **1991**, *138*, 606–622. [CrossRef]
12. Hermanutz, L. Outcrossing in the weed, *Solanum ptycanthum* (Solanaceae): A comparison of agrestal and ruderal populations. *Am. J. Bot.* **1991**, *78*, 638–646. [CrossRef]
13. Hao, K.; Tian, Z.X.; Wang, Z.C.; Huang, S.Q. Pollen grain size associated with pollinator feeding strategy. *Proc. R. Soc. Lond. B Biol. Sci.* **2020**, *287*, 20201191. [CrossRef] [PubMed]
14. Larson, B.; Barrett, S.C.H. A comparative analysis of pollen limitation in flowering plants. *Biol. J. Linn. Soc.* **2000**, *69*, 503–520. [CrossRef]
15. Preston, C.; Powles, S.B. Evolution of herbicide resistance in weeds: Initial frequency of target site-based resistance to acetolactate synthase-inhibiting herbicides in *Lolium rigidum*. *Heredity* **2002**, *88*, 8–13. [CrossRef] [PubMed]
16. Conner, A.J.; Glare, T.R.; Nap, J.P. The release of genetically modified crops into the environment: Part II. Overview of ecological risk assessment. *Plant J.* **2003**, *33*, 19–46. [CrossRef] [PubMed]
17. Gibson, R.H.; Nelson, I.L.; Hopkins, G.W.; Hamlett, B.J.; Memmott, J. Pollinators webs, plant communities and the conservation of rare plants: Arable weeds as a case study. *J. App. Ecol.* **2006**, *43*, 246–257. [CrossRef]
18. Benvenuti, S. Weed dynamics in the Mediterranean urban ecosystem: Ecology, biodiversity and management. *Weed Res.* **2004**, *5*, 341–354. [CrossRef]
19. Krenn, H.W.; Plant, J.D.; Szucsich, N.U. Mouthparts of flower visiting insect. *Arthropod Struct. Dev.* **2005**, *34*, 1–40. [CrossRef]
20. Pacini, E.; Nepi, M.; Vesprini, J.L. Nectar biodiversity: A short review. *Plant Syst. Evol.* **2003**, *238*, 7–21. [CrossRef]
21. Jürgens, A. Comparative floral morphometrics in day-flowering, night-flowering and self-pollinated Caryophylloideae (Agrostemma, Dianthus, Saponaria, Silene, and Vaccaria). *Plant Syst.Evol* **2006**, *257*, 233–250. [CrossRef]
22. Albre, J.; Quilichini, A.; Gibernau, M. Pollination ecology of *Arum italicum* (Araceae). *Bot. J. Linn. Soc.* **2003**, *141*, 205–214. [CrossRef]
23. Kephart, S.; Theiss, K. Pollinator-mediated isolation in sympatric milkweeds (*Asclepias*): Do floral morphology and insects behavior influence species boundaries? *New Phytol.* **2003**, *161*, 265–277. [CrossRef]
24. Benvenuti, S. Soil texture involvement in wildflower strip ecosystem services delivery in Mediterranean agro-environment. *Eur. J. Agron.* **2023**, *145*, 126793. [CrossRef]
25. Brys, R.; Jacquemyn, H. 2011 Variation in the functioning of autonomous self-pollination, pollinator services and floral traits in three *Centaurium* species. *Ann. Bot.* **2011**, *107*, 917–925. [CrossRef] [PubMed]
26. Benvenuti, S.; Mazzoncini, M.; Cioni, P.L.; Flamini, G. Wildflower-pollinator interactions: Which phytochemicals are involved? *Basic Appl. Ecol.* **2020**, *45*, 62–75. [CrossRef]
27. Haaland, C.; Gyllin, M. Butterflies and bumblebees in greenways and sown wildflower strips in southern Sweden. *J. Insect Conserv.* **2010**, *14*, 125–132. [CrossRef]
28. Larson, D.L.; Royer, R.A.; Royer, M.R. Insect visitation and pollen deposition in an invaded prairie plant community. *Biol. Conserv.* **2006**, *130*, 148–159. [CrossRef]
29. Newman, D.A.; Thomson, J.D. Interactions among nectar robbing, floral herbivory, and ant protection in *Linaria vulgaris*. *Oikos* **2005**, *110*, 497–506. [CrossRef]
30. Kawaano, S.; Odaki, M.; Yamaoka, R.; Odatanabe, M.; Takeuchi, M.; Kawano, N. Pollination biology of *Cenotera* (Onagraceae). The interplay between floral UV absorbancy patterns and floral volatiles as signals to nocturnal insects. *Plant Spec. Biol.* **1995**, *10*, 31–35. [CrossRef]
31. Young, H.J. Diurnal and nocturnal pollination of *Silene alba* (Caryophyllaceae). *Am. J. Bot.* **2002**, *89*, 433–440. [CrossRef] [PubMed]
32. Kay, Q.O.N.; Lack, A.J.; Bamber, F.C.; Davies, C.R. Differences between sexes in floral morphology, nectar production and insect visit in a dioecious species, *Silene dioica*. *New Phytol.* **1984**, *98*, 515–519. [CrossRef]
33. Folke, S.H.; Delph, L.F. Environmental and physiological effects on pistillate flower production in *Silene noctiflora* L. (Caryophyllaceae). *Int. J. Plant Sci.* **1997**, *158*, 501–509. [CrossRef]
34. Beattie, A.J. Pollination mechanism in *Viola*. *New Phytol.* **1971**, *70*, 343–346. [CrossRef]

35. Petanidou, T.; Lamborn, E. A land for flowers and bees: Studying pollination ecology in Mediterranean communities. *Plant Biosyst.* **2005**, *139*, 279–294. [CrossRef]
36. Rasheed, S.A.; Harder, L.D. Economic motivation for plant species preferences of pollen-collecting bumble bees. *Ecol. Entomol.* **1997**, *22*, 209–219. [CrossRef]
37. Thomas, S.G.; Frankin-Tong, E.F. Self-incompatibility triggers programmed cell death in *Papaver* pollen. *Nature* **2004**, *429*, 305–309. [CrossRef] [PubMed]
38. Dafni, A. Autumnal and winter pollination adaptations under Mediterranean conditions. *Bocconea* **1996**, *5*, 171–181.
39. Petanidou, T.; Vokou, D. Pollination ecology of Labiatae in a phryganic (East Mediterranean) ecosystem. *Am. J. Bot.* **1993**, *80*, 892–899. [CrossRef]
40. Levin, D.A.; Anderson, W.W. Competition for pollinators between simultaneous flowering species. *Am. Nat.* **1970**, *104*, 455–467. [CrossRef]
41. Westwood, J.H.; Tominaga, T.; Weller, S.C. Characterization and breakdown of self-incompatibility in field bindweed (*Convolvulus arvensis* L.). *J. Hered.* **1997**, *88*, 459–465. [CrossRef]
42. Motten, A.F. Determinants of outcrossing rate in a predominantly self-fertilizing weed, *Datura stramonium* (Solanaceae). *Am. J. Bot.* **1992**, *79*, 419–427. [CrossRef]
43. Larson, B.; Kevan, P.; Inouye, D. Flies and flowers: Taxonomic diversity of anthophiles and pollinators. *Canad. Entomol.* **2001**, *133*, 439–465. [CrossRef]
44. Kearns, C.A.; Inouye, D.W.; Waser, N.M. Endangered mutualisms: The Conservation of Plant-Pollinator Interactions. *Ann. Rev. Ecol. Syst.* **1998**, *29*, 83–112. [CrossRef]
45. Jennersten, O. Flower visitation and pollination efficiency of some North European butterfly. *Oecologia* **1984**, *63*, 80–89. [CrossRef]
46. Sakai, S. A review of brood-site pollination mutualism: Plants providing breeding sites for their pollinators. *J. Plant Res.* **2002**, *115*, 161–168. [CrossRef]
47. Bosch, J.; Retana, J.; Cerdá, X. Flowering phenology, floral traits and pollinator composition on a herbaceous Mediterranean plant community. *Oecologia* **1997**, *101*, 583–591. [CrossRef]
48. Beattie, A.J.; Turnbull, C.L.; Knox, R.B.; Williams, E.G. Ant inhibition of pollen function: A possible reason why ant pollination is rare. *Am. J. Bot.* **1984**, *71*, 421–426. [CrossRef]
49. Junker, R.; Chung, A.Y.C.; Blüthgen, N. Interaction between flowers, ants and pollinators: Additional evidence for floral repellence against ants. *Ecol. Res.* **2007**, *22*, 665–670. [CrossRef]
50. Stout, J.C.; Allen, J.A.; Goulson, D. Nectar robbing, forager efficiency and seed set: Bumblebees foraging on the self incompatible plant *Linaria vulgaris* (Scrophulariaceae). *Acta Oecol.* **2000**, *21*, 277–283. [CrossRef]
51. Maloof, J.E.; Inouye, D.W. Is nectar robbers' cheaters or mutualists? *Ecology* **2000**, *81*, 2651–2661. [CrossRef]
52. Bourgeois, B.; Munoz, F.; Fried, G.; Mahaut, L.; Armengot, L.; Denelle, P.; Storkey, J.; Gaba, S.; Violle, C. What makes a weed a weed? A large-scale evaluation of arable weeds through a functional lens. *Am. J. Bot.* **2019**, *106*, 90–100. [CrossRef] [PubMed]
53. Storkey, J.; Westbury, D.B. Managing arable weeds for biodiversity. *Pest Manag. Sci.* **2007**, *63*, 517–523. [CrossRef] [PubMed]
54. Johnson, S.D.; Steiner, K.E. Generalization versus specialization in plant pollination systems. *Trends Ecol. Evol.* **2000**, *15*, 140–143. [CrossRef] [PubMed]
55. Huang, S.Q. Debates enrich our understanding of pollination biology. *Trends Ecol. Evol.* **2006**, *21*, 233–234. [CrossRef]
56. Sutcliffe, O.L.; Kay, Q.O.N. Changes in the arable flora of central southern England since the 1960s. *Biol. Conserv.* **2000**, *93*, 1–8. [CrossRef]
57. Aigner, P.A. Optimality modelling and fitness trade-offs: When should plants become pollinator specialist? *Oikos* **2001**, *95*, 177–184. [CrossRef]
58. Giurfa, M.; Dafni, A.; Neal, P.R. Floral symmetry and its role in plant-pollinator systems. *Inter. J. Plant Sci.* **1999**, *160* (Suppl. S6), 541–550. [CrossRef] [PubMed]
59. Nilson, L.A. Deep flowers for long tongues. *Trends Ecol. Evol.* **1998**, *13*, 259–260. [CrossRef]
60. Fenster, C.B.; Armbruster, W.S.; Wilson, P.; Dudash, M.R.; Thomson, J.D. Pollination syndromes and floral specialization. *Ann. Rev. Ecol. Syst.* **2004**, *35*, 375–403. [CrossRef]
61. Konzmann, S.; Kluth, M.; Karadana, D.; Lunau, K. Pollinator effectiveness of a specialist bee exploiting a generalist plant-tracking pollen transfer by *Heriades truncorum* with quantum dots. *Apidologie* **2020**, *51*, 201–211. [CrossRef]
62. Klumpers, S.G.; Stang, M.; Klinkhamer, P.G. Foraging efficiency and size matching in a plant–pollinator community: The importance of sugar content and tongue length. *Ecol. Lett.* **2019**, *22*, 469–479. [CrossRef]
63. Armbruster, W.S. Evolution of floral form: Electrostatic forces, pollination, and adaptative compromise. *New Phytol.* **2001**, *152*, 181–183. [CrossRef]
64. Fenner, M. The phenology of growth and reproduction in plants. *Perspect. Plant Ecol. Evol. Syst.* **1998**, *1*, 78–91. [CrossRef]
65. Herrera, C.M. Floral biology, microclimate, and pollination by ectothermic bees in an early-blooming herb. *Ecology* **1995**, *76*, 221–228. [CrossRef]
66. Blionis, G.J.; Vokou, D. Pollination ecology of *Campanula* species on Mt Olympus, Greece. *Ecography* **2001**, *24*, 287–297. [CrossRef]
67. Elzinga, J.A.; Atlan, A.; Biere, A.; Gigord, L.; Weis, A.E.; Bernasconi, G. Time after time: Flowering phenology and biotic interactions. *Trends Ecol. Evol.* **2007**, *22*, 432–439. [CrossRef] [PubMed]

68. Davis, S.L.; Delph, L.F. Prior selfing and gynomonoecy in *Silene noctiflora* L. (Caryophyllaceae): Opportunities for enhanced outcrossing and reproductive assurance. *Int. J. Plant Sci.* **2005**, *166*, 475–480. [CrossRef]
69. Barthelmess, E.L.; Richards, C.M.; McCauley, D.E. Relative effects of nocturnal vs diurnal pollinators and distance on gene flow in small *Silene alba* populations. *New Phytol.* **2006**, *169*, 689–698. [CrossRef]
70. Corbet, A.A. Nectar sugar content: Estimating standing crop and secretion in the field. *Apidologie* **2003**, *34*, 1–10. [CrossRef]
71. Dupont, Y.L.; Hansen, D.M.; Rasmussen, J.T.; Olsen, J.M. Evolutionary changes in nectar sugar composition with switches bird and insect pollination: The Canarian bird-flower element revisited. *Funct. Ecol.* **2004**, *18*, 670–676. [CrossRef]
72. Petanidou, T.; Van Laere, A.; Ellis, W.N.; Smets, E. What shapes amino acid and sugar composition in Mediterranean floral nectars? *Oikos* **2006**, *115*, 155–169. [CrossRef]
73. Gardener, M.C.; Gillman, P.M. The taste of nectar—A neglected area of pollination ecology. *Oikos* **2002**, *98*, 552–557. [CrossRef]
74. Gardener, M.C.; Gillman, P.M. The effects of soil fertilizer on amino acids in the floral nectar of corncockle, *Agrostemma githago* (Caryophyllaceae). *Oikos* **2001**, *92*, 101–106. [CrossRef]
75. Gange, A.C.; Smith, A.K. Arbuscolar mycorrhyzal fungi influence visitation rates of pollinating insects. *Ecol. Entomol.* **2005**, *30*, 600–606. [CrossRef]
76. Adler, L.S. The ecological significance of toxic nectar. *Oikos* **2000**, *91*, 409–420. [CrossRef]
77. Narbona, E.; Arista, M.; Whittall, J.B.; Camargo, M.G.G.; Shrestha, M. The role of flower color in angiosperm evolution. *Front. Plant Sci.* **2021**, *12*, 736998. [CrossRef]
78. Menzel, R.; Shmida, A. The ecology of flower colours and the natural colour vision of insect pollinators: The Israeli flora as a case study. *Biol. Rev.* **1993**, *68*, 81–120. [CrossRef]
79. Yoshioka, Y.; Horisaki, A.; Kobaiyashi, K.; Syfaruddin, S.; Niikura, S.; Ninomiya, S.; Ohsawa, R. Intraspecific variation in the ultraviolet colour proportion of flowers in *Brassica rapa* L. *Plant Breed.* **2005**, *124*, 551–556. [CrossRef]
80. Gray, L.A.; Varga, S.; Soulsbury, C.D. Floral ultraviolet absorbance area responds plastically to ultraviolet irradiance in *Brassica rapa*. *Plant-Environ. Interact.* **2022**, *3*, 203–211. [CrossRef]
81. Nakano, C.; Washitani, I. Variability and specialization of plant-pollinator systems in a nothern maritime grassland. *Ecol. Res.* **2003**, *18*, 221–246. [CrossRef]
82. Goulson, D. Are insects flower constant because they use search images to find flowers? *Oikos* **2000**, *88*, 547–552. [CrossRef]
83. Johnson, S.D.; Dafni, A. Response of bee-flies to the shape and pattern of model flowers: Implications for floral evolution in a Mediterranean herb. *Funct. Ecol.* **1998**, *12*, 289–297. [CrossRef]
84. Montgomery, B.R. Effect of introduced *Euphorbia esula* on the pollination of *Viola pedatifida*. *Botany* **2009**, *87*, 283–292.
85. Van Schie, C.C.N.; Haring, M.A.; Schuurink, R.C. Regulation of terpenoid and benzenoid production in flowers. *Curr. Opin. Plant Biol.* **2006**, *9*, 203–206. [CrossRef] [PubMed]
86. Van Kleunen, M.; Manning, J.C.; Pasqualetto, V.; Johnson, S.D. Phylogenetically independent associations between autonomous self-fertilization and plant invasiveness. *Am. Nat.* **2008**, *171*, 195–201. [CrossRef] [PubMed]
87. Booth, B.D.; Swanton, C.J. Assembly theory applied to weed communities. *Weed Sci.* **2002**, *50*, 2–13. [CrossRef]
88. Snell, R.; Aarssen, L.W. Life history traits in selfing versus outcrossing annuals: Exploring the 'time-limitation' hypothesis for the fitness benefit of self-pollination. *BMC Ecol.* **2005**, *5*, 1–14. [CrossRef] [PubMed]
89. Brenner, D.M.; Baltensperger, D.D.; Kulakow, P.A.; Lehmann, J.W.; Myers, R.L.; Slabbert, M.M.; Sleugh, B.B. Genetic resources and breeding of *Amaranthus*. *Plant Breed. Rev.* **2000**, *19*, 227–285.
90. Costea, M.; Weaver, S.E.; Tardif, F.J. The biology of Canadian weeds. 130. *Amaranthus retroflexus* L., *A. powellii* S. Watson and *A. hybridus* L. *Can. J. Plant Sci.* **2004**, *130*, 631–668. [CrossRef]
91. Prokop, P. Urban environment decreases pollinator availability, fertility, and prolongs anthesis in the field bindweed (*Convolvulus arvensis* Linnaeus, 1753). *Plant Sign. Behav.* **2024**, *19*, 2325225. [CrossRef]
92. Verkleij, J.A.C.; De Boer, A.M.; Lugtenborg, T.F. On the ecogenetics of *Stellaria media* (L.) Vill. And *Stellaria pallida* (Dum.) pire from abandoned arable field. *Oecologia* **1980**, *46*, 354–359. [CrossRef] [PubMed]
93. Zimmerman, C.A. A comparison of breeding systems and seed physiologies in three species of *Portulaca* L. *Ecology* **1977**, *58*, 860–868. [CrossRef]
94. Gibbs, P.E.; Talavera, S. Breeding system studies with three species of *Anagallis* (Primulaceae): Self-incompatibility and reduced female fertility in *A. monelli* L. *Ann. Bot.* **2001**, *88*, 139–144. [CrossRef]
95. Jiménez-Lobato, V.; Martínez-Borda, E.; Núñez-Farfán, J.; Valverde, P.L.; Cruz, L.L.; López-Velázquez, A.; Santos-Gally, R.; Arroyo, J. Changes in floral biology and inbreeding depression in native and invaded regions of *Datura stramonium*. *Plant Biol.* **2018**, *20*, 214–223. [CrossRef]
96. Motten, A.F.; Stone, J.L. Heritability of stigma position and the effect of stigma-anther separation on outcrossing in a predominantly self-fertilizing weed, *Datura stramonium* (Solanaceae). *Am. J. Bot.* **2000**, *87*, 339–347. [CrossRef] [PubMed]
97. Boucher, J.J.; Ireland, H.S.; Wang, R.; David, K.M.; Schaffer, R.J. The genetic control of herkogamy. *Funct. Plant Biol.* **2024**, *51*, 5. [CrossRef] [PubMed]
98. Wheeler, M.J.; Armstrong, S.A.; Franklin-Tong, V.E.; Franklin, F.C.H. Genomic organization of the *Papaver rhoeas* self-incompatibility S 1 locus. *J. Exp. Bot.* **2023**, *54*, 131–139. [CrossRef] [PubMed]
99. Lundquist, A. The self-incompatibility system in *Ranunculus repens* (Ranunculaceae). *Hereditas* **1994**, *120*, 151–157. [CrossRef]
100. Kay, Q.O.N. Biological flora of the British Isles: *Anthemis cotula* L. *J. Ecol.* **1971**, *59*, 637–648. [CrossRef]

101. Stevens, J.P.; Kay, Q.O.N. The number, dominance relationships and frequencies of self-incompatibility alleles in a natural population of *Sinapis arvensis* L. in South Wales. *Heredity* **1989**, *62*, 199–205. [CrossRef]
102. Sampson, D.R. Frequency and distribution of self-incompatibility alleles in *Raphanus raphanistrum*. *Genetics* **1967**, *56*, 241–251. [CrossRef]
103. Benvenuti, S.; Mazzoncini, M. Entomogamy in wildflowers: What level of pollinator biodiversity is required? *Acta Oecol.* **2021**, *111*, 103737. [CrossRef]
104. Stanton, M.L.; Snow, A.A.; Handel, S.N.; Bereczky, J. The impact of a flower-color polymorphism on mating patterns in experimental populations of Wild Radish (*Raphanus raphanistrum* L.). *Evolution* **1989**, *43*, 335–346. [PubMed]
105. Devaux, C.; Lepers, C.; Porcher, E. Constraints imposed by pollinator behaviour on the ecology and evolution of plant mating systems. *J. Evol. Biol.* **2014**, *27*, 1413–1430. [CrossRef] [PubMed]
106. Eckert, C.G. Contributions of autogamy and geitonogamy to self-fertilization in a mass-flowering, clonal plant. *Ecology* **2000**, *81*, 532–542. [CrossRef]
107. De Jong, T.J. Geitonogamy: The neglected side of selfing. *Trends Ecol. Evol.* **1993**, *8*, 321–325. [CrossRef]
108. Koul, P.; Koul, A.K.; Hamal, I.A. Reproductive biology of wild and cultivated carrot (*Daucus carota* L.). *New Phytol.* **1989**, *112*, 437–443. [CrossRef]
109. Lamborn, E.; Ollerton, J. Experimental assessment of the functional morphology of inflorescences of *Daucus carota* (Apiaceae): Testing the "fly catcher effect". *Funct. Ecol.* **2000**, *14*, 445–454. [CrossRef]
110. De Jong, T.J. From pollen dynamics to adaptative dynamics. *Plant Spec. Biol.* **2000**, *15*, 31–41. [CrossRef]
111. Ishii, H.S.; Harder, L.D. The size of individual *Delphinum* flowers and the opportunity for geitonogamous pollination. *Funct. Ecol.* **2006**, *20*, 1115–1123. [CrossRef]
112. Klinkhamer, P.G.L.; De Jong, T.L. Effects of plant density and sex differential reward visitation in the protandrous *Echium vulgare* (Boraginaceae). *Oikos* **1990**, *57*, 399–405. [CrossRef]
113. Leiss, K.A.; Klinkhammer, G.L. Genotype by environment interactions in the nectar production of *Echium vulgare*. *Funct. Ecol.* **2005**, *19*, 454–459. [CrossRef]
114. Westerbergh, A.; Saura, A. Gene flow and pollinator behaviour in *Silene dioica* populations. *Oikos* **1994**, *71*, 215–224. [CrossRef]
115. Blair, A.C.; Wolfe, L.M. The evolution of an invasive plant: An experimental study with *Silene latifolia*. *Ecology* **2004**, *85*, 3035–3042. [CrossRef]
116. Petanidou, T. Sugars in Mediterranean floral nectars: An ecological and evolutionary approach. *J. Chem. Ecol.* **2005**, *31*, 1065–1088. [CrossRef]
117. Fahn, A. Secretory tissues in vascular plants. *New Phytol.* **1988**, *108*, 229–257. [CrossRef]
118. Albrecht, H.; Mattheis, A. The effect of organic and integrated farming on rare arable weeds on the Forschungsverbund Agrarökosysteme München (FAM) research station in southern Bavaria. *Biol. Conserv.* **1998**, *86*, 347–356. [CrossRef]
119. Levin, D.A. Competition for pollinator service: A stimulus for the evolution of the autogamy. *Evolution* **1971**, *26*, 668–674. [CrossRef]
120. Marshall, E.J.P.; Brown, V.K.; Boatman, N.D.; Lutman, P.J.W.; Squire, G.R.; Ward, L.K. The role of weeds in supporting biological diversity within crop fields. *Weed Res.* **2003**, *43*, 77–89. [CrossRef]
121. Robinson, R.A.; Sutherland, W.J. Post-war changes in arable farming and biodiversity in Great Britain. *J. Appl. Ecol.* **2002**, *39*, 157–176. [CrossRef]
122. Munoz, F.; Fried, G.; Armengot, L.; Bourgeois, B.; Bretagnolle, V.; Chadoeuf, J.; Mahaut, L.; Plumejeaud, C.; Storkey, J.; Violle, C.; et al. Ecological specialization and rarity of arable weeds: Insights from a comprehensive survey in France. *Plants* **2020**, *9*, 824. [CrossRef] [PubMed]
123. Biesmeijer, J.C.; Roberts, S.P.M.; Reemer, M.; Ohlemüller, R.; Edwards, M.; Peeters, T.; Schaffers, A.P.; Potts, S.G.; Kleukers, R.; Thomas, C.D.; et al. Parallel Declines in Pollinators and Insect-Pollinated Plants in Britain and the Netherlands. *Science* **2006**, *313*, 351–354. [CrossRef]
124. Lázaro, A.; Jakobsson, A.; Totland, Ø. How do pollinator visitation rate and seed set relate to species' floral traits and community context? *Oecologia* **2013**, *173*, 881–893. [CrossRef] [PubMed]
125. Feldman, T.S. Pollinator aggregative and functional responses to flower density: Does pollinator response to patches of plants accelerate at low densities? *Oikos* **2006**, *115*, 128–140. [CrossRef]
126. Pontin, D.R.; Wade, M.R.; Kehrli, P.; Wratten, S.D. Attractiveness of single and multiple species flower patches to beneficial insects in agroecosystems. *Ann. Appl. Biol.* **2005**, *148*, 39–47. [CrossRef]
127. Chancellor, R.J. A preliminary survey of arable weeds in Britain. *Weed Res.* **1977**, *17*, 283–289. [CrossRef]
128. Chancellor, R.J. Decline of arable weeds during 20 years in soil under grass on the periodicity of seedling emergence after cultivation. *J. Appl. Ecol.* **1983**, *23*, 631–637. [CrossRef]
129. Baessler, C.; Klotz, S. Effects of changes in agricultural land-use on landscape structure and arable weed vegetation over the last 50 years. *Agric. Ecosyst. Environ.* **2006**, *115*, 43–50. [CrossRef]
130. Andreasen, C.; Stryhn, H.; Streibig, J.C. Decline of the flora in Danish arable field. *J. Appl. Ecol.* **1996**, *33*, 619–626. [CrossRef]
131. Chamorro, L.; Masalles, R.M.; Sans, F.X. Arable weed decline in Northeast Spain: Does organic farming recover functional biodiversity? *Agric. Ecosys. Environ.* **2016**, *223*, 1–9. [CrossRef]

132. Dutoit, T.; Buisson, E.; Roche, P.; Didier, A. Land history and botanical changes in the calcareous hillsides of Upper-Normandy (north-western France): New implications for their conservation management. *Biol. Conserv.* **2004**, *115*, 1–19. [CrossRef]
133. Gu, H.; Cao, A.; Walter, G.H. Host selection and utilisation of *Sonchus oleraceus* (Asteraceae) by *Helicoverpa armigera* (Lepidoptera: Noctuidae): A genetic analysis. *Ann. Appl. Biol.* **2001**, *138*, 293–299. [CrossRef]
134. Witt, T.; Jürgens, A.; Gottsberger, G. Nectar sugar composition of European Caryophylloideae (Caryophyllaceae) in relation to flower length, pollination biology and phylogeny. *J. Evol. Biol* **2013**, *26*, 2244–2259. [CrossRef]
135. Shapiro, A.M. The California urban butterfly is dependent on alien plants. *Diver. Distrib.* **2002**, *8*, 31–40. [CrossRef]
136. Wiklund, C. Generalist vs. specialist oviposition behaviour in *Papilio machaon* (Lepidoptera) and functional aspects of the hierarchy of oviposition preferences. *Oikos* **1981**, *36*, 163–170. [CrossRef]
137. Alam, S.M.; Bashar, A.; Aich, U.; Akand, S.; Rahman, S. Colonization and strategic stratification of butterfly *Pachliopta aristolochiae* (Lepidoptera: Papilionidae) and its relationship with the host plant *Aristolochia indica* (Piperales: Aristolochiaceae). *J. Biodiv. Conserv. Biores. Manag.* **2017**, *3*, 45–54. [CrossRef]
138. Konvicka, M.; Novak, J.; Benes, J.; Fric, Z.; Bradley, J.; Keil, P.; Hrcek, J.; Chobot, K.; Marhoul, P. The last population of the Woodland Brown butterfly (*Lopinga achine*) in the Czech Republic: Habitat use, demography and site management. *J. Insect Conserv.* **2008**, *12*, 549–560. [CrossRef]
139. Courtney, S.P. Coevolution of pierid butterflies and their cruciferous foodplants IV. Crucifer apparency and *Anthocharis cardamines* (L.) oviposition. *Oecologia* **1982**, *52*, 258–265. [CrossRef]
140. Novak, I.; Severa, F. *Impariamo a Conoscere le Farfalle*; Istituto Geografico De Agostini: Novara, Italy, 1980; 352p.
141. Bowers, M.D. The role of iridoid glycosides in host-plant specificity of checkerspot butterflies. *J. Chem. Ecol.* **1983**, *9*, 475–493. [CrossRef]
142. Warren, M.S. The ecology and conservation of the heath fritillary butterfly, *Mellicta hatalia* I. host selection and phenology. *J. Appl. Ecol.* **1987**, *24*, 467–482. [CrossRef]
143. Dennis, R.L.H.; Hodgson, J.G.; Grenyer, R.; Shreeve, T.G.; Roy, D.B. Host plants and butterfly biology. *Ecol. Entomol.* **2004**, *29*, 12–26. [CrossRef]
144. Koh, L.P.; Sodhi, N.S.; Brook, B.W. Co-extinctions of tropical butterflies and their host plants. *Biotropica* **2004**, *36*, 272–274.
145. DeVries, P.J.; Baker, I. Butterfly exploitation of a plant–ant mutualism: Adding insult to herbivory. *J. N. Y. Entomol. Soc.* **1989**, *97*, 332–340.
146. Fiedler, K. Ants that associate with Lycaeninae butterfly larvae: Diversity, ecology and biogeography. *Div. Distrib.* **2001**, *7*, 45–60. [CrossRef]
147. Kells, A.R.; Goulson, D. Preferred nesting sites of bumblebee queens (Hymenoptera: Apidae) in agroecosystems in the UK. *Biol. Conserv.* **2003**, *109*, 165–174. [CrossRef]
148. Potts, S.G.; Petanidou, T.; Roberts, S.; O'Toole, C.; Hulbert, A.; Willmer, P. Plant-pollinator biodiversity and pollination services in a complex Mediterranean landscape. *Biol. Conserv.* **2006**, *129*, 519–529. [CrossRef]
149. Ferguson, D.J.; Li, X.; Yeates, D.K. Silent, underground warfare of flies: An endoparasitic bee fly (Diptera: Bombyliidae) larva parasitising a predatory assassin fly (Diptera: Asilidae) larva. *Austral Entomol.* **2020**, *59*, 582–592. [CrossRef]
150. Kephart, S. Pollination mutualisms in Caryophyllaceae. *New Phytol.* **2006**, *169*, 637. [CrossRef] [PubMed]
151. Jones, R.E. Behavioural evolution in the cabbage butterfly (*Pieris rapae*). *Oecologia* **1987**, *72*, 69–76. [CrossRef]
152. Dufaÿ, M.; Anstett, M.C. Conflicts between plants and pollinators that reproduce within inflorescences: Evolutionary variations on a theme. *Oikos* **2003**, *100*, 3–14. [CrossRef]
153. Rundlölf, M.; Smith, H.G. The effect of organic farming on butterfly diversity depends on landscape context. *J. Appl. Ecol.* **2006**, *43*, 1121–1127. [CrossRef]
154. Heigland, S.J.; Boeke, L. Relationships between the density and diversity of floral resources and flower visitor activity in a temperate grassland community. *Ecol. Entomol.* **2006**, *31*, 532–538. [CrossRef]
155. Dukas, R.; Morse, D.H. Crab spiders affect flower visitation by bees. *Oikos* **2003**, *101*, 157–163. [CrossRef]
156. Suttle, K.B. Pollinators as mediators of top-down effects on plants. *Ecol. Lett.* **2003**, *6*, 688–694. [CrossRef]
157. Su, Q.; Qi, L.; Zhang, W.; Yun, Y.; Zhao, Y.; Peng, Y. Biodiversity survey of flower-visiting spiders based on literature review and field study. *Environ. Entomol.* **2020**, *49*, 673–682. [CrossRef] [PubMed]
158. Gabriel, D.; Tscharntke, T. Local diversity of arable weeds increases with landscape complexity. *Agric. Ecosyst. Environ.* **2005**, *7*, 85–93. [CrossRef]
159. Gerowitt, B. Development and control of weeds in arable farming systems. *Agric. Ecosyst. Environ.* **2003**, *98*. 247–254. [CrossRef]
160. Bengtsson, J.; Ahnström, J.; Weitbull, A.C. The effects of organic agriculture on biodiversity and abundance: A meta-analysis. *J. Appl. Ecol.* **2005**, *42*, 261–269. [CrossRef]
161. Barbault, R. Biodiversity dynamics: From population and community ecology approaches to a landscape ecology point of view. *Landsc. Urban Plan.* **1995**, *31*, 89–98. [CrossRef]
162. Gabriel, D.; Tscharntke, T. Insect pollinated plants benefit from organic farming. *Agric. Ecosyst. Environ.* **2007**, *118*, 43–48. [CrossRef]
163. Rookwood, P. Landscape planning for biodiversity. *Landsc. Urban Plan.* **1995**, *31*, 379–385. [CrossRef]
164. Kevan, P.G.; Baker, H.G. Insects as Flower Visitors and Pollinators. *Ann. Rev. Entomol.* **1989**, *28*, 407–453. [CrossRef]

165. Banaszak, J. Strategy for conservation of wild bees in an agricultural landscape. *Agric. Ecosyst. Environ.* **1992**, *40*, 179–192. [CrossRef]
166. Zamora, J.; Verdù, J.R.; Galante, E. Species richness in Mediterranean agroecosystems: Spatial and temporal analysis for biodiversity conservation. *Biol. Conserv.* **2007**, *134*, 113–121. [CrossRef]
167. Vulliamy, B.; Potts, S.G.; Willmer, P.G. The effects of cattle grazing on plant-pollinator communities in a fragmented Mediterranean landscape. *Oikos* **2006**, *114*, 529–543. [CrossRef]
168. Tschanz, P.; Vogel, S.; Walter, A.; Keller, T.; Albrecht, M. Nesting of ground-nesting bees in arable fields is not associated with tillage system per se, but with distance to field edge, crop cover, soil and landscape context. *J. Appl. Ecol.* **2023**, *60*, 158–169. [CrossRef]
169. Atkins, E.L.; Kellum, D. Comparative morphogenic and toxicity studies on the effect of pesticides on honeybee brood. *J. Apic. Res.* **1986**, *25*, 242–255. [CrossRef]
170. Marshall, E.J.P.; Arnold, G.M. Factors affecting field weed and field margin flora on a farm in Essex, UK. *Landsc. Urban Plan.* **1995**, *31*, 205–216. [CrossRef]
171. Ma, M.; Tarmi, S.; Helenius, J. Revisiting the species-area relationship in a semi-natural habitat: Floral richness in agricultural buffer zones in Finland. *Agric. Ecosyst. Environ.* **2002**, *89*, 137–148. [CrossRef]
172. Steffan-Dewenter, I.; Tscharntke, T. Early succession of butterfly and plant communities on set-aside fields. *Oecologia* **1997**, *109*, 294–302. [CrossRef]
173. Graves, S.D.; Shapiro, A.M. Exotic as host plants of the Calofornia butterfly fauna. *Biol. Conserv.* **2003**, *110*, 413–433. [CrossRef]
174. Saarinen, K.A. comparison of butterfly communities along field margins under traditional and intensive management in SE Finland. *Agric. Ecosyst. Environ.* **2002**, *90*, 59–65. [CrossRef]
175. Carvell, C.; Meek, W.R.; Pywell RFGouldson, D.; Nowakoski, M. Comparing the efficacy of agri-environment schemes to enhance bumble bee abundance and diversity on arable field margins. *J. Appl. Ecol.* **2007**, *44*, 29–40. [CrossRef]
176. Debnam, S.; Saez, A.; Aizen, M.A.; Callaway, R.M. Exotic insect pollinators and native pollination systems. *Plant Ecol.* **2021**, *222*, 1075–1088. [CrossRef]
177. Ghisbain, G.; Gérard, M.; Wood, T.J.; Hines, H.M.; Michez, D. Expanding insect pollinators in the Anthropocene. *Biol. Rev.* **2021**, *96*, 2755–2770. [CrossRef] [PubMed]
178. Bjerknes, A.L.; Totland, Ø.; Hegland, S.J.; Nielsen, A. Do alien plant invasions really affect pollination success in native plant species? *Biol. Conserv.* **2007**, *138*, 1–12. [CrossRef]
179. Hanley, M.E.; Goulsonm, D. Introduced weeds pollinated by introduced bees: Cause or effect? *Weed Biol. Manag.* **2003**, *3*, 204–212. [CrossRef]
180. Lau, J.A.; Funk, J.L. How ecological and evolutionary theory expanded the 'ideal weed' concept. *Oecologia* **2023**, *203*, 251–266. [CrossRef] [PubMed]
181. Fijen, T.P.; Read, S.F.; Walker, M.K.; Gee, M.; Nelson, W.R.; Howlett, B.G. Different landscape features within a simplified agroecosystem support diverse pollinators and their service to crop plants. *Landsc. Ecol.* **2022**, *37*, 1787–1799. [CrossRef]
182. Benvenuti, S. Wildflowers-pollinator-crab spider predator food-web as indicator of the agroecosystem biodiversity. *Ecol. Indic.* **2022**, *143*, 109272. [CrossRef]

plants

MDPI

Article

From Leaves to Reproductive Organs: Chemodiversity and Chemophenetics of Essential Oils as Important Tools to Evaluate *Piper mollicomum* Kunth Chemical Ecology Relevance in the Neotropics

Daniel de Brito Machado [1,2,*], Jéssica Sales Felisberto [1,2], George Azevedo de Queiroz [3], Elsie Franklin Guimarães [2], Ygor Jessé Ramos [1,4] and Davyson de Lima Moreira [1,2,5,*]

1 Graduate Program in Plant Biology, Institute of Biology, State University of Rio de Janeiro, Maracanã, Rio de Janeiro 20550-013, Brazil; jessickasalles@gmail.com (J.S.F.); ygorjesse@gmail.com (Y.J.R.)
2 Rio de Janeiro Botanical Garden Research Institute, Botanical Garden of Rio de Janeiro, Rio de Janeiro 22460-030, Brazil; eguimar@jbrj.gov.br
3 West Zone Campus, State University of Rio de Janeiro, Rua Manuel Caldeira de Alvarenga, Rio de Janeiro 23070-200, Brazil; georgeazevedo08@gmail.com
4 Earth's Pharmacy Laboratory, School of Pharmacy, Federal University of Bahia, Salvador 40170-215, Brazil
5 Oswaldo Cruz Foundation, Farmanguinhos, Manguinhos, Rio de Janeiro 21041-250, Brazil
* Correspondence: danielmachado.cedae@gmail.com (D.d.B.M.); davysonmoreira@hotmail.com (D.d.L.M)

Abstract: *Piper mollicomum* Kunth (Piperaceae) plays a vital role in the preservation of the Brazilian Atlantic Forest by contributing to the regeneration of deforested areas. Recent scientific investigations have analyzed the chemical constituents and seasonal dynamics of essential oils (EO) from various *Piper* L. species, highlighting the need to elucidate their chemical–ecological interactions. This study aims to expand the chemical–ecological knowledge of this important taxon in neotropical forests, using *P. mollicomum* as a model. The methodologies employed include the collection of plant material, EO extraction by hydrodistillation, analysis of EO by gas chromatography–mass spectrometry (GC–MS) and gas chromatography–flame ionization detector (GC–FID), recording the frequency of visits by potential pollinators and microclimatic variables, and by conducting calculations of chemodiversity and chemophenetic indices. Chemical analyses indicated that the diversity of EO and environmental factors are linked to the activities of potential pollinators. In the Tijuca Forest, *P. mollicomum* revealed significant interactions between its volatile constituents and microclimatic variables, showing that the chemodiversity of the leaves and reproductive organs correlates with pollinator visitation. Additionally, a notable difference in chemical evenness was observed between these vegetative structures. The chemophenetic indices by Ramos and Moreira also revealed correlations with chemical diversity.

Keywords: Piperaceae; Atlantic Forest; essential oils; monoterpenes; sesquiterpenes

1. Introduction

A brief recreational stroll through a garden, or even exhaustive fieldwork, can offer much more than just a visual experience, as, in these environments, our senses are stimulated in diverse ways. For example, while walking along a path adorned with woods or herbs, we may encounter anything from the fresh and inviting aroma of mint (*Mentha piperita* L.) to the unpleasant and repulsive odor of titan arum (*Amorphophallus titanum* (Becc.) Becc. ex Arcang, 1878). These aromas are not only perceived by humans; they are samples of a complex and elegant set of chemical languages used in plant–plant, plant–insect, and plant–microorganism communication, essential for the integration of plant species into their different niches [1–11]. Informed by these insightful understandings, can we infer that these constituents have always been tasked with the same ecological activities? If the answer is yes, what has contributed, retrospectively, to the current patterns of disparity, regularity, occurrence, and chemical diversity of these metabolites? Regardless, we know there are some approaches on the subject at hand, as will be discussed below.

Chemical diversity refers to the array of constituents present at different levels of organization in a specific ecosystem [12]. It is worth noting that tropical forests, for instance, have played a pivotal role as "living pharmacies", offering a rich source of novel substances [12,13]. Therefore, understanding the dynamics of metabolites present in these areas is crucial for advancing the field of human health [14]. Basic research in natural products, with an emphasis on plant studies, is highly relevant because the wide range of plant constituents play important roles in the therapeutic process [15,16].

Chemodiversity is a metric used to quantify the chemical diversity of a population, and it is subject to a range of influences, from environmental conditions (abiotic) and genetic factors to biotic interactions [11,12]. For a long time, the influence of these chemical constituents on the attraction of pollinating insects was considered only in specific cases of coevolution or treated in a stochastic manner. Today it is known that these volatiles directly and indirectly influence the network of ecological interactions to which plant species are subjected [11,16–21]. Therefore, it is crucial to study taxa that are influenced by these interactions to understand trophic and other ecological interactions in their habitat. In this context, the genus *Piper* L. stands out.

The Atlantic Forest, a coastal biome stretching from Northeastern Brazil to Southeastern Argentina, stands out for its phytophysiognomic diversity, adapted to variations in climate, soil, and topography. Despite harboring rich biodiversity, with over 15,000 plant species, including the Piperaceae family, it faces severe threats from uncontrolled urbanization, fragmentation, deforestation, and pollution. With less than 10% of its original cover remaining, competition for resources and edge effects compromise its stability. Nevertheless, the Atlantic Forest plays a crucial role in climate regulation, water conservation, and disaster mitigation, making it essential for Brazil's socio-economic sustainability [11,16].

Piper mollicomum Kunth is a plant species that allocates a significant portion of its energy to the biosynthesis of volatile building blocks for the establishment of their ecological interactions [11,22–27]. This species occurs in various tropical forests across Brazilian territory, ranging from high-light environments to shaded, humid locations [27–30]. In addition to their bioactive effects [31–34], these plants play vital ecological roles in ecosystems, for example, as pioneer plants in secondary forests [11,35,36]. Additionally, their inflorescences and infructescences constitute important nutritional hotspots, scarce in forests degraded by human activities, and are essential for maintaining the population density of generalist pollinators and native seed dispersers [11,35–37].

Ramos and collaborators published a comprehensive analysis of the chemical composition of essential oils (EO) from different organs of *P. mollicomum* from different regions of Brazil [30]. Their study showed the relevant presence of arylpropanoids in the roots, while mono- and sesquiterpenes predominated in the aerial parts of the plant. It is worth noting that this research marked the beginning of a pioneering investigation into the spatial and temporal patterns of compound biosynthesis in *Piper* L species. Additionally, aspects of the volatile chemistry of *P. mollicomum* and *Piper aduncum* L. were elaborated in a spatial-temporal scales and intra-plant context, which demonstrated substantial disparities in relation to collection sites, time (circadian cycle and seasonality), and organs of these species [30]. In addition to these two species, the study of the ecophysiology of *Piper gaudichaudianum* Kunth has been the central focus of recent research. For example, it has been recorded that phenotypic plasticity in the chemical composition of its EO, on a spatiotemporal scale, is directly associated with the level of oxidation of these volatile mixtures [26].

Continuing these research efforts, in 2022 De Brito-Machado and collaborators published the results of their investigation into the seasonal variation of the EO in leaves, inflorescences, and infructescences of *P. mollicomum* during its reproductive period. Their results provide important insights into the chemical variability among these different plant organs, but no significant correlations were found between the major volatile constituents and the attraction of potential pollinators [11].

This research, unlike previous studies, focuses on the influence of the chemical diversity of EO on the attraction of potential pollinators and how abiotic variables can modify this interaction. It introduces a scientific innovation by suggesting that, as the volatile mixture becomes richer and more diverse, it tends to become proportionally more oxidized. To expand chemical–ecological knowledge about this important taxon in neotropical forests, it is crucial to continue these investigations. Therefore, it is essential to assess the correlations between microclimatic variations and the chemodiversity of EO from different organs of *P. mollicomum*, as well as to gather more information on the influence of the chemical profile on pollinator visitation frequency.

2. Results

2.1. Chemical Composition of the Essential Oils of *P. mollicomum*

The main objective of this research was to investigate whether differences in the chemodiversity of the EO from the leaves and four distinct stages of the reproductive organ in *P. mollicomum* (Figure S1) are related to the attraction of potential pollinators. The results reveal variations in the yields of the EO across the four stages, ranging from 0.01% to 1.12% (*w*/*w*) (Tables S1 and S2). Additionally, variability in the chemical profile of the volatile constituents was observed during the reproductive period of the plant. Over a period of months, there was a predominance of non-oxygenated sesquiterpenes in the leaves and oxygenated monoterpenes at the different stages of the reproductive organ.

During the early months of flowering (September and October), the monoterpene linalool recorded the highest relative percentage in the leaves (44.84% and 27.20%, respectively), as well as in most stages of the reproductive organs (29.24% to 73.14%). Another notable monoterpene was limonene, which was found in a relatively high percentage (3.12% to 17.08%). In November, the predominant constituents in the leaves were identified as α-terpineol (11.41%) and β-elemene (19.95%), while 1,8-cineole showed the highest relative percentage at the different stages of the reproductive organs (26.07% to 58.77%). In December, the major constituents in the leaves were α-pinene (11.30%) and β-pinene (6.49%), as well as linalool (11.28%) and β-elemene (8.62%). On the other hand, in the reproductive organ stages, the main constituent was 1,8-cineole (22.96% to 32.01%), along with the monoterpenes α-pinene (3.50% to 14.70%) and linalool (0.00% to 10.46%). In January, the last month of the research, 1,8-cineole continued to be the constituent with the highest relative percentages in the organs, both in the leaves and at the different stages (8.62% to 49.32%). β-elemene also showed a high relative percentage content in the leaf organ (11.55%). Tables S1 and S2 present the chemical constituents that, in any month of the analysis, recorded a relative percentage content greater than or equal to 5%. However, all chemicals of the EO from the leaves and different reproductive stages are included in Table S3. It is worth noting that, for the calculation of chemodiversity and Ramos and Moreira (R&M) indices, all identified compounds were used, regardless of their relative percentage content.

2.2. Chemodiversity Variability of the Organs of *P. mollicomum*: A Multifaceted Analysis of the Leaves and Reproductive Organs

Figure 1 highlights the mean variations between the chemodiversity α indices in the mentioned leaves and reproductive organs. The graphical representation shows that the leaves exhibited significantly higher values for chemodiversity α, considering the Shannon index ($p < 0.05$), a relevant chemical richness in its EO compared with the different stages of the reproductive organs. The other chemodiversity indices did not show significant differences ($p > 0.05$) when compared between leaves and reproductive organs.

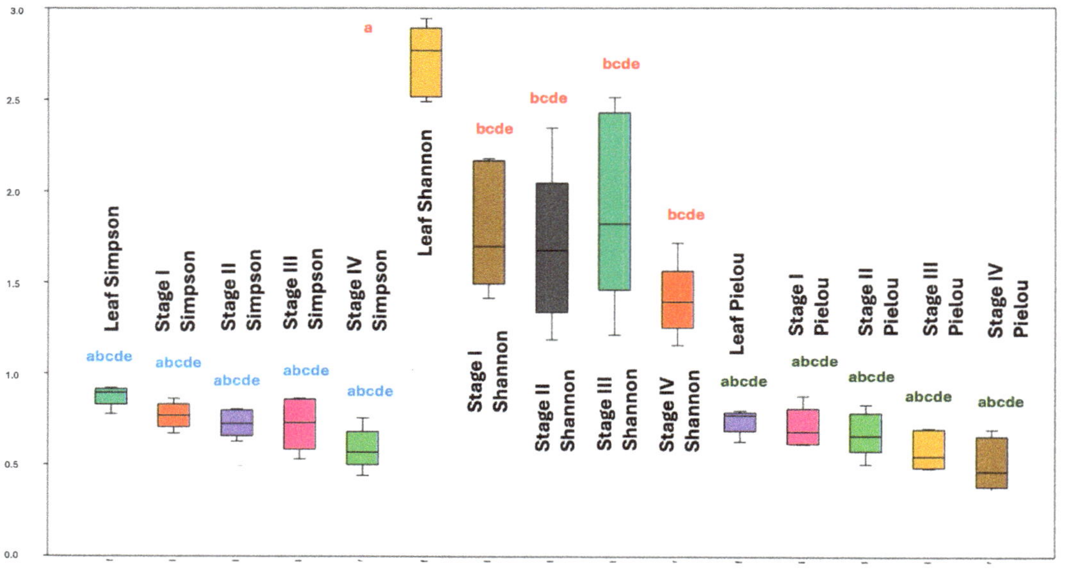

Figure 1. Comparison between means (analysis of variance) of the chemodiversity α indices (Shannon, Simpson and Pielou) between leaves and the different stages of the reproductive organ of *P. mollicomum*. Leg. Same letter indicates no statistical difference ($p > 0.05$).

As there was a significant difference in chemodiversity between the leaves and reproductive organs, considering the Shannon index by ANOVA (Figure 1), we applied the Tukey Post-Hoc test to compare these different compartments (Table S4 and Figure 2). These analyses revealed that the chemodiversity of the EO from the leaves and stage III (anthesis) showed significant similarity ($p = 0.103$). However, there is a significant difference ($p < 0.05$) between leaves and the other stages of the reproductive organ (I, II, and IV), with the EO from the leaves being the most chemodiverse compared with these stages. Additionally, the EO from stage IV (which presents immature drupes) has the lowest richness in constituents. The average richness given by the Shannon index of constituents between stages I to IV was not statistically different ($p > 0.05$) (Table S4 and Figure 2).

Although the richness of the samples may be comparatively equal, how equal, in terms of chemical constituents, are they actually? To answer this question, the Jaccard index was used, aiming to deepen the study of chemodiversity concerning the dichotomy of the chemical profile among the samples studied. This index was used because it represents chemodiversity β and it was applied to the reproductive structures to compare the similarity between chemical constituents [38]. The results confirm the low similarity between stages I and II (53%), and between stages I and III (50%). Between stages I and IV, the similarity was even lower (36%), indicating that the development of the reproductive organ leads to chemical phenotypic diversity (Figure 3). The data from this analysis indicate chemical similarity between one stage and the next, as evidenced by the Jaccard indices between stages II and III (55%) and III and IV (55%). However, there was a gradual decrease in similarity between subsequent stages (stage II and IV—41%). This suggests that, as development progresses, the chemical similarity between stages tends to decrease.

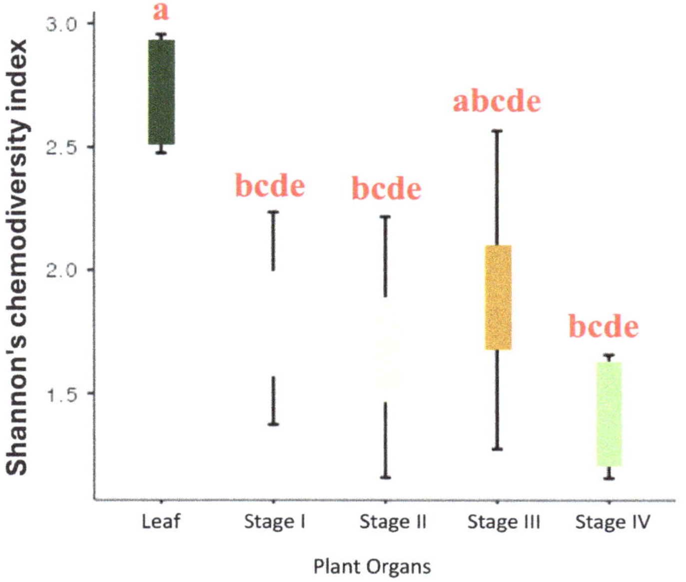

Figure 2. Tukey test (ANOVA) of the means of Shannon's chemodiversity index for leaves and reproductive organ stages of *P mollicomum*. Leg. Same letter indicates no statistical difference ($p > 0.05$).

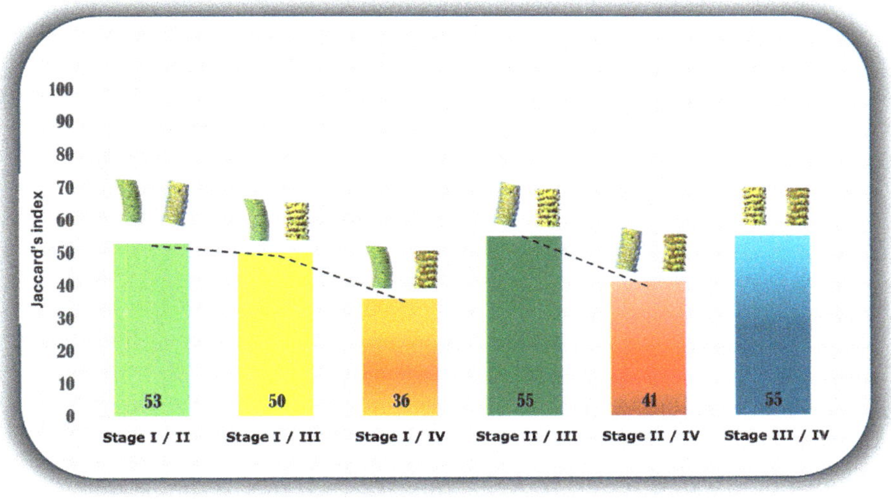

Figure 3. Jaccard index for assessing chemodiversity β among the distinct stages of the reproductive organ of *P. mollicomum*. Leg. numbers in the bars indicate similarity percentual.

With the aim of deepening the analyses and gaining comprehensive insights into the relationship between the gradual decrease in chemodiversity from the initial stages of inflorescence maturation to fruiting, this research proposed an innovative scatterplot to represent the interaction between richness (Shannon index) and evenness (Pielou index), as illustrated in Figure 4. This analysis confirmed that the leaves were the only organ to exhibit high levels of richness and chemical uniformity, mainly as the most prominent when compared with all other reproductive organs.

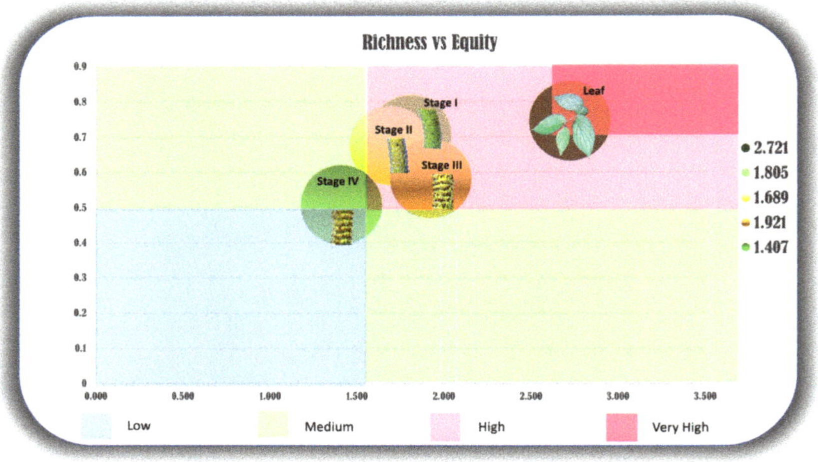

Figure 4. Scatter plot representing two chemodiversity α indices: Shannon (richness) and Pielou (evenness). Leg. The y-axis represents the Shannon index and the x-axis represents the Pielou index.

2.3. Chemical Ecology of P. mollicomum: Interactions between Chemical Diversity, Potential Pollinators, Volatile Compounds, and Microclimatic Variables

The data from potential pollinator observations show that bees and flies were the most frequently observed insect groups that visited *P. mollicomum* blooming inflorescences in 4332 instances (Figure 5). Bees were the most incident visitors, showing high visits on all surveyed days. The most frequent floral visitor, *Tetragonisca angustula* Latreille, 1811, was recorded as having a prominent number of visits, totaling 3042 interactions. Another Hymenoptera, belonging to the Colletidae insect family, stood out with a considerable 820 visits. Two other bee species belonging to the family Halictidae were recorded with frequencies of 238 and 67, respectively. Dipteran insects, on the other hand, showed distinct visitation rates, as one species of the Syrphidae family was the fourth most frequent insect, with 164 visits while another was observed only once on the blooming inflorescences, at the beginning of flowering (Figure 5).

Figure 6 shows the Pearson analyses carried out to assess the potential correlations between the variability in microclimate data and the frequency of visits by potential pollinators versus chemodiversity and the Ramos and Moreira indices (R&M). For instance, a strong positive correlation was identified between the R&M of stage III and the chemodiversity α indices of the leaves ($p < 0.05$). As the diversity, richness, and evenness of volatile constituents increased, the EO compounds of the flowering inflorescence at anthesis (stage III), as well as that of the leaves, tend to become more oxidized. A positive correlation between the R&M in the early stages of inflorescences (I and II) and the chemodiversity α indices of stage IV was also identified. This relationship remained consistent throughout the different stages of the analyzed inflorescences, suggesting that, as the volatile mixture becomes richer and more diversified, it becomes concurrently and proportionally more oxidized (Figure 7).

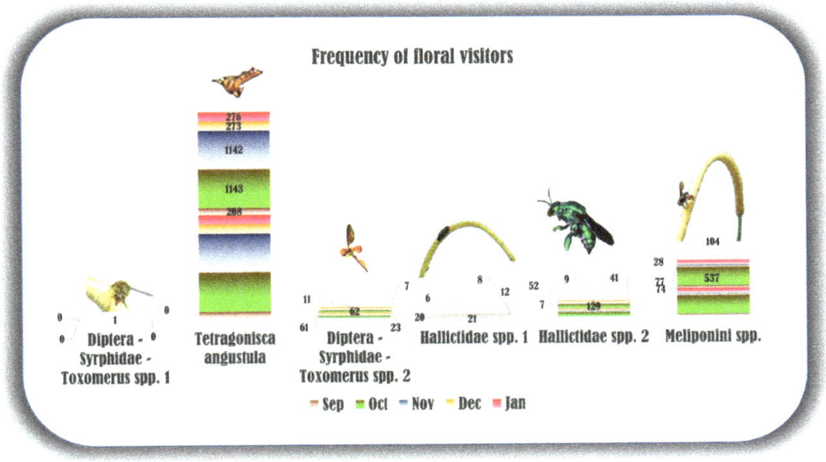

Figure 5. Visit frequencies of potential pollinators of *P. mollicomum* from Tijuca Forest/RJ (September 2020 to January 2021). Leg. The x-axis represents the visit frequencies of potential pollinators, and the y-axis represents the months under analysis. Sep—September, Oct—October, Nov—November, Dec—December, Jan—January.

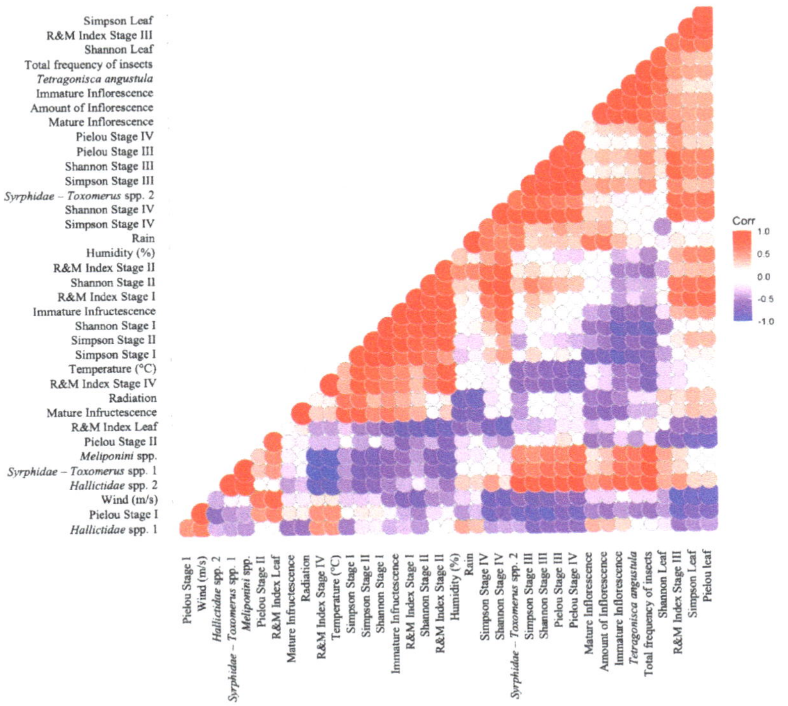

Figure 6. Pearson correlation analyses regarding the following variables: insect visit frequency, chemodiversity α indices, Ramos and Moreira redox indices, phenology (quantity of reproductive organs), and microclimatic factors. Leg. Corr—Pearson correlation; R&M—Ramos and Moreira index.

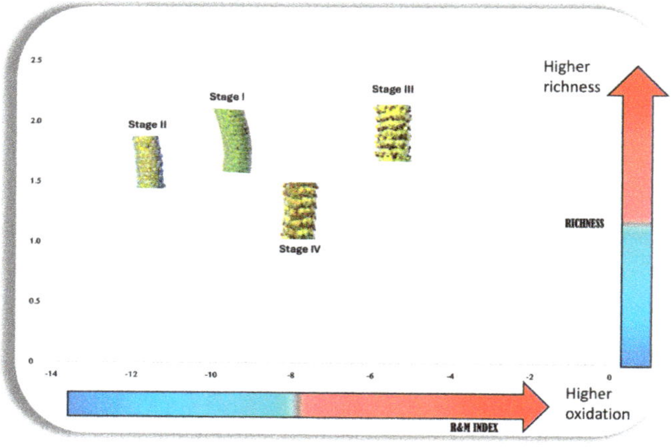

Figure 7. Scatter plot depicting the relationship between the Shannon index (Chemical richness) and the Ramos and Moreira index (R&M, redox pattern) of the essential oils from different stages of the reproductive organ of *P. mollicomum*.

Furthermore, the analyses indicated a positive correlation between an increase in the Shannon's chemodiversity of the leaves (richness) and an increase in the frequency of visits by potential pollinators to flowering inflorescences at anthesis (stage III). Additionally, these results reveal a positive correlation between the number of visits by these harmonious insects and the total number of inflorescences.

Another relevant correlation found in the EO of the early stages of the reproductive organ (stages I and II) was found between the microclimatic variable temperature and the chemodiversity α indices. According to these results, the increased temperature resulted in greater richness, abundance, and homogeneity of the EO of these organs.

The analyses also showed a negative correlation between the chemodiversity α indices of the leaves and the microclimatic variable wind. Continuing to regard negative correlations, an inversely proportional relationship was identified between the Simpson's chemodiversity index (diversity) in the early stages of inflorescence maturation and the total visits of all insects, especially *T. angustula*.

Finally, a negative correlation was also observed between the frequency of visits from some potential pollinators, such as *Meliponini* spp., one species of Syrphidae, and others of Halictidae, and the R&M in stage IV. This suggests that the more oxidized the mixture of volatile compounds at that stage, the lower the frequency of insect visits. Similarly, when the metabolite mixture of the EO is reduced further, there is an increase in the visitation of potential pollinators to the flowering inflorescences. Additionally, the analyses revealed negative correlations between the chemodiversity α indices (leaves and stage II).

3. Discussion

3.1. Chemical Composition of P. mollicomum Essential Oils

To understand the chemical mediation between *P. mollicomum* and its natural enemies in the Tijuca Forest (Rio de Janeiro), we investigated the chemical composition of EO from leaves and the developmental stages of the reproductive organs of this plant and the correlation between these aspects with visitors and microclimatic environmental conditions. During the early flowering stage, when most specimens studied had immature inflorescences, the predominance of linalool and limonene biosynthesis in these reproductive structures was observed. In this period, the plants were in a phase prior to the proper production of pollen grains, which likely results in pollination limitation. Therefore, we hypothesize that chemical defense of these immature organs is crucial to maintain the reproductive viability of their specimens. Our results are corroborated by previous studies which have suggested that plant species emit toxic volatiles more intensely during the early stages of flowering to protect their pubescent organs against herbivory [39,40]. Additionally, research has shown that genes contributing to the protection of reproductive organs are primarily stimulated during the early stages of the development of young tissues [41]. Previous studies have confirmed limonene's ability to repel pest organisms, such as aphids, or to attract ladybugs, forming a tritrophic interaction between plant and herbivore and carnivore [42–44]. Examining the importance of the constituent linalool, many studies have also highlighted its significant role in attracting insect pollinators [11,24,45–48]. Additionally, this monoterpene may also have defensive effects against herbivores [49–52].

The analyses of the final stages of fruiting revealed a greater amount of α-terpineol (11.41%), 1,8-cineole (8.00% to 58.77%), α-pinene (11.30%), β-pinene (6.49%), linalool (11.28%), β-elemene (8.62%), and eupatoriochromene (32.92%). In this stage of development, mature inflorescences stood out, suggesting that the biosynthesis of these constituents is ecologically important, either directly or indirectly, to facilitate reproduction. Previous studies have demonstrated that some of these compounds can play prominent ecological roles, such as in the development of reproductive structures and in protection against herbivores [53–56]. It is also important to note that the biosynthesis of sesquiterpenes, such as *E*-nerolidol, β-elemene, and germacrene D, was recorded in senescent leaves of *P. mollicomum*, suggesting a protective role for these plant parts [27]. The presence of these constituents in the EO of reproductive organs may also be related to the need to attract the visitation of pollinators and seed dispersers [11,57,58].

For example, the presence of 1,8-cineole in specimens of *P. mollicomum* from the Tijuca Forest has been found for several years [11,30]. Studies have revealed that this monoterpene may possess toxic effects and alter the intestinal microbiota of herbivores and frugivores, thus compromising the digestibility and nutrient absorption of these consumers [59–61]. These and other analyses suggest that this major constituent may play an important ecological role in defense against herbivores, increasing in the early stages of inflorescence development and persisting throughout the flowering period [11,30]. Our findings suggest that these volatiles, despite having distinct ecological roles, at the beginning and end of flowering period in *P. mollicomum*, may offer similar functional expression in the chemical communication of this species, perhaps acting synergistically when biosynthesized concomitantly. These hypotheses need to be specifically investigated.

The innovative aspect of this research is its consideration of the way that the ecological roles of these compounds, predominant during the reproductive period in *P. mollicomum*, have often been studied in isolation, disregarding the relevance of minor compounds [11,14,26,30,37]. To improve this discussion, we can draw an analogy between volatiles and soccer players. One star player may have a pivotal role in their team. Then, just as a soccer club typically invests its financial efforts to acquire a valued player, plants also allocate their resources to biosynthesize major and ecologically important substances. However, this player cannot act alone; they depend on other members to form an efficient team. Similarly, "supporting compounds" can have equally crucial importance, acting synergistically with the major compounds and contributing to the survival, acclimatization, and adaptation of plants in their niches; especially when considering that resources, being so scarce, cannot be spent on synthesizing compounds without any function for these plants [38,62–64].

3.2. Chemodiversity Analysis of P. mollicomum: Ecophysiological and Ecological Interconnections of Volatile Metabolites along Developmental Stages

Small variations in volatile chemodiversity occur over time and space, mainly because EO are directly influenced by biotic and abiotic environmental conditions [38,62–64]. For instance, infections caused by insects on vegetative structures can induce the production of new volatile constituents, both locally and systemically; moreover, neighboring plants under herbivore attack may release volatile emissions triggering changes in the EO of adjacent plants [38,62–64]. Therefore, studying the volatile chemodiversity and its role as a protagonist in ecological activities is of utmost importance for a comprehensive understanding of its ecological expression.

During the 1980s, O. R. Gottlieb initiated his studies on the evolution of chemical diversity and the oxidation state of compounds in plant species [65,66]. Studies conducted by his group revealed that such phytochemicals play an essential role in the adaptive capacity of organisms to adverse environmental conditions, ranging from high UV radiation to oxygen-rich atmospheres, all factors that may have been crucial for the adaptation of early plants to the terrestrial environment [16,65,66]. From these initial functional traits, the richness of chemodiversity emerged as an intricate phenotypic expression of intra- and interspecific interactions, which are under genetic and environmental control [10,11,16]. These associations among plants of the same and different species promote the biosynthesis of essential metabolites [11,14,16]. Numerous studies have corroborated the importance of these biotic and abiotic factors, highlighting the occurrence of specific chemical variability in response to abrupt environmental changes, such as light radiation [67], temperature [68], herbivory and pollination [37,69], allelopathic influence [70], latitude and longitude [71]; and precipitation [11,30].

Some results of our study reveal the important relationship between temperature, the chemical profile of the reproductive organs, and the chemodiversity of leaves. The increase in temperature was a determining factor for the increase in richness, abundance, and homogeneity of the EO mixture of leaves and reproductive organs. This suggests that thermal conditions may play a crucial role in the biosynthesis and accumulation of volatiles in these plant parts. Our findings corroborate previous studies, which provided specific evidence of how temperature variations can influence the concentration of some terpenes, demonstrating the degradation and expressive biosynthesis of some constituents due to photo-oxidation at high temperatures [72,73]. On the other hand, the results also indicate a negative correlation between the chemodiversity α indices of leaves and the wind microclimatic variable. This relationship suggests that an increase in wind intensity may result in greater volatilization of the chemicals, leading to a reduction in the richness of the EO.

It is known that some enzymes responsible for terpene biosynthesis are influenced by light, such as for 1,8-cineole synthase and linalool synthase. An in vitro study on volatile constituents at different developmental stages of *Vitis vinifera* L. (Vitaceae) revealed that increased UVB radiation results in a significant increase in the proportion of oxygenated cyclic monoterpenes, such as 1,8-cineole [67]. This observation may also explain our results regarding the decrease in chemodiversity related to the increased volatilization of some constituents, also previously observed [24].

These findings contribute to understanding the underlying mechanisms of production and variation of volatile constituents in plants, considering the influence of abiotic factors. They may have significant implications in the chemical ecology of these plants, as well as in practical applications, such as agricultural management, the production of bioactive substances, and ecosystem studies. However, further research is needed to deepen our understanding of the genetic and/or epigenetic influence on the molecular mechanisms involved in these ecological responses, as well as the potential impact of these changes on plant–environment interaction [62].

This distinction in EO chemodiversity suggests the presence of unique biosynthetic mechanisms regulated by "promiscuous enzymes" in metabolic pathways of each plant organ [4,74–76]. As a result, a relevant variety of odoriferous substances is biosynthesized. This complex system is notoriously understood to be controlled by widely studied enzyme superfamilies called terpene synthases (TPS) [4,74–76]. Through their activity, an intricate network of mono- and sesquiterpenes can be formed, generating a universe of aromas that play critical roles in plant communication. TPS are even distinguished at phylogenetic levels, being a major theme in chemotaxonomy [4,37,74–77], and are subjected to years of pressures, selective bottlenecks, and adaptations and mediated by various biotic and abiotic factors [4,37,74–77].

In the context of plant–pollinator ecology, our results provide valuable insights into the importance of chemodiversity in the specificity of ecological activities. This phenomenon may act by attracting pollinators to locate, recognize, and determine the quality of the reward offered by the plant [4,11,24,27,30,74–76]. These findings corroborate the data analyzed in this research, which shows strong positive correlations between the chemodiversity α indices of different plant parts, presenting a fine association between the frequency of visits by potential pollinators and responses to environmental variables. For example, our investigations found that leaves and inflorescences in anthesis (stage III) exhibited the highest chemodiversity indices. These analyses may suggest that the plant possibly directs its allocation of chemical resources with the purpose of defending its still immature organs against herbivores, as well as to attract and reward its mutualistic agents [11,69].

Conversely, infructescences showed low chemodiversity indices, possibly indicating a shift in resource redirection towards the biosynthesis of carbohydrates to attract seed dispersers [36,78–82]. These results suggest the possibility of a drain effect between vegetative and reproductive organs [83], where resource allocation may influence the chemodiversity of different plant organs, consequently characterizing each organ as a specific functional trait [11,69].

Furthermore, an increase in chemodiversity by Shannon index, directly related to the increase in the frequency of visits by potential pollinators to inflorescences in anthesis (stage III), was found in our research. It is also important to note that the highest frequency of visits was recorded in the months of October and November, a period during which a greater number of blooming inflorescences (stage III) were quantitatively observed [11]. This finding may suggest the hypothesis that the richness of volatiles present in the EO may positively influence the interaction between plants and their mutualistic agents [11,30]. Also observed, concurrently, was an inversely proportional relationship between the chemodiversity by Simpson index in the early stages of inflorescence maturation and the total visits of all insects, especially *T. angustula* (jataí). The Simpson index represents the diversity of one or more volatile constituents in samples of these stages [12,16,26,38,84]. This latter analysis suggests that the diversity of substances, typically found in the EO of *P. mollicomum* may not be sufficient for insect attraction [11,30,69].

Our studies infer that EO synthesized with higher richness (Shannon index) and evenness (Pielou index) in *P. mollicomum* may be more attractive to insects. Leaves were the most "Equidiverse" * organs (* a term proposed by the authors to express the idea of an abundant and uniform chemical profile). Stages I, II, and III also exhibited high equidiversity, indicating that, at different stages of reproductive organ maturation, they may present a volatile composition with high diversity and uniformity. This is so for two vital reasons: protection of these developing structures (stages I and II), and attraction of insects for the dispersal of their generative whorls (stage III) [11,26,30]. On the other hand, stage IV showed low equidiversity, which allows us to infer that there is a channeling of adaptive metabolism towards survival in these structures, which corroborates the perspectives of the drain effect [83]. For example, a recent investigation, which focused on the analysis of the chemical composition of EO from *Satureja hortensis* L., revealed the presence of a notable diversity and uniformity of volatile constituents in this plant. The research emphasized the synergistic effects demonstrated by the substances, which showed relevant insecticidal properties in combating pests [85].

We can infer that this "phytochemical socialism" is probably crucial, both to attract and/or to repel chemically mediated interactions. The "volatile fog" [11] present in the leaves and in different stages of the reproductive organs, may function as an innate chemical barrier or attractant. If an herbivore manages to overcome the first protective barrier on the leaves, it can find other volatiles, rich and evenly distributed in "chemical weapons", which act together to minimize the herbivory [11,85,86]. However, as mentioned above, reproductive stage IV showed low equidiversity, indicating that the richness of secondary metabolites may not be a fundamental factor for attracting seed dispersers [36,78,79,81,82]. These results are consistent with previous studies that have investigated the allocation of substances at different stages of plant development, as well as the importance of carbohydrate biosynthesis in attracting dispersers [36,78–82]. In conclusion, the analyses conducted have provided important insights into chemodiversity in different structures and stages of development of *P. mollicomum*.

3.3. Similarity between the Chemical Profiles of the Reproductive Organs of P. mollicomum Using the Jaccard Index

The exploration of the distinct facets of chemodiversity, the focus of our study, has allowed a deeper investigation into the potential dichotomy in the chemical profile of the samples, with emphasis on the different stages of the reproductive organs of *P. mollicomum*. The assessment of chemodiversity β, by Jaccard index, was employed to examine the chemical congruence among the different developmental stages [16,38]. Our results reveal interesting nuances in this context. A similarity of 53% was found between stages I and II, and 50% between stages I and III, as well as 36% between stages I and IV. Other results show stages II and III (55%), II and IV (41%) and, III and IV (55%). These findings indicate that the chemical profile affinity among the different stages is considerably reduced, highlighting the chemical uniqueness of each germinative structure, even if they retain anatomical similarities. This discovery reinforces the notion that different developmental stages of the reproductive organ of *P. mollicomum* possess unique chemical profiles, which may be related to specific physiological and systemic functions at each maturation stage.

At an individual level, the chemodiversity of a specimen depends on the amount of information present in its vicinity. It is known that the less interference from other biotic interactions, the lower the chemical diversity [12,30,87]. Previous studies conducted by the group have shown that chemodiversity depletes in plant populations removed from their niche [88]. Therefore, the indices quantified in this work play a fundamental role in inferring the characterization of chemical variations, mainly concerning their geographical (different niches), spatial (distinct anatomical structures), and temporal (circadian rhythm) distribution [11,16,27,30].

Thus, understanding and assessing different levels of chemodiversity is essential for comprehending the complex interactions between organisms in their niches. The proper application of these indices provides a valuable quantitative insight into chemical diversity and its role in species ecology and acclimatization, offering a solid subsidy for future ecological and biotechnological studies. When addressing the intricate networks of interactions that permeate the production of volatile constituents, it is essential to consider the potential impact of biotic and microclimatic factors in this process. These elements play a crucial role in the biosynthesis, accumulation, and release of essential metabolites, consequently influencing the attractive response of pollinators or the repellent response of disharmonic insects [11,30]. Understanding these complex "chemo-signaling networks" is of utmost importance for a holistic view of the ecology of ecological activities provided by *P. mollicomum* in its niche. Furthermore, it is well established that these interactions can influence the dynamics of oxidative-reductive patterns present in the EO of these plants [14,16,26].

3.4. Ramos and Moreira Index: A Redox Evaluation

It has been recorded in this study that the richer and more diversified the mixture of volatile components, the higher the R&M. The mixture becomes more oxidized when it is chemically more diverse. This was observed in the EO of both leaves and reproductive organ stages, except for stage IV. Dr. Gottlieb's hypothesis, mentioned above in this text, presumes that plant secondary metabolites function to capture free radicals [16,65,66,89]. The intense enzymatic activity (peroxidases, terpene synthases) in leaves and reproductive stages, especially stage III (anthesis), aims to generate richer and more diversified volatile mixtures, which would also lead to increased production of these free radicals. Consequently, these free radicals (mainly reactive oxygen species (ROS)) are eliminated by mixtures of compounds that include EO, resulting in increased oxidation of the mixture, then reflecting on R&M [16,65,66,89].

4. Conclusions

In summary, analyses of the chemical composition of *P. mollicomum* in the neotropical Tijuca Forest revealed important insights into the interactions among volatile constituents, environmental factors, and their potential pollinators. The results indicate that variations in the chemodiversity of the volatile chemical profiles of leaves and different stages of the reproductive organ showed correlations between microclimatic variables and activities of potential pollinator visits, along with significant differences in chemical evenness among these different vegetative structures. These findings reveal the importance of understanding the complex interactions among plant metabolism, its environment, and ecological relationships in its niche, and provides a comprehensive view of the ecology and evolution of ecological activities provided by this specie. We prove that, by inserting chemodiversity indices in the analysis, these interactions can be quantified for better understanding.

5. Materials and Methods

5.1. Study Area

The area chosen for this research was the Tijuca National Park (TNP) (43°14′29.64″ W, 22°58′9.80″ S), located in the South Zone of Rio de Janeiro city. According to the Köppen–Geiger classification, the TNP has a tropical monsoon climate [90]. Six adult specimens of *P. mollicomum* Kunth, with an average height of 1.65 m, were selected in an open area with elevations ranging from 68 to 127 m. The license permission for conducting these investigations was obtained from the Biodiversity Authorization and Information System (SISBIO—number 57296-1; authentication code 47749568). The experimental plot was surrounded by native vegetation. The fertile specimens were previously identified by the taxonomist Dr. George Azevedo de Queiroz at the Rio de Janeiro Botanical Garden Research Institute (JBRJ), and the samples were deposited in the Herbarium of the State University of Rio de Janeiro—Maracanã Campus (HRJ/UERJ) (Table S5).

5.2. Collection of Microclimatic Data

The averages of the data regarding microclimatic variables were collected weekly and recorded by manual measuring instruments [11]. A digital windmeter anemometer (SIN2919025384—Brazil) was used to measure wind speed, temperature, and relative humidity. Additionally, a luxmeter (INSTRUTEMP, 1712268—Brazil) was employed to measure local luminosity. An infrared thermometer (EXBOM—TDI 330) was utilized to record the surface temperatures of leaves and inflorescences during anthesis. These measurements were taken once at each interval corresponding to the frequency of visits by potential pollinators. Climatological data related to rainfall indices were obtained from the Meteorological Station of Forte de Copacabana, under the jurisdiction of the State of Rio de Janeiro, during the corresponding weeks of observation periods. This institution is affiliated with the Brazilian Meteorological Institute (INMET), responsible for managing climatic information. The analyzed meteorological variables included air temperature (°C), inflorescence temperature (°C), leaf temperature (°C), relative air humidity (%), wind speed (m/s), rainfall (mm), and light intensity (kJ/m^2).

5.3. Evaluation of Pollinators Visit Frequency

The observations regarding insect visit frequency were conducted weekly from September 2020 to January 2021, from 8:00 a.m. to 5:00 p.m. (with 30 min of surveillance and 30 min of break), totaling 128 h. This period was chosen due to the presence of inflorescences in anthesis (stage III) on the *P. mollicomum* individuals. Frequency counting was performed whenever insects were foraging on the inflorescences during this period [91,92]. The collection of these ecosystemic agents was carried out using an entomological net with a reach of 3 m (basket with approximately 35 cm in diameter, 80 cm in depth, and 3 mm mesh).

After capture, the animals were promptly anesthetized in vials containing cotton slightly dampened with 70% (*v*/*v*) hydroalcoholic solution for subsequent identification [93]. Insect descriptions were made using identification keys and entomological specimens from the Reproductive Biology and Pollination Laboratory of the Research Institute of the Botanical Garden of Rio de Janeiro (JBRJ).

5.4. Reproductive and Vegetative Phenological Study

Quantitative studies on the reproductive phenology of *P. mollicomum* were conducted throughout the observation of the phenological events: quantities of immature and mature inflorescences, and the total quantity of reproductive organs. Due to the small size of the flowers and fruits, it was necessary to use a manual lens with a magnification of 60× to determine which flowers were in anthesis [94]. The intensity of Fournier (IF) methodology [95] was employed for the quantification of phenophases, which involves creating a scale that classifies the analyzed phenological patterns in plants, previously described in [11].

5.5. Essential Oil Obtention and Analysis

The collection of the EO from the leaves (150 g) and the four stages of the reproductive organ (40 g) occurred monthly, between 9:00 a.m. and 10:00 a.m. The samples were comminuted and placed in a 2 L round-bottom flask containing 700 mL of distilled water and subjected to the hydrodistillation method using a modified Clevenger-type apparatus. The process lasted for 2 h. Upon completion, the pure EO was separated from the aqueous phase, dried with anhydrous sodium sulfate, and placed in amber vials for storage in a freezer at −20 °C. The four stages were described by taxonomists Dr. Elsie Franklin Guimarães and Dr. George Azevedo de Queiroz, and they are illustrated in Figure S1 and previously characterized [11].

The obtained EO was solubilized in spectroscopic grade dichloromethane from Tedia (Brazil), to reach a final concentration of approximately 1000 μg/mL. These diluted solutions were then subjected to gas chromatography–mass spectrometry (GC–MS) analysis using an HP Agilent GC 6890—MS 5973N instrument to identify volatile constituents by their respective mass spectra. In turn, for the determination of relative percentage parameters and calculation of retention index (RI), the analyses were conducted by gas chromatography–flame ionization detector (GC–FID). The conditions employed in the GC–MS analyses involved the use of an analytical capillary column, HP-5MS (30 m × 0.25 mm i.d. × 0.25 μm, film thickness), with a temperature ramp from 60 °C to 240 °C, increment of 3 °C/min, and helium as carrier gas (~99.99%) at a constant flow rate of 1.0 mL/min. Additionally, the mass scan range (*m*/*z*) ranged from 40 to 600 atomic mass units (u), with an electron impact energy of 70 (eV), operating in positive mode. A 1 μL sample of the EO solution was injected splitless, with the injector temperature set at 270 °C. GC–FID analyses were conducted under the same conditions, though hydrogen (mobile phase with a flow rate of 1 mL/min) and synthetic air were used to produce the flame. Retention times (tR) were measured in minutes, without correction, and the relative percentage of each identified substance was determined based on the signal area. RIs were calculated from the results of the analysis of a homologous series of saturated aliphatic hydrocarbons (C_8–C_{28}), from Sigma-Aldrich, Rio de Janeiro, Brazil, using the same column and conditions employed in the GC–FID analyses. The identification of constituents was performed by comparing the calculated RIs and obtained mass spectra with information available in the literature [96]. GC–FID analyses were conducted in triplicate [11]. Constituents were identified by comparison of their calculated LRIs with those in the literature, and by comparison of the mass spectrum with those recorded by the National Institute of Standards and Technology (NIST) library "NIST14" and Wiley (ChemStation data system) "WILEY7n." [14]. Additionally, authentic pattern co-injection was performed whenever possible [14].

5.6. Chemodiversity Indices Calculation

To quantify the dynamics of the chemical profile of the EO from the leaves and different stages of the reproductive organ, the α and β chemodiversity indices were calculated [12,16,26,38]. It is relevant to mention that these indices were initially developed to assess species diversity on a spatial scale [26,84,97–105] and currently have been adapted to measure chemical phenotypic variability [12,16,26,38].

To evaluate chemodiversity α, the following indices were used: Shannon [99], Simpson [84], and Pielou [100]. For chemodiversity β, the Jaccard index was applied to assess the similarity of samples from different stages of the reproductive organ [97]. These levels are classified based on intra- and interspecific chemodiversity categories, as well as richness, abundance, and presence/absence of chemical constituents, and adapted for chemical diversity analyses in different reproductive organs. These quantifications are described below:

Chemodiversity α: This index was adapted to infer the chemical diversity of an individual. The diversity of chemical profiles can be evaluated at different temporal scales or even in a specific tissue [16,38]. This variability can characterize an adaptation or acclimatization of the individual or organ to ecological complexity influenced by positive or negative biotic and abiotic interactions [12,16,38].

The equations for chemodiversity α indices are presented below:

$$\text{Shannon Index} = \Sigma(\text{Pi} \times \ln(\text{Pi})) \quad (1)$$

$$\text{Simpson Diversity Index:} = \Sigma(\text{Pi} \times \ln(\text{Pi} - 1)) \quad (2)$$

$$\text{Pielou Index:} = -\Sigma(\text{Pi} \times \ln(\text{Pi}/\ln(\text{S})) \quad (3)$$

Pi represents the proportional abundance of identified substances, obtained by dividing the relative abundance (or yield of the identified compound) by the total number of determined metabolites, as well as the number of substances in the sample. S is equal to the total number of substances. The natural logarithm (ln) is the logarithm to the base "e" (Euler's number, approximately 2.71828) × n.

Chemodiversity β: This index was adapted to infer the chemical diversity of different individuals of the same species in a geographically segregated population. Moore and colleagues (2014), from a contemplative perspective, presented the idea that intraspecific chemical changes occur at different levels of diversity, synthesized by distinct metabolic pathways influenced by many variables [38]. Therefore, in this research, this index was developed to evaluate the similarity of chemical profiles among different stages of the reproductive organ of *P. mollicomum*.

To assess chemodiversity β, the Jaccard index was used [4], as described below:

$$\text{Jaccard Index: } (c/(a + b - c)) \quad (4)$$

In this equation, "a" represents the number of substances in one sample, "b" the number of substances in another sample, and "c" the number of common substances in both samples.

5.7. Ramos and Moreira Index (R&M)

In this research, the Ramos and Moreira index (R&M), a chemophenetic index, was also applied. This index offers a quantitative approach to evaluate oxidation-reduction (redox) patterns in mixtures, allowing for a deeper understanding of redox mechanisms in a context that can be applied to chemical ecology studies [16,26].

For the calculations, redox indices were initially used to analyze substances in relation to their oxidation number (NOX), following rules established by Hendrickson–Cram–

Hammond [16,26,106], to determine the sum of the oxidation states of each atom in the molecule [16,26,107].

To calculate the weighted mean oxidation-reduction pattern (SRO), the oxidation number (NOX) of the constituent of interest was multiplied by the relative percentage content obtained in the sample analysis (Q%), and this value was subsequently divided by the number of carbon atoms in the molecular skeleton (n) [16,26]. The equation is described below (5):

$$SRO = (NOX\ substance \times Q\%)/n \quad (5)$$

This equation provides a weighted average value of the oxidation of the carbon atoms of the mixture constituents (in this case, EO), and represents an intermediate step to obtain the Ramos and Moreira Index (also called General Mixture Redox Index (GMRO)), as defined by Equation (6) [16,26]. To calculate the GMRO, it is necessary to sum the SRO values of all constituents of the mixture and divide by the number of substances identified in the sample (NCI) [16,26]:

$$R\&M = (\Sigma\ SRO)/NCI) \quad (6)$$

5.8. Statistical Analysis

Statistical analyses were conducted using the statistical software Jamovi 2.2.5. To explore the data descriptively, measures of central tendency and dispersion were employed, providing a more precise visualization of the results.

For the final analysis, we developed an innovative graph representing the relationship between richness and evenness. For this, the data obtained from the Shannon index (x-axis) and Pielou index (y-axis) were used, representing respectively the richness and uniformity of the chemical profile of the samples. Based on these data, four distinct ranges were established for categorizing the values "very high" for values between 80% and 90%, "high" for values between 80% and 60%, "medium" for values between 60% and 50%, and "low" for values below 50%. These ranges were referenced from the literature of prominent ecologists such as Robert E. Ricklefs and Relyea Rick in "The Economy of Nature", and Peroni and Hernandez in "Ecology of Populations and Communities" [108,109].

Additionally, the Tukey post-hoc test was performed to obtain a better interpretation of the differences between the means of the compared groups. This test allows for multiple comparisons between samples, identifying which are statistically different from each other. The interpretation of the Tukey test is based on the significance values (p-value) obtained for each pairwise comparison. If the significance value is less than a certain pre-established level (usually 0.05), this indicates that there is a statistically significant difference between the compared groups [110].

Pearson correlation tests were also conducted to assess relationships between abiotic factors, volatile constituents present in different stages of reproductive organs, and frequency of visitation by potential pollinators. The correlation coefficient of this inspection can be positive (directly proportional) or negative (inversely proportional), assimilating quantitative values that suggest whether the relationship between variables is strong, moderate, or weak, as follows: 0.19 (very weak), 0.20 to 0.39 (weak), 0.40 to 0.69 (moderate), 0.70 to 0.89 (strong), and 0.90 to 1.00 (very strong)

These analyses were conducted in the R program and designed to evaluate the multivariate relationship between the different data, across different collection periods [111].

Supplementary Materials: The following supporting information can be downloaded at https://www.mdpi.com/article/10.3390/plants13172497/s1, Figure S1: Developmental stages of the reproductive organ of *Piper mollicomum* from Tijuca Forest, Rio de Janeiro city, Rio de Janeiro State. (A1–A4) Different inflorescence stages. Source: Author's own work, 2021. Table S1: Major chemical constituents of the essential oils from the leaves and stages I and II of the reproductive organ of *Piper mollicomum* Kunth during the reproductive period; Table S2: Main chemical constituents of the essential oils from stages III and IV of the reproductive organ of *Piper mollicomum* Kunth; Table S3: Chemical constituents of the essential oils from the leaves and stages I—IV of the reproductive organ of *Piper mollicomum* Kunth; Table S4: Turkey's post-hoc test for comparing the means of the Shannon index (α-chemodiversity) of leaves and reproductive organ stages of *Piper mollicomum* Kunth; Table S5: Characterization of the collection sites of *Piper mollicomum* Kunth accessions in the Tijuca National Park (TNP)—RJ.

Author Contributions: D.d.B.M., D.d.L.M. and Y.J.R. designed the survey; D.d.B.M., D.d.L.M., Y.J.R. and J.S.F. analyzed the results and contributed to the writing of the article. Y.J.R. and D.d.L.M. contributed to the methodology and statistical treatments. E.F.G. and G.A.d.Q. contributed to the descriptions and characterizations of reproductive organs. All authors have read and agreed to the published version of the manuscript.

Funding: This research was funded by Conselho Nacional de Pesquisas e Desenvolvimento Científico e Tecnológico e Inovação (CNPq)—Brazil (PROEP no. 407845/2017) and the Fundação de Amparo à Pesquisa do Estado do Rio de Janeiro (FAPERJ)—Brazil (APQ1 no. NE26/210.245/2019 and CNE no. 201.211/2022).

Data Availability Statement: Data is contained within the article and Supplementary Material.

Conflicts of Interest: The authors declare no conflict of interest.

References

1. Tsahar, E.; Friedman, J.; Izhaki, I. Impact on fruit removal and seed predation of a secondary metabolite, emodin, in *Rhamnus alaternus* fruit pulp. *Oikos* **2002**, *99*, 290–299. [CrossRef]
2. Irwin, R.E.; Adler, L.S.; Brody, A.K. The dual role of floral traits: Pollinator attraction and plant defense. *Ecology* **2004**, *85*, 1503–1511. [CrossRef]
3. Cipollini, M.L.; Paulk, E.; Mink, K.; Vaughn, K.; Fischer, T. Defense Tradeoffs in Fleshy Fruits: Effects of Resource Variation on Growth, Reproduction, and Fruit Secondary Chemistry in *Solanum carolinense*. *J. Chem. Ecol.* **2004**, *30*, 1–17. [CrossRef] [PubMed]
4. Tholl, D. Terpene synthases and the regulation, diversity and biological roles of terpene metabolism. *Curr. Opin. Plant Biol.* **2006**, *9*, 297–304. [CrossRef]
5. Cazetta, E.; Schaefer, H.M.; Galetti, M. Does attraction to frugivores or defense against pathogens shape fruit pulp composition? *Oecologia* **2007**, *155*, 277–286. [CrossRef]
6. Kessler, A.; Halitschke, R. Testing the potential for conflicting selection on floral chemical traits by pollinators and herbivores: Predictions and case study. *Funct. Ecol.* **2009**, *23*, 901–912. [CrossRef]
7. McCall, A.C.; Fordyce, J.A. Can optimal defence theory be used to predict the distribution of plant chemical defences? *J. Ecol.* **2010**, *98*, 985–992. [CrossRef]
8. Giuliani, C.; Ascrizzi, R.; Lupi, D.; Tassera, G.; Santagostini, L.; Giovanetti, M.; Flamini, G.; Fico, G. *Salvia verticillata*: Linking glandular trichomes, volatiles and pollinators. *Phytochemistry* **2018**, *155*, 53–60. [CrossRef]
9. Nazem, V.; Sabzalian, M.R.; Saeidi, G.; Rahimmalek, M. Essential oil yield and composition and secondary metabolites in self- and open-pollinated populations of mint (*Mentha* spp.). *Ind. Crops Prod.* **2019**, *130*, 332–340. [CrossRef]
10. Whitehead, S.R.; Schneider, G.F.; Dybzinski, R.; Nelson, A.S.; Gelambi, M.; Jos, E.; Beckman, N.G. Fruits, frugivores, and the evolution of phytochemical diversity. *Oikos* **2022**, *2022*, e08332. [CrossRef]
11. De Brito-Machado, D.; Ramos, Y.J.; Defaveri, A.C.A.E.; Queiroz, G.A.; Guimarães, E.F.; Moreira, D.L. Volatile Chemical Variation of Essential Oils and Their Correlation with Insects, Phenology, Ontogeny and Microclimate: *Piper mollicomum* Kunth, a Case of Study. *Plants* **2022**, *11*, 3535. [CrossRef]
12. Kessler, A.; Kalske, A. Plant secondary metabolite diversity and species interactions. *Annu. Rev. Ecol. Evol. Syst.* **2018**, *49*, 115–138. [CrossRef]
13. Harvey, A.L.; Edrada-Ebel, R.; Quinn, R.J. The reemergence of natural products for drug discovery in the genomics era. *Nat. Rev. Drug Discov.* **2015**, *14*, 111–129. [CrossRef] [PubMed]
14. Costa-Oliveira, C.; Gouvêa-Silva, J.G.; Brito-Machado, D.; Felisberto, J.R.S.; Queiroz, G.A.; Guimarães, E.F.; Ramos, Y.J.; Moreira, D.L. Chemical Diversity and Redox Values Change as a Function of Temporal Variations of the Essential Oil of a Tropical Forest Shrub. *Diversity* **2023**, *15*, 715. [CrossRef]
15. Theodoridis, S.; Drakou, E.G.; Hickler, T.; Thines, M.; Nogues-Bravo, D. Evaluating natural medicinal resources and their exposure to global change. *Lancet Planet Health* **2023**, *7*, e155–e163. [CrossRef] [PubMed]
16. Ramos, Y.J.; Gouvêa-Silva, J.G.; de Brito, M.D.; Felisberto, J.S.; Pereira, R.C.; Sadgrove, N.J.; de Lima Moreira, D. Chemophenetic and Chemodiversity Approaches: New Insights on Modern Study of Plant Secondary Metabolite Diversity at Different Spatiotemporal and Organizational Scales. *Rev. Bras. Farmacogn.* **2023**, *33*, 49–72. [CrossRef]
17. Baker, H.G.; Baker, I. Studies of nectar-constitution and pollinator-plant coevolution. In *Coevolution of Animals and Plants: Symposium V, Proceedings of the First International Congress of Systematic and Evolutionary Biology, Boulder, CO, USA, 4–12 August 1973*; Gilbert, L.E., Raven, P.H., Eds.; University of Texas Press: New York, NY, USA, 1975; pp. 100–140.
18. Detzel, A.; Wink, M. Attraction, deterrence or intoxication of bees (*Apis mellifera*) by plant allelochemicals. *Chemoecology* **1993**, *4*, 8–18. [CrossRef]

19. Mačukanović-Jocić, M.; Zora, D.S.; Mića, M.; Gojko, J. Flower morphophysiology of selected Lamiaceae species in relation to pollinator attraction. *J. Apic. Res.* **2011**, *50*, 89–101. [CrossRef]
20. Milet-Pinheiro, P.; Daniela, M.A.F.; Stefan, D.; Airton, T.C.; Carlos, E.P.; Manfred, A.; Clemens, S. Pollination biology in the dioecious orchid *Catasetum uncatum*: How does floral scent influence the behaviour of pollinators? *Phytochemistry* **2015**, *116*, 149–161. [CrossRef]
21. Tölke, E.D.; Capelli, N.d.V.; Pastori, T.; Alencar, A.C.; Cole, T.C.H.; Demarco, D. Diversity of Floral Glands and Their Secretions in Pollinator Attraction. In *Co-Evolution of Secondary Metabolites*; Mérillon, J.M., Ramawat, K., Eds.; Reference Series in Phytochemistry; Springer: Cham, Switzerland, 2020.
22. Regnault-Roger, C. The potential of botanical essential oils for insect pest control. *Integr. Pest Manag. Rev.* **1997**, *2*, 25–34. [CrossRef]
23. Wheeler, D.A.; Isman, M.B. Antifeedant and toxic activity of *Trichilia americana* extract against the larvae of *Spodoptera litura*. *Entomol. Exp. Appl.* **2001**, *98*, 9–16. [CrossRef]
24. Barros, F.; Zambarda, E.; Heinzmann, B. Variabilidade sazonal e biossíntese de terpenoides presentes no óleo essencial de *Lippia alba* (Mill.) n.e. brown (Verbenaceae). *Quim. Nova* **2009**, *32*, 861–867. [CrossRef]
25. Togashi, K.; Goto, M.; Rim, H.; Hattori, S.; Ozawa, R.; Arimura, G. Mint companion plants attract the predatory mite *Phytoseiulus persimilis*. *Sci. Rep.* **2019**, *9*, 1704. [CrossRef] [PubMed]
26. Ramos, Y.J.; Costa-Oliveira, C.D.; Candido-Fonseca, I.; Queiroz, G.A.D.; Guimarães, E.F.; Defaveri, A.C.A.; Sadgrove, N.J.; Moreira, D.L. Advanced Chemophenetic Analysis of Essential Oil from Leaves of *Piper gaudichaudianum* Kunth (Piperaceae) Using a New Reduction-Oxidation Index to Explore Seasonal and Circadian Rhythms. *Plants* **2021**, *10*, 2116. [CrossRef] [PubMed]
27. Machado, D.B.; Ramos, Y.J.; Queiroz, G.A.; Defaveri, A.C.A.; Gobatto, A.A.; Moreira, D.L. Study of volatile chemical constituents and insect-plant interaction in *Piper mollicomum* Kunth (Piperaceae) from Tijuca Forest, Rio de Janeiro—RJ, Brazil. *Rev. Virtual Quím.* **2021**, *13*, 1216–1225. [CrossRef]
28. Queiroz, G.A.; Guimarães, E.F. *Piper* L. (Piperaceae) Do Leste Metropolitano, RJ, Brasil/*Piper* L. (Piperaceae) from Eastern Metropolitan, RJ, Brazil. *Braz. J. Dev.* **2020**, *6*, 93597–93634. [CrossRef]
29. Flora do Brasil. Jardim Botânico do Rio de Janeiro. 2020. Available online: http://floradobrasil.jbrj.gov.br/ (accessed on 7 April 2022).
30. Ramos, Y.J.; Machado, D.d.B.; de Queiroz, G.A.; Guimarães, E.F.; Defaveri, A.C.A.; Moreira, D.L. Chemical composition of the essential oils of circadian rhythm and of different vegetative parts from *Piper mollicomum* Kunth—A medicinal plant from Brazil. *Biochem. Syst. Ecol.* **2020**, *92*, 104116. [CrossRef]
31. Lago, J.H.G.; Ito, A.T.; Fernandes, C.M.; Young, M.C.M.; Kato, M.J. Secondary metabolites isolated from *Piper chimonantifolium* and their antifungal activity. *Nat. Prod. Res.* **2011**, *26*, 770–773. [CrossRef]
32. Potrich, F.B.; Baggio, C.H.; Freitas, C.S.; Mayer, B.; Santos, A.C.; Twardowschy, A.; Guedes, A.; Marques, M.C.A. Ação de extratos de plantas medicinais sobre a motilidade do trato gastrointestinal. *Rev. Bras. Plantas Med.* **2014**, *16*, 750–754. [CrossRef]
33. Stefnia, P.S.; Simone, S.V.; Nathalia, F.C.; Andrea, S.C.; Keila, S.C.L.; Valber, S.F.; Antonio, L.S.L. Chemical composition and antinociceptive activity of the essential oil of *Piper mollicomum* and *Piper rivinoides*. *J. Med. Plants Res.* **2014**, *8*, 788–793. [CrossRef]
34. Messias, M.C.T.B.; Menegatto, M.F.; Prado, A.C.C.; Santos, B.R.; Guimarães, M.F.M. Popular use of medicinal plants and the socioeconomic profile of the users: A study in the urban area of Ouro Preto, Minas Gerais, Brazil. *Rev. Bras. Plantas Med.* **2015**, *17*, 76–104. [CrossRef]
35. Thies, W.; Kalko, E.K.V. Phenology of neotropical pepper plants (Piperaceae) and their association with their main dispersers, two short-tailed fruit bats, *Carollia perspicillata* and *C. castanea* (Phyllostomidae). *Oikos* **2004**, *104*, 362–376. [CrossRef]
36. Maynard, L.D.; Ananda, A.; Sides, M.F.; Burk, H.; Whitehead, S.R. Dietary resource overlap among three species of frugivorous bat in Costa Rica. *J. Trop. Ecol.* **2019**, *35*, 165–172. [CrossRef]
37. Vargas, R.D.L.; Vieira, M.F. Sex expression, breeding system and pollinators of *Piper caldense* (piperaceae) in the brazilian atlantic forest. *Acta Biol. Colomb.* **2017**, *22*, 370–376.
38. Moore, B.D.; Andrew, R.L.; Külheim, C.; Foley, W.J. Explaining intraspecific diversity in plant secondary metabolites in an ecological context. *New Phytol* **2014**, *201*, 733–750. [CrossRef]
39. Meena, R.K.; Jangra, S.; Wadhwa, Z.; Leela-Wati, M. Role of Plant Volatiles in Defense and Communication. *Int. J. Curr. Microbiol. Appl. Sci.* **2017**, *6*, 300–313. [CrossRef]
40. Zheng, L.; Jingrui, L.I.; Yan, D.; Wen, Z.; Hongtong, B.; Shu, L.; Su, W.; Hui, L.; Lei, S. Gene co-expression modulating terpene metabolism is associated with plant anti-herbivore defence during initial flowering stages. *Authorea* **2020**. [CrossRef]
41. Fujita, Y.; Koeduka, T.; Aida, M.; Suzuki, H.; Iijima, Y.; Matsui, K. Biosynthesis of volatile terpenes that accumulate in the secretory cavities of young leaves of Japanese pepper (*Zanthoxylum piperitum*): Isolation and functional characterization of monoterpene and sesquiterpene synthase genes. *Plant Biotechnol.* **2017**, *34*, 17–28. [CrossRef]
42. Zhu, J.; Park, K.-C. Methyl Salicylate, a Soybean Aphid-Induced Plant Volatile Attractive to the Predator *Coccinella septempunctata*. *J. Chem. Ecol.* **2005**, *31*, 1733–1746. [CrossRef]
43. Picard, I.; Hollingsworth, R.G.; Salmieri, S.; Lacroix, M. Repellency of Essential Oils to *Frankliniella occidentalis* (Thysanoptera: Thripidae) as Affected by Type of Oil and Polymer Release. *J. Econ. Entomol.* **2012**, *105*, 1238–1247. [CrossRef]
44. Smith, G.H.; Roberts, J.M.; Pope, T.W. Terpene based biopesticides as potential alternatives to synthetic insecticides for control of aphid pests on protected ornamentals. *Crop Prot.* **2018**, *110*, 125–130. [CrossRef]

45. Reisenman, C.E.; Riffell, J.A.; Bernays, E.A.; Hildebrand, J.G. Antagonistic effects of floral scent in an insect–plant interaction. *Proc. R. Soc. B Biol. Sci.* **2010**, *277*, 2371–2379. [CrossRef] [PubMed]
46. Glinwood, R.; Blande, J.D. Deciphering Chemical Language of Plant Communication: Synthesis and Future Research Directions. *Signal. Commun. Plants* **2016**, *1*, 319–326.
47. Ma, Q.; Ma, R.; Su, P.; Jin, B.; Guo, J.; Tang, J.; Chen, T.; Zeng, W.; Lai, C.; Ling, F.; et al. Elucidation of the essential oil biosynthetic pathways in *Cinnamomum burmannii* through identification of six terpene synthases. *Plant Sci.* **2022**, *317*, 111203. [CrossRef]
48. Bergström, G.; Tengö, J. Linalool in mandibular gland secretion of *Colletes* bees (Hymenoptera: Apoidea). *J. Chem. Ecol.* **1978**, *4*, 437–449. [CrossRef]
49. Wang, Q.; Hillwig, M.L.; Okada, K.; Yamazaki, K.; Wu, Y.; Swaminathan, S.; Yamane, H.; Peters, R.J. Characterization of CYP76M5–8 Indicates Metabolic Plasticity within a Plant Biosynthetic Gene Cluster. *J. Biol. Chem.* **2012**, *287*, 6159–6168. [CrossRef] [PubMed]
50. Wu, Y.; Wang, Q.; Hillwig, M.L.; Peters, R.J. Picking sides: Distinct roles for CYP76M6 and CYP76M8 in rice oryzalexin biosynthesis. *Biochem. J.* **2013**, *454*, 209–216. [CrossRef] [PubMed]
51. Boachon, B.; Junker, R.R.; Miesch, L.; Bassard, J.-E.; Höfer, R.; Caillieaudeaux, R.; Seidel, D.E.; Lesot, A.; Heinrich, C.; Ginglinger, J.F.; et al. CYP76C1 (Cytochrome P450)-Mediated Linalool Metabolism and the Formation of Volatile and Soluble Linalool Oxides in Arabidopsis Flowers: A Strategy for Defense against Floral Antagonists. *Plant Cell* **2015**, *27*, 2972–2990. [CrossRef]
52. Shilpashree, H.B.; Sudharshan, S.J.; Shasany, A.K.; Nagegowda, D.A. Molecular characterization of three CYP450 genes reveals their role in withanolides formation and defense in *Withania somnifera*, the Indian Ginseng. *Sci. Rep.* **2022**, *12*, 1602. [CrossRef]
53. Miehe-Steier, A.; Roscher, C.; Reichelt, M.; Gershenzon, J.; Unsicker, S.B. Light and Nutrient Dependent Responses in Secondary Metabolites of Plantago lanceolata Offspring Are Due to Phenotypic Plasticity in Experimental Grasslands. *PLoS ONE* **2015**, *10*, e0136073. [CrossRef]
54. Nafea, H.M.; Kawaz, A.M.N.A. Synthesis, Characterization, Antimicrobial, DNA Cleavage and Fluorescent Activity of Metal ion (II) Coordinate with 2H-Chromene Azo Novel Ligand. 2020. Available online: https://ijrps.com/home/article/view/1038 (accessed on 19 April 2022).
55. Moraes, M.M.; Kato, M.J. Biosynthesis of Pellucidin A in *Peperomia pellucida* (L.) HBK. *Front. Plant Sci.* **2021**, *12*, 641717. [CrossRef] [PubMed]
56. Gaia, A.M.; Yamaguchi, L.F.; Guerrero-Perilla, C.; Kato, M.J. Ontogenetic Changes in the Chemical Profiles of *Piper* Species. *Plants* **2021**, *10*, 1085. [CrossRef]
57. Bianconi, G.V.; Mikich, S.B.; Teixeira, S.D.; Maia, B.H.L.N.S. Attraction of Fruit-Eating Bats with Essential Oils of Fruits: A Potential Tool for Forest Restoration. *Biotropica* **2007**, *39*, 136–140. [CrossRef]
58. Bianconi, G.V.; Suckow, U.M.S.; Cruz-Neto, A.P.; Mikich, S.B. Use of Fruit Essential Oils to Assist Forest Regeneration by Bats. *Restor. Ecol.* **2010**, *20*, 211–217. [CrossRef]
59. Wiggins, N.L.; McArthur, C.; McLean, S.; Boyle, R. Effects of Two Plant Secondary Metabolites, Cineole and Gallic Acid, on Nightly Feeding Patterns of the Common Brushtail Possum. *J. Chem. Ecol.* **2003**, *29*, 1447–1464. [CrossRef] [PubMed]
60. Marsh, K.J.; Wallis, I.R.; Mclean, S.; Sorensen, J.S.; Foley, W.J. Conflicting demands on detoxification pathways influence how common brushtail possums choose their diets. *Ecology* **2006**, *87*, 2103–2112. [CrossRef] [PubMed]
61. Yearsley, J.M.; Villalba, J.J.; Gordon, I.J.; Kyriazakis, I.; Speakman, J.R.; Tolkamp, B.J.; Illius, A.W.; Duncan, A.J. A Theory of Associating Food Types with Their Postingestive Consequences. *Am. Nat.* **2006**, *167*, 705–716. [CrossRef]
62. Dirk, S.; Maik, K. Influencing the product quality by deliberately applying drought stress during the cultivation of medicinal plants. *Ind. Crops Prod.* **2013**, *42*, 558–566.
63. De Macêdo, D.; Souza, M.M.A.; Morais-Braga, M.F.B.; Coutinho, H.D.M.; Dos Santos, A.T.L.; da Cruz, R.P.; da Costa, J.G.M.; Rodrigues, F.F.G.; Quintans-Junior, L.J.; da Silva, A.J.R.G.; et al. Effect of seasonality on chemical profile and antifungal activity of essential oil isolated from leaves *Psidium salutare* (Kunth). *PeerJ–Life Environ.* **2018**, *6*, e5476. [CrossRef]
64. Sikron-Persi, N.; Granot, G.; Batushansky, A.; Toubiana, D.; Grafi, G.; Fait, A. Mass spectrometry-based metabolite profiling reveals functional seasonal shifts in the metabolome of Zygophyllum dumosum Boiss and its relation to environmental conditions. *Planta* **2023**, *258*, 10. [CrossRef]
65. Gottlieb, O.R. The role of oxygen in phytochemical evolution towards diversity. *Phytochemistry* **1989**, *28*, 2545–2558. [CrossRef]
66. Gottlieb, O.R.; Kaplan, M.A.C. Phytochemical Evolution: The Redox Theory. *Nat. Prod. Lett.* **1993**, *2*, 171–176. [CrossRef]
67. Gil, M.; Bottini, R.; Berli, F.; Pontin, M.; Silva, M.F.; Piccoli, P. Volatile organic compounds characterized from grapevine (*Vitis vinifera* L. cv. Malbec) berries increase at pre-harvest and in response to UV-B radiation. *Phytochemistry* **2013**, *96*, 148–157. [CrossRef] [PubMed]
68. Schrader, W.; Geiger, J.; Klockow, D.; Korte, E.H. Degradation of α-Pinene on Tenax during Sample Storage: Effects of Daylight Radiation and Temperature. *Environ. Sci. Technol.* **2001**, *35*, 2717–2720. [CrossRef] [PubMed]
69. Vieira, E.A.; Arruda, R.; Massuda, K.F.; Cardoso-Gustavson, P.; Guimarães, E.F.; Trigo, J.R. Volatiles released by damaged leaves of *Piper mollicomum* (Piperaceae) act as cues for predaceous wasps: Evidence using plasticine dummies as herbivore model. *Arthropod-Plant Interact.* **2019**, *13*, 593–601. [CrossRef]
70. Benchaa, S.; Hazzit, M.; Abdelkrim, H. Allelopathic Effect of Eucalyptus citriodora Essential Oil and Its Potential Use as Bioherbicide. *Chem. Biodivers.* **2018**, *15*, e1800202. [CrossRef]

71. Zhang, T.; Zheng, Y.; Fu, C.; Yang, H.; Liu, X.; Qiu, F.; Wang, X.; Wang, Z. Chemical Variation and Environmental Influence on Essential Oil of *Cinnamomum camphora*. *Molecules* **2023**, *28*, 973. [CrossRef]
72. Presto, A.A.; Huff-Hartz, K.E.; Donahue, N.M. Secondary Organic Aerosol Production from Terpene Ozonolysis. Effect of UV Radiation. *Environ. Sci. Technol.* **2005**, *39*, 7036–7045. [CrossRef]
73. Babar, Z.B.; Park, J.H.; Lim, H.J. Influence of NH_3 on secondary organic aerosols from the ozonolysis and photooxidation of α-pinene in a flow reactor. *Atmos. Environ.* **2017**, *164*, 71–84. [CrossRef]
74. Muhlemann, J.K.; Klempien, A.; Dudareva, N. Floral volatiles. *Plant Cell Environ.* **2014**, *37*, 1936–1949. [CrossRef]
75. Beran, F.; Köllner, T.G.; Gershenzon, J.; Tholl, D. Chemical convergence between plants and insects: Biosynthetic origins and functions of common secondary metabolites. *New Phytol.* **2019**, *223*, 52–67. [CrossRef] [PubMed]
76. Ramya, M.; Jang, S.; An, H.-R.; Lee, S.-Y.; Park, P.-M.; Park, P.H. Volatile Organic Compounds from Orchids: From Synthesis and Function to Gene Regulation. *Int. J. Mol. Sci.* **2020**, *21*, 1160. [CrossRef] [PubMed]
77. Valentin-Silva, A.; Coelho, V.P.M.; Ventrella, M.C.; Vieira, M.F. Timing of pollen release and stigma receptivity period of *Piper vicosanum*: New insights into sexual reproduction of the genus. *Am. J. Bot.* **2015**, *102*, 626–633. [CrossRef] [PubMed]
78. Fleming, T.H. Fecundity, fruiting pattern, and seed dispersal in *Piper amalago* (Piperaceae), a bat-dispersed tropical shrub. *Oecologia* **1981**, *51*, 42–46. [CrossRef] [PubMed]
79. Piper, J.K. Seasonality of Fruit Characters and Seed Removal by Birds. *Oikos* **1986**, *46*, 303–310. [CrossRef]
80. Rodríguez, A.; Alquézar, B.; Peña, L. Fruit aromas in mature fleshy fruits as signals of readiness for predation and seed dispersal. *New Phytol.* **2012**, *197*, 36–48. [CrossRef]
81. Rocha, V.J.; Barbosa, G.P.; Rossi, H.R.S.; Sekiama, M.L. Chiropteran richness and diversity (Chiroptera; Mammalia) in Permanent Preservation Areas from UFSCar-Araras campus (SP). *Ciência Tecnol. Ambiente* **2018**, *8*, 21–29. [CrossRef]
82. Maynard, L.D.; Slinn, H.L.; Glassmire, A.E.; Matarrita-Carranza, B.; Dodson, C.D.; Nguyen, T.T.; Whitehead, S.R. Secondary metabolites in a neotropical shrub: Spatiotemporal allocation and role in fruit defense and dispersal. *Ecology* **2020**, *101*, e03192. [CrossRef]
83. Silber, A.; Goldberg, T.; Shapira, O.; Hochberg, U. Nitrogen up take and Macronutrients Distribution in Mango (*Mangifera indica* L. Cv. Keitt) Trees. *Plant Physiol. Biochem* **2022**, *181*, 23–32. [CrossRef]
84. Simpson, E.H. Measurement of diversity. *Nature* **1949**, *163*, 688. [CrossRef]
85. Ebadollahi, A. Estragole-rich essential oil of summer savory (*Satureja hortensis* L.) as an eco-friendly alternative to the synthetic insecticides in management of two stored-products insect pests. *Acta Agric. Slov.* **2020**, *115*, 307–314. [CrossRef]
86. Amaral, W.; Deschamps, C.; Biasi, A.L.; Bizzo, R.H.; Machado, M.P.; Silva, L.E. Yield and chemical composition of the essential oil of species of the *Asteraceae* family from Atlantic Forest, South of Brazil. *J. Essent. Oil Res.* **2018**, *30*, 278–284. [CrossRef]
87. Schiestl, F.P. Ecology and evolution of floral volatile-mediated information transfer in plants. *New Phytol.* **2015**, *206*, 571–577. [CrossRef] [PubMed]
88. Ramos, Y.J. Identificação de Possíveis Quimiotipos de *Piper aduncum* L. e *Piper mollicomum* Kunth (Piperaceae) com Base no Estudo dos Componentes Químicos de Óleos Essenciais. Master's Thesis, Universidade do Estado do Rio de Janeiro, Rio de Janeiro, Brazil, 2018; 246p. Available online: https://www.bdtd.uerj.br:8443/handle/1/17559 (accessed on 29 April 2024).
89. Kawano, T. Roles of the reactive oxygen species-generating peroxidase reactions in plant defense and growth induction. *Plant Cell Rep.* **2003**, *21*, 829–837. [CrossRef] [PubMed]
90. Koppen, W.; Geiger, G.C. *Das geographisca System der Klimate, Handbuch der Klimatologie*; Borntraeger: Berlin, Germany, 1936; pp. 1–44.
91. Sakagami, S.F.; Laroca, S.; Moure, J.S. Wild bee biocoenotics in São José dos Pinhais (PR), South Brazil. *J. Fac. Sci. Hokkaido Univ. Ser. VI Zool.* **1967**, *16*, 253–291.
92. Campos, G.P.A.; Barros, C.T.; Carneiro, L.T.; Santa-Martinez, E.; Oliveira, M.M.; Castro, C.C. Pollinator efficiency in openly grown eggplants: Can non-vibrating bees produce high-quality fruits? *Arthropod-Plant Interact.* **2022**, *16*, 159–170. [CrossRef]
93. Polizel, A.L.; Nanka, S.; Conte, H. Insetos dípteras como polinizadores em Orchidaceae. *Rev. Uningá* **2015**, *46*, 11–15.
94. Valentin-Silva, A.; Staggemeier, V.G.; Batalha, M.A.; Guimarães, E. What factors can influence the reproductive phenology of Neotropical *Piper* species (Piperaceae) in a semi-deciduous seasonal forest? *Botany* **2018**, *96*, 675–684. [CrossRef]
95. Fournier, L.A. Un método cuantitativo para la medición de características fenológicas en árboles. *Turrialba* **1974**, *24*, 422–423.
96. Adams, R.P. *Identification of Essential Oil Components by Gas Chromatography/Mass Spectroscopy*, 4th ed.; Allured Business: Carol Stream, IL, USA, 2009.
97. Jaccard, P. Distribution de la flore alpine dans le bassin des dranses et dans quelques régions voisines. *Bull. Société Vaudoise Sci. Nat.* **1901**, *37*, 241–272.
98. Sørensen, T. *A Method of Establishing Groups of Equal Amplitude in Plant Sociology Based on Similarity of Species Content and Its Application to Analyses of the Vegetation on Danish Commons*; Biologiske Skrifter/Kongelige Danske Videnskabernes Selskab: Copenhagen, Denmark, 1948; pp. 1–34.
99. Shannon, C.E. A mathematical theory of communication. *Bell Syst. Tech. J.* **1948**, *27*, 379–423. [CrossRef]
100. Pielou, E. Species-diversity and pattern-diversity in the study of ecological succession. *J. Theor. Biol.* **1966**, *10*, 370–383. [CrossRef] [PubMed]
101. Cody, M.L. Towards a theory of continental species diversities. In *Ecology and Evolution of Communities*; Cody, M.L., MacArthur, R.H., Diamond, J.M., Eds.; Harvard University Press: Cambridge, MA, USA, 1975.

102. Gouyon, P.H.; Vernet, P.H.; Guillerm, J.L.; Valdeyron, G. Polymorphisms and environment: The adaptive value of the oil polymorphisms in *Thymus vulgaris* L. *Heredity* **1986**, *57*, 59–66. [CrossRef]
103. Iason, G.R.; Lennon, J.J.; Pakeman, R.J.; Thoss, V.; Beaton, J.K.; Sim, D.A.; Elston, D.A. Does chemical composition of individual scots pine trees determine the biodiversity of their associated ground vegetation? *Ecol. Lett.* **2005**, *8*, 364–369. [CrossRef]
104. Salazar, D.; Jaramillo, A.; Marquis, R.J. The impact of plant chemical diversity on plant–herbivore interactions at the community level. *Oecologia* **2016**, *181*, 1199–1208. [CrossRef]
105. Feng, X.; Zhang, W.; Wu, W.; Bai, R.; Kuang, S.; Shi, B.; Li, D. Chemical composition and diversity of the essential oils of Juniperus rigida along the elevations in Helan and Changbai Mountains and correlation with the soil characteristics. *Ind. Crops Prod.* **2021**, *159*, 113032. [CrossRef]
106. Hendrickson, J.B.; Cram, D.J.; Hammond, G.S. *Organic Chemistry*, 3rd ed.; McGraw-Hill: New York, NY, USA, 1970.
107. Emerenciano, V.P.; Cabrol-Bass, D.; Ferreira, M.J.; Alvarenga, S.A.; Brant, A.J.; Scotti, M.T.; Barbosa, K.O. Chemical evolution in the asteraceae. The oxidation-reduction mechanism and production of secondary metabolites. *Nat. Prod. Commun.* **2006**, *1*, 495–507. [CrossRef]
108. Peroni, N.; Hernández, M.I.M. *Ecologia de Populações e Comunidades*, 1st ed.; CCB/EAD/UFSC: Florianópolis, Brazil, 2011; ISBN 978-85-61485-39-9.
109. Ricklefs, R.; Relyea, R. *Economia da Natureza*, 7th ed.; Guanabara Koogan: Rio de Janeiro, Brazil, 2018; ISBN 9788527728768.
110. Kevser, M.; Doğan, M. Islamic Financial Literacy and Its Determinants: A Field Study on Turkey. *J. Transit. Stud. Rev.* **2021**, *28*, 91–120.
111. Ainsley, M.; Kate, P. Anxiety around Learning R in First Year Undergraduate Students: Mathematics versus Biomedical Sciences Students. *J. Stat. Data Sci. Educ.* **2024**, *32*, 47–53.

plants

MDPI

Article

Impact of Cassava Cultivars on Stylet Penetration Behavior and Settling of *Bemisia tabaci* Gennadius (Hemiptera: Aleyrodidae)

Sudarat Pimkornburee [1], Supawadee Pombud [1], Kumri Buensanteai [1], Weravart Namanusart [2], Sukanya Aiamla-or [1] and Jariya Roddee [1,*]

[1] School of Crop Production Technology, Institute of Agricultural Technology, Suranaree University of Technology, Nakhon Ratchasima 30000, Thailand; m6500016@g.sut.ac.th (S.P.); supawadeepb@gmail.com (S.P.); kumrai@sut.ac.th (K.B.); sukanya.aia@sut.ac.th (S.A.-o.)

[2] Department of Plant Science, Faculty of Agricultural Innovation and Technology, Raja Mangala University of Technology Isan, Nakhon Ratchasima 30000, Thailand; weravart.na@rmuti.ac.th

* Correspondence: jariyaroddee@g.sut.ac.th

Abstract: This study investigates the settling preferences and feeding behavior of the *Bemisia tabaci* whitefly on six cassava cultivars using electrical penetration graph techniques. Six distinct electrical penetration graph waveforms—non-probing, stylet pathway, phloem salivation, phloem ingestion, intracellular puncture, and xylem feeding—were identified and analyzed. Significant differences in the frequency and duration of these waveforms were observed among the cassava cultivars. The whiteflies spent the majority of their time in the non-probing phase, particularly on the Huaybong 80, Kasetsart 50, Rayong 9, and Rayong 72 cultivars. CMR-89 cultivar exhibited higher total probe durations in the phloem salivation and ingestion waveforms, suggesting a greater potential for transmission of the Sri Lankan cassava mosaic virus. The study also examined trichome density and size across the cassava cultivars, revealing that CMR-89 had the highest density and small trichomes, while Huaybong 80 had the lowest density. Trichome characteristics significantly impacted whitefly behavior: larger trichomes were negatively correlated with whitefly settling, whereas higher trichome density was positively correlated with longer settling durations. These findings indicate that trichome-based resistance mechanisms are crucial in whitefly deterrence. Overall, the results suggest that cultivars with lower trichome density and larger trichomes are more resistant to whitefly infestation and subsequent Sri Lankan cassava mosaic virus transmission. These insights are valuable for cassava breeding programs focused on enhancing pest resistance, highlighting the importance of trichome characteristics in developing more resilient cassava varieties.

Keywords: *Bemisia tabaci*; cassava mosaic virus; electrical penetration graph; host preference; trichome

1. Introduction

Cassava (*Manihot esculenta* Crantz, family Euphorbiaceae) is an important root crop in Africa, Asia, and South America. It provides a livelihood to over 500 million people and is among the most important food staples worldwide [1]. Cassava is the world's third-largest source of carbohydrates for human food and raw material for starch-based industries. However, its cultivation faces significant challenges from various pests and diseases, including over 20 viral diseases associated with vegetative propagation. Another factor increasingly affecting its production is cassava mosaic disease (CMD), resulting in losses estimated at USD 1.6 billion [2–4]. CMD, caused by cassava mosaic geminiviruses (family Geminiviridae: genus Begomovirus), is a major threat to global agriculture. Cassava mosaic geminiviruses are one of the top 10 viruses that affect economically important crops, significantly influencing cassava production in Africa and the Indian subcontinent [3,5–7]. The devastating impact of CMD was first observed in East Africa in 1894, and the disease has since spread across Africa, India, and Sri Lanka, largely attributed to geminiviruses and their vector, *Bemisia tabaci* [8–10]. The Sri Lankan cassava mosaic virus (SLCMV) was identified in Sri Lanka and linked to CMD in India [9].

The whitefly species predominantly spread CMD, particularly *B. tabaci* (Gennadius) (Hemiptera: Aleyrodidae). The *B. tabaci* whitefly is known to vector cassava mosaic begomoviruses and cassava brown streak ipomovirus, the causative agents of CMD, the cassava brown streak disease, and the SLCMV [7,9,11]. *B. tabaci*, the only vector of cassava geminiviruses, is spread predominantly by virus-infected cuttings. In Africa, these vectors have been shown to reduce cassava yields by significantly 35–60%. While nine CMV species have been reported in Africa and the Indian Ocean islands, only two are found in Asia: Indian cassava mosaic virus and the SLCMV, with the latter reported exclusively in Southeast Asia. The SLCMV has caused devastating yield losses in Thailand, with reductions of up to 80% [12–14].

Whiteflies, specifically *B. tabaci*, are tiny pests commonly found in tropical and subtropical areas [15]. These sap-sucking insects act as vectors for viruses, significantly harming various host plants. They significantly threaten cassava production in tropical regions [16]. Besides transmitting viruses, whiteflies directly harm cassava plants by feeding on the phloem of their leaves, resulting in leaf discoloration, shedding, and a potential yield reduction of up to 50% in vulnerable varieties [17]. Furthermore, the sugary substance (honeydew) excreted by whiteflies promotes the growth of a sooty mold, further impeding the photosynthetic capability of cassava plants.

CMD can be managed through the multiplication and distribution of disease-free stem cuttings. The main management for the SLCMV is host plant resistance against viruses through CMD used throughout and introduced to Thailand. However, producing cassava resistant to the SLCMV via conventional breeding methods is challenging due to the high heterozygosity and inbreeding depression in elite cultivars [18]. Additionally, the excessive use of chemical pesticides for whitefly control harms the ecosystem and raises production costs, proving uneconomical for small-scale farmers [19]. Consequently, developing host plant resistance to insect vectors emerges as one of the most effective strategies for controlling vector-borne viral diseases [20]. The successful transmission of plant viruses by insect vectors hinges on their behavior and dispersal capabilities. Whiteflies are particularly pivotal in virus transmission, making their host plant selection process critically important. Insect herbivores rely on sensory mechanisms, such as olfactory receptors for plant volatiles and gustatory and mechanoreceptive sensillae for feeding and oviposition, to identify suitable host plants [21–23]. Plant defenses, including trichome density and leaf surface characteristics, play essential roles in repelling pests [24–26]. As hemipteran vectors of plant viruses, whiteflies adopt steps to locate and identify suitable host plants for settling and feeding, engaging in sustained phloem-sap ingestion once they deem the host plant appropriate [27].

Whitefly-stylet activities—particularly penetration into phloem sieve elements—are closely associated with transmitting phloem-restricted persistent viruses. Most hemipterans prefer settling on the abaxial side of the leaf upon landing [28]. For instance, *Myzus persicae* aphids tend to settle on the abaxial leaf surface, and so do *B. tabaci* crawlers when given a choice [29]. After touching the plant surface, *B. tabaci* evaluates host plant quality through labial dabbing and probing with its piercing mouthparts. The feeding behavior of piercing–sucking insects such as whiteflies can be closely monitored using the electrical penetration graph (EPG) technique. This method, pioneered by McLean et al., [30] and refined by Tjallingii [31,32], has proven instrumental in understanding the feeding behavior of whiteflies such as *Trialeurodes vaporariorum* and *B. tabaci* [33–40]. The waveform patterns of *B. tabaci* were categorized by amplitude, relative voltage level, R/emf origin, frequency, and the context of the waveform as non-probing (Np), stylet pathway (C), phloem salivation (E1), phloem ingestion (E2), intracellular puncture—potential drop (Pd), and xylem feeding (G). The EPG technique involves creating an electric circuit that includes the insect and the plant, where voltage fluctuations are associated with specific stylet activities. This technique helps pinpoint the tissues in which plant resistance factors operate [41].

Understanding whitefly preferences for feeding on different cassava cultivars is crucial for developing resistant varieties and mitigating the impact of viral diseases on cassava production. This study examines the feeding behavior and settling preferences of *B. tabaci* across diverse cassava cultivars and evaluates whitefly spread among these cultivars. The insights gained will contribute to developing whitefly-resistant cassava varieties, ultimately bolstering cassava production and mitigating the impact of viral diseases on agricultural livelihoods.

2. Results

2.1. EPG Waveform Characteristics and Feeding Behavior of Bemisia Tabaci Whiteflies on Different Cassava Cultivars

2.1.1. EPG Waveform Characteristics

The six probing waveforms identified in this study were Np, C, E1, E2, Pd, and G, which provide a comprehensive overview of whitefly feeding behavior. The waveform patterns, types, and characteristics are depicted in Figure 1. These waveform characteristics align with the findings of previous studies and reaffirm the complexity of whitefly interactions with host plants [36–38,40,42–45].

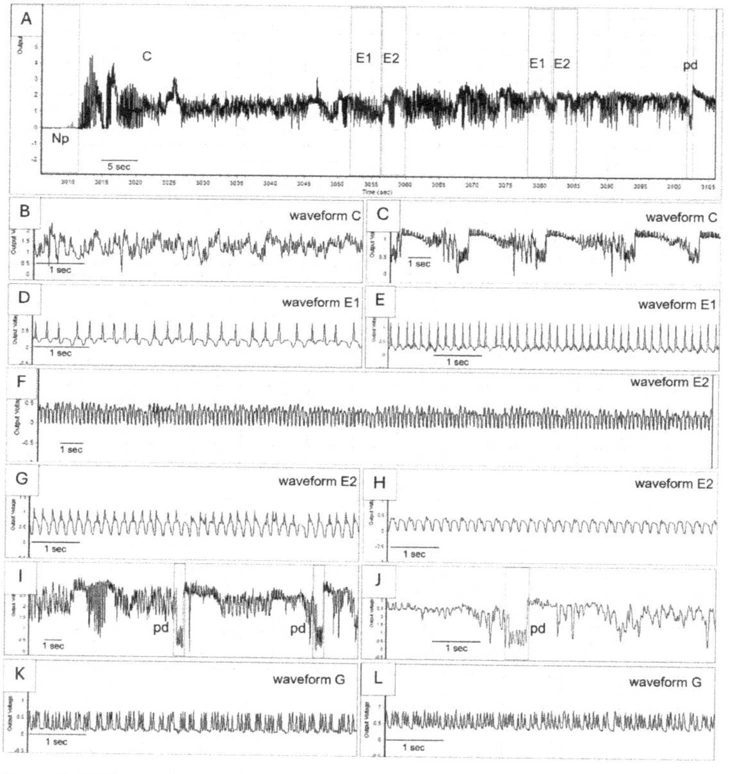

Figure 1. EPG waveforms of *Bemisia tabaci* whitefly were observed in the study. (**A**) The overview of EPG waveforms and details for 100 s with non-probing (Np), stylet pathway (**C**), phloem salivation (E1), phloem ingestion (E2), and potential drop (Pd) waveforms. Waveform C, (**B**,**C**). Waveform E1 (**D**,**E**). Waveform E2 (**F**–**H**). Waveform pd (**I**,**J**). Waveform G (**K**,**L**).

2.1.2. Feeding Behavior of *Bemisia tabaci* Whiteflies on Different Cassava Cultivars

The feeding behavior of *B. tabaci* on six cassava cultivars was analyzed using the EPG technique. The frequency and duration of the six probing waveforms—Np, C, E1, E2, Pd, and G—were compared among the cultivars.

The TPD and TWD of adult whiteflies were calculated and compared between the six cassava cultivars (Figure 2). The data indicated that the adult whiteflies spent most of their time in the Np phase on the cassava leaf tissue, followed by the E1 waveform. Notably, the percentage of time spent in the Np phase was significantly longer on Huaybong 80, Kasetsart 50, Rayong 9, and Rayong 72 cultivars, averaging around 70%, compared to Rayong 5 and CMR 89 cultivars. The TPD for adult whiteflies revealed that the whiteflies spent more than 50% of the recording time on CMR 89 cultivars in probing. Additionally, CMR 89 showed higher total probe durations in the E1 and E2 waveforms (Figure 2).

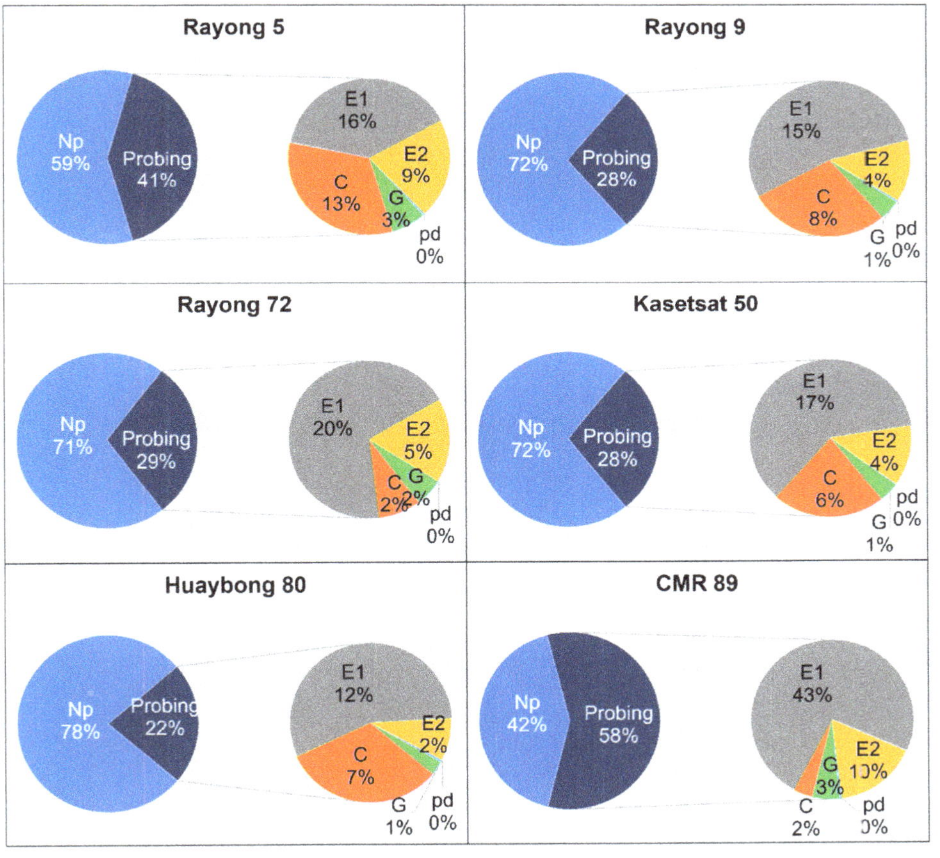

Figure 2. The total waveform duration (TWD) and the total probe duration (TPD) of adult *B. tabaci* whiteflies on different cassava cultivars during 3 h recording. Np, C, E1, E2, Pd waveform, and G waveform. Np: non-probing, waveform C: stylet pathway, waveform E1: phloem salivation, waveform E2: phloem ingestion, waveform Pd: intracellular puncture—potential drop, and waveform G: xylem feeding.

2.1.3. Number of Waveform Events per Insect (NWEI)

The NWEI for the Pd waveform did not show significant differences among the six cassava cultivars. However, the number of Np (F5,182 = 1.504, p = 0.023), C (F5,276 = 1.458, p = 0.024), E1 (F5,287 = 1.530, p = 0.022), E2 (F5,195 = 1.640, p = 0.057), and G (F5,105 = 1.788, p = 0.018) waveform events per insect revealed significant differences among the six cassava cultivars. Adult *B. tabaci* whiteflies spent more time on the E1 (phloem salivation) waveform in Huaybong 80, Rayong 9, and CMR-89 cultivars. They also spent more time on the E2 (phloem ingestion) waveform in Rayong 5, Huaybong 80, CMR-89, and Rayong 9 cultivars, similar to the G waveform (Figure 3).

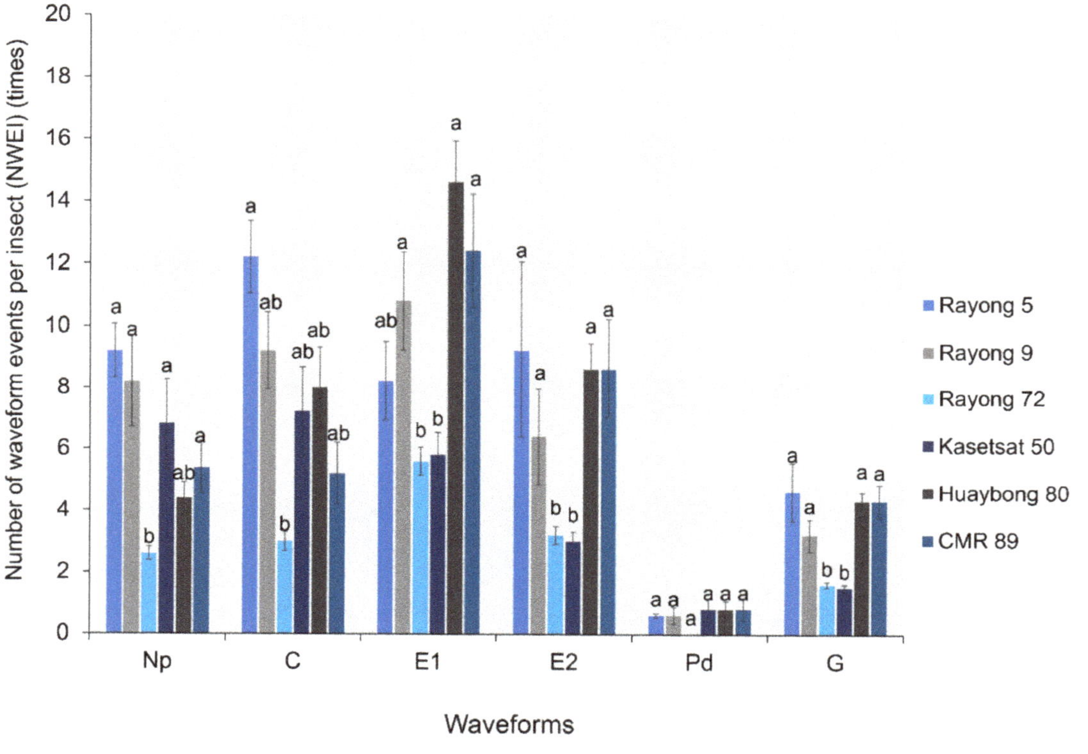

Figure 3. The number of waveform events per insect (NWEI) for all waveforms on six cassava cultivars. Bars (mean ± SE) with the same letter at the top within a waveform category are not significantly different at p = 0.05 (Tukey's test): Np = non-probe; C = pathway phase; E1 = salivation phase; E2 = ingestion phase; Pd = potential drops; and G = xylem feeding.

2.1.4. Waveform Duration per Event per Insect (WDEI)

The WDEI for the C, Pd, and G waveforms did not show significant differences among the six cassava cultivars. However, the duration of the Np, E1, and E2 waveforms differed significantly (Figure 4). Adult *B. tabaci* whiteflies spent a longer duration per event on the Np waveform when fed on Huaybong 80 (3981.04 ± 148.53 s) and Kasetsart 50 (3818.58 ± 149.98 s). Additionally, the WDEI for E1 and E2 waveforms was longer in CMR-89 (625.61 ± 109.68 s for E1; 302.04 ± 12.50 s for E2), Rayong 72 (554.73 ± 144.1 s for E1; 160.87 ± 10.09 s for E2), and Rayong 5 (480.96 ± 104.73 s for E1; 133.54 ± 73.76 s for E2) than in Rayong 9, Kasetsart 50, and Huaybong 80 (Figure 4).

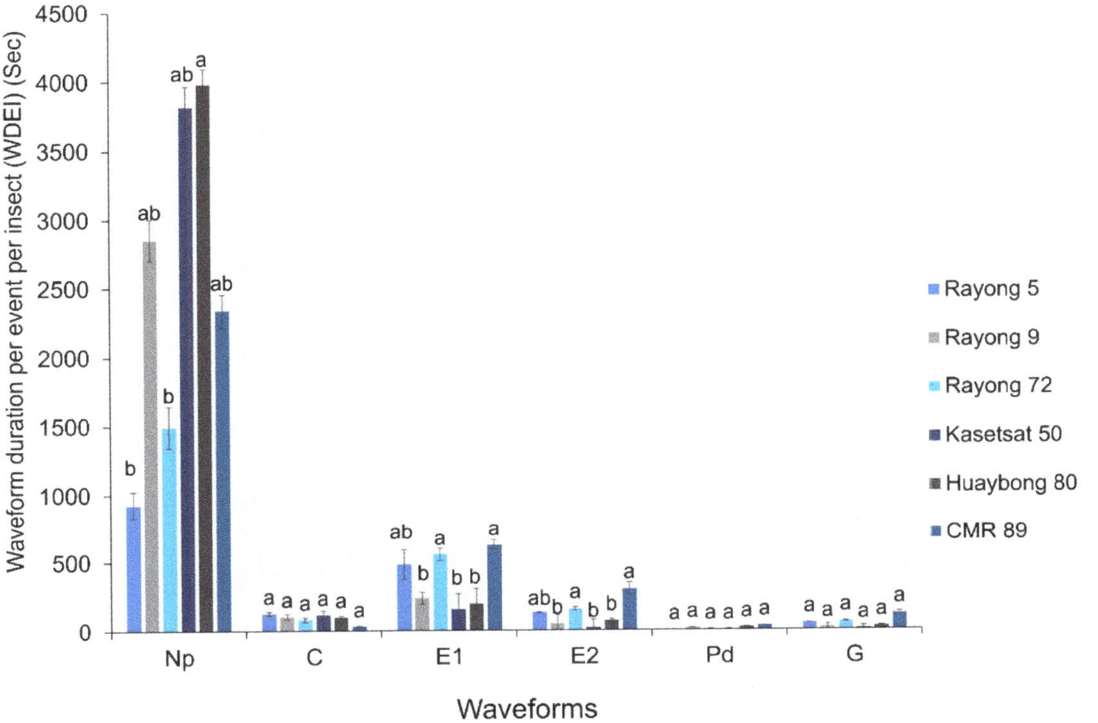

Figure 4. The waveform duration per event per insect (WDEI) for all waveforms on six cassava cultivars. Bars (mean ± SEM) with the same letter at the top within a waveform category are not significantly different at $p = 0.05$ (Tukey's test): Np = non-probe; C = pathway phase; E1 = salivation phase; E2 = ingestion phase; Pd = potential drops; and G = xylem feeding.

2.1.5. Waveform Duration per Insect (WDI)

The WDI for the Pd and G waveforms did not show significant differences among the six cassava cultivars. However, the WDI for the Np, C, E1, and E2 waveforms were significantly different ($F_{5,182} = 2.841$, $p = 0.037$ for waveform Np; $F_{5,276} = 2.026$, $p = 0.011$ for waveform C; $F_{5,287} = 3.972$, $p = 0.009$ for waveform E1; and $F_{5,195} = 1.981$, $p = 0.045$ for waveform E2). The Np waveform in adult *B. tabaci* whiteflies had a shorter duration in CMR-89 (4671.56 ± 357.59 s) than in other cassava cultivars. However, the E1 and E2 waveforms had longer durations in CMR-89 (4752.75 ± 199.02 s for E1; 1068.16 ± 139.95 s for E2) than in other cassava cultivars (Figure 5).

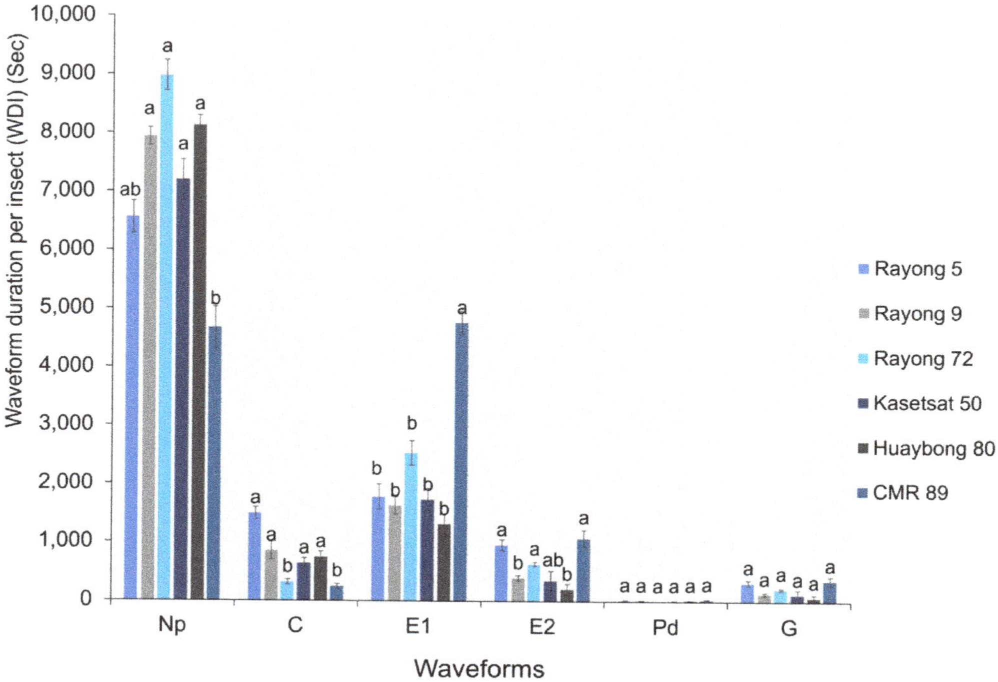

Figure 5. The WDI for all waveforms on six cassava cultivars. Bars (mean ± SEM) with the same letter at the top within a waveform category are not significantly different at $p = 0.05$ (Tukey's): Np = non-probe; C = pathway phase; E1 = salivation phase; E2 = ingestion phase; Pd = potential drops; and G = xylem feeding.

2.2. *Adult Bemisia tabaci Whitefly Settling Under Free Choice*

An analysis of variance was conducted to assess the settling preferences of the *B. tabaci* whiteflies on different cassava cultivars under field conditions for nine weeks. The results indicated significant differences in the settling behavior of the whiteflies among the cultivars ($p \leq 0.05$) during the first week. Specifically, Rayong 9 and Huaybong 80 had the minimum number of adult whiteflies, while CMR-89 and Rayong 5 had the maximum ($F5,30 = 16.140$, $p < 0.001$). In the second week, the minimum number of adult whiteflies varied significantly among cultivars, particularly on Kasetsart 50 and Huaybong 80 ($F5,30 = 1.0533$, $p = 0.043$). From weeks 3 to 9, the minimum number of adult whiteflies remained significantly lower on Kasetsart 50 and Huaybong 80 compared to other cultivars (Figure 6). Conversely, CMR-89 and Rayong 5 consistently had the maximum number of adult whiteflies during weeks 3, 4, 5, 7, and 8. Specifically, significant differences were observed in week 3 ($F5,30 = 3.933$, $p = 0.024$), week 4 ($F5,30 = 12.582$, $p < 0.001$), week 5 ($F5,30 = 6.060$, $p = 0.005$), week 7 ($F5,30 = 13.147$, $p < 0.001$), and week 8 ($F5,30 = 7.793$, $p = 0.002$) (Figure 6). After four months, the CMR-89 cultivar showed moderate to severe SLCMV disease symptoms, which were severe in 30% of total cassava recorded. This suggests a correlation between the high whitefly settling preference on CMR-89 and the increased severity of the SLCMV in this cultivar.

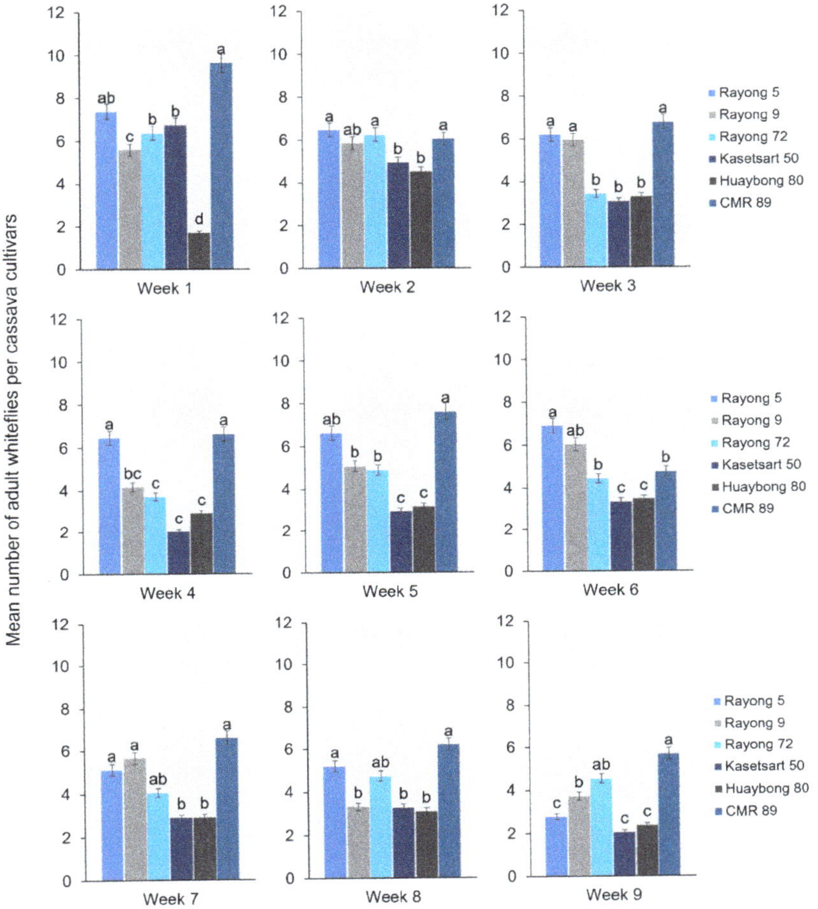

Figure 6. The mean number of adult whiteflies settled on the cassava cultivar was 1 to 9 weeks after release. Bars represent the mean percentage of whiteflies settled (mean ± SEM). Different letters above the bars indicate significant differences.

2.3. Trichome Density and Size

The morphology of leaves of different cassava cultivars was observed by an SEM, which revealed that each cultivar's characteristic was non-glandular trichomes (Figure 7). The trichome numbers per microscopic length scale of 50 and 100 μm^2 of the SEM for the six cassava cultivars (Figure 7, Table 1) differed significantly ($F5,59 = 56.89$; $p = < 0.01$). The CMR 89 cultivar had the highest number of trichomes (256 ± 12.08 per 100 μm^2), while the Huaybong 80 had the lowest (128.50 ± 14.19 per 100 μm^2). For 50 μm of the SEM, the six cassava cultivars differed significantly ($F5,59 = 34.669$; $p \leq 0.01$). The CMR 89 and Rayong 5 cultivars had the highest number of trichome density at 75.00 ± 5.29 and 50.83 ± 3.46 per 50 μm^2, respectively, while the Huaybong 80 had the lowest (33.25 ± 4.66 per 50 μm^2). The average size of each cassava cultivar leaf's trichomes was significantly different ($p \leq 0.01$). The Kasetsart 50 cultivar had the largest trichomes (11.34 ± 0.29 μm^2), followed by the Huaybong 80 and Rayong 9 cultivars, which were 10.39 ± 0.25, and 10.02 ± 0.20 per 50 μm^2 ($F5,59 = 26.236$, $p \leq 0.01$), respectively (Table 1). The study highlights the variation in trichome characteristics among different cassava cultivars, indicating potential differences in their physiological and ecological traits.

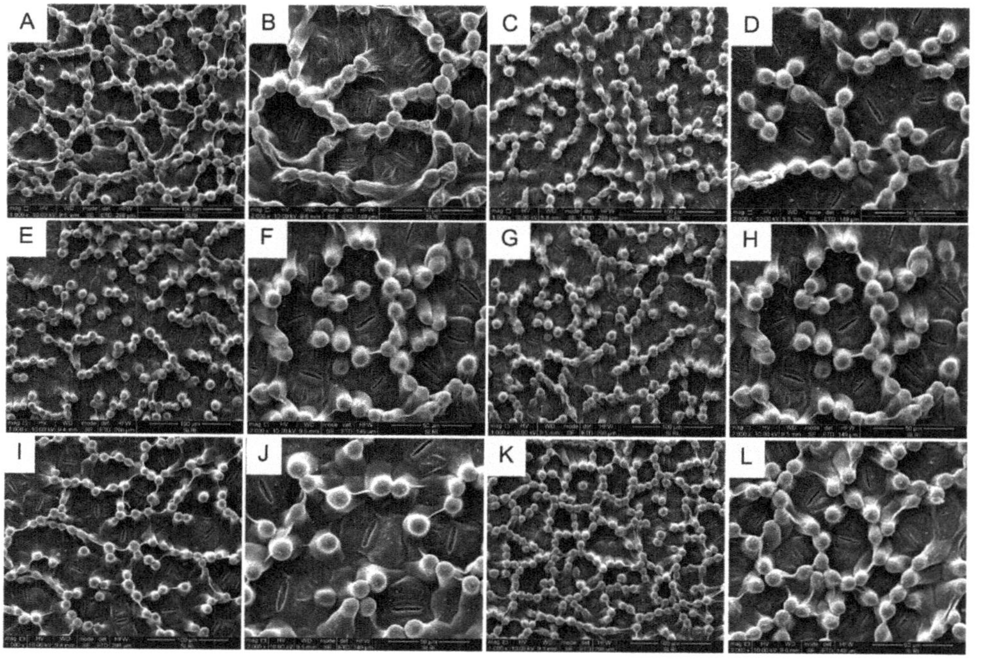

Figure 7. Trichomes of six cassava cultivar leaves with scanning electron micrograph of cassava leaf (**A**,**B**) Rayong 5, (**C**,**D**) Rayong 9, (**E**,**F**) Rayong 72, (**G**,**H**) Kasetsart 50, (**I**,**J**) Huaybong 80, and (**K**,**L**) CMR-89. (**A**,**C**,**E**,**G**,**I**,**K**): SEM of cassava leaves with 100 μm^2, (**B**,**D**,**F**,**H**,**J**,**L**): SEM of cassava left with 50 μm^2.

Table 1. Mean ± SEM trichome density and trichome size of six cassava cultivars.

Cassava Cultivars	Trichome Density per 100 μm^2				Trichome Density per 50 μm^2				Size of Trichome (μm)			
Rayong 5	187.33	±	18.16	b [1/]	50.83	±	3.46	b	9.76	±	0.27	c
Rayong 9	170.33	±	12.19	b	48.67	±	0.88	b	10.02	±	0.20	b
Rayong 72	180.67	±	15.61	b	47.00	±	3.06	b	9.80	±	0.26	c
Kasetsat 50	161.00	±	25.01	b	46.50	±	3.50	b	11.34	±	0.29	a
Huaybong 80	128.50	±	14.19	c	33.25	±	4.66	c	10.39	±	0.25	b
CMR 89	256.00	±	12.08	a	75.00	±	5.29	a	9.01	±	0.17	c

[1/] Different letters indicate significant differences.

2.4. The Correlation Analysis Between the EPG Parameters, Settling Preference of Whitefly and Trichome Density and Size

The correlation analysis conducted in this study aimed to explore the relationship between various parameters of the EPG activity of an adult *B. tabaci* whitefly and the settling preference on cassava cultivars in conjunction with trichome density and size (Table 2).

Table 2. Correlation coefficients (Pearson correlation) and significance levels of the EPG parameters, the settling preference of whiteflies, and trichome density and size among six cassava varieties.

Parameters	Trichome per 100 μm	Trichome per 50 μm	Size of Trichome	Settling Preference of Adult Whiteflies on Cassava Cultivars After Different Infection Times								
				Week1	Week2	Week3	Week4	Week5	Week6	Week7	Week8	Week9
Number of Np waveform events per insect	0.04	0.11	0.10	0.23	0.21	0.55	0.33	0.24	0.67	0.32	−0.05	−0.35
Number of C waveform events per insect	−0.27	−0.23	0.15	−0.16	0.02	0.37	0.24	0.06	0.60	0.07	−0.20	−0.57
Number of E1 waveform events per insect	0.00	0.04	−0.33	−0.38	−0.37	0.31	0.22	0.13	−0.05	0.23	−0.04	0.12
Number of E2 waveform events per insect	0.23	0.22	−0.55	−0.04	0.10	0.64	0.68	0.52	0.45	0.46	0.35	0.10
Number of Pd waveform events per insect	0.01	0.13	0.22	−0.05	−0.54	0.22	0.03	−0.07	−0.16	0.02	−0.16	−0.33
Number of G waveform events per insect	0.23	0.22	−0.55	−0.04	0.10	0.64	0.68	0.52	0.45	0.46	0.35	0.10
Waveform Np duration per event per insect	−0.50	−0.39	0.64	−0.55	−0.95 **	−0.50	−0.72	−0.73	−0.74	−0.53	−0.72	−0.44
Waveform C duration per event per insect	−0.73	−0.72	0.65	−0.44	−0.17	−0.31	−0.37	−0.50	0.23	−0.52	−0.61	−0.87 *
Waveform E1 duration per event per insect	0.81 *	0.71	−0.86 *	0.68	0.80	0.52	0.78	0.84 *	0.37	0.65	0.95 **	0.81
Waveform E2 duration per event per insect	0.88 *	0.81 *	−0.87 *	0.64	0.55	0.56	0.76	0.83 *	0.17	0.68	0.93 **	0.86 *
Waveform Pd duration per event per insect	0.42	0.47	−0.49	0.05	−0.19	0.41	0.29	0.32	−0.16	0.48	0.22	0.55
Waveform G duration per event per insect	0.88 *	0.81 *	−0.87 *	0.64	0.55	0.56	0.76	0.83	0.17	0.68	0.93	0.86
Waveform Np duration per insect	−0.79	−0.83	0.45	−0.70	−0.23	−0.70	−0.68	−0.68	−0.21	−0.63	−0.67	−0.39
Waveform C duration per insect	−0.34	−0.34	0.16	−0.18	0.14	0.25	0.22	0.04	0.64	−0.03	−0.16	−0.60
Waveform E1 duration per insect	0.93 **	0.91 **	−0.71	0.75	0.42	0.51	0.61	0.72	0.01	0.67	0.83	0.87
Waveform E2 duration per insect	0.88 *	0.82 *	−0.78	0.84	0.80	0.74	0.91 **	0.94 **	0.55	0.75	0.97 **	0.64
Waveform Pd duration per insect	0.65	0.67	−0.64	0.43	0.25	0.83	0.81	0.74	0.41	0.72	0.61	0.38
Waveform G duration per insect	0.88 *	0.82 *	−0.78	0.84 *	0.80	0.74	0.91 **	0.94 **	0.55	0.75	0.97 **	0.64
Trichome density per 100 μm		0.98 **	−0.75	0.90 *	0.65	0.73	0.78	0.88 *	0.32	0.83	0.90 *	0.83
Trichome density per 50 μm			−0.67	0.90 *	0.57	0.73	0.73	0.83	0.29	0.83	0.83	0.78
Size of trichome				−0.47	−0.70	−0.75	−0.87 *	−0.90 *	−0.52	−0.84	−0.82	−0.85 *

Asterisks indicate significant difference (* $p < 0.05$, ** $p < 0.01$).

Conversely, the duration of settling preference and trichome density were positively correlated. Specifically, weeks 1, 5, 8, and 9 exhibited strong positive correlations with trichome density per 100 µm (r = 0.90*, r = 0.88*, r = 0.90*, r = 0.83, respectively) and per 50 µm (r = 0.90*, r = 0.83*, r = 0.83*, r = 0.78, respectively). Additionally, positive correlations were observed between settling preference and certain EPG waveform parameters, such as E1 and E2 duration per event per insect and E2 and G duration per insect. Additionally, the EPG waveform positively correlated with the trichome density. These findings suggest that higher trichome density, particularly in the CMR 89 cultivar, may not deter the feeding and settlement of adult whiteflies.

3. Discussion

3.1. EPG Waveform Characteristics and Feeding Behavior of *Bemisia tabaci* Whiteflies on Different Cassava Cultivars

Trichome density, size, and EPG waveform characteristics significantly influence the interaction between cassava plants and whiteflies (*B. tabaci*). Through the analysis of EPG waveform characteristics, six distinct probing waveforms were identified, including Np, C, E1, E2, Pd, and G, with the findings of previous studies and reaffirmed the complexity of whitefly interactions with host plants [36–38,40,42–45]. The waveforms provide insights into the feeding behavior of whiteflies on different cassava cultivars. The study revealed that adult whiteflies predominantly spent their time in the Np phase, followed by the E1 waveform, indicating a pattern of probing and feeding behavior. Significant differences were observed in the TPD and TWD among the six cassava cultivars. Specifically, CMR 89 cultivars exhibited longer total probe durations in the E1 and E2 waveforms than other cultivars, suggesting variations in whitefly feeding behavior across cultivars. Moreover, the NWEI and the WDEI varied significantly among the cassava cultivars. While the number of Pd waveform events did not show significant differences, other waveforms such as Np, C, E1, E2, and G exhibited significant variations. These differences reflect the distinct responses of whiteflies to the cassava cultivars, potentially influenced by factors such as leaf morphology and chemical composition.

The analysis of feeding behavior revealed that adult whiteflies predominantly remained in the Np phase, especially on Huaybong 80, Kasetsart 50, Rayong 9, and Rayong 72 cultivars. This finding suggests that these cultivars may possess certain deterrent properties or structural defenses that inhibit whitefly probing and feeding. In contrast, the CMR 89 and Rayong 5 cultivars exhibited higher probing activities, indicating a lower resistance to whitefly feeding. Moreover, the significant differences in the NWEI (Np, C, E1, E2, G) among the six cassava cultivars underscore the variability in whitefly feeding behavior influenced by plant genotype. Notably, the higher frequency of E1 and E2 waveforms in certain cultivars, such as CMR 89, indicates a preference for these plants, possibly due to easier access to phloem sap. In CMR 89, the shorter duration of the Np waveform and longer durations for E1 and E2 waveforms reinforce the susceptibility of this cultivar to whitefly feeding. The significant differences in WDI for the Np, C, E1, and E2 waveforms among the cultivars suggest that certain cassava genotypes are more conducive to whitefly feeding, which has implications for virus transmission dynamics [46,47]. The EPG technique has been used to study insect-feeding behavior in relation to the transmission of pathogens [48,49]. The prolonged duration in the non-probing phase on specific cultivars indicates a potential resistance mechanism, making these cultivars less suitable for whitefly infestation. This finding is significant for cassava breeding programs aiming to enhance pest resistance.

3.2. Adult *Bemisia tabaci* Whitefly Settling Under Free Choice Tested

During the first week, Rayong 9 and Huaybong 80 exhibited the lowest number of adult whiteflies, indicating an initial deterrence effect. In contrast, CMR-89 and Rayong 5 attracted the highest number of whiteflies. These differences were statistically significant, highlighting that the cultivars' inherent characteristics influence whiteflies' initial selection of host plants. In the second week, Kasetsart 50 and Huaybong 80 continued to show a significantly lower number of whiteflies. From weeks 3 to 9, Kasetsart 50 and Huaybong 80 consistently had fewer whiteflies, indicating sustained deterrence. The persistent lower whitefly numbers on these cultivars suggest they have long-term resistance traits, making them less attractive or hospitable to whiteflies.

Conversely, CMR-89 and Rayong 5 consistently had the highest number of whiteflies during the same period, particularly in weeks 3, 4, 5, 7, and 8. The significant differences observed in these weeks indicate that these cultivars are more attractive to whiteflies, possibly due to higher phloem availability or fewer physical or chemical deterrents. The high whitefly settling preference on CMR-89 notably correlates with severe SLCMV disease expression in 30% of the plants after four months. This correlation highlights the significant role of whitefly preference in the epidemiology of SLCMV disease. However, whiteflies transmitted infections were less severe than those originating from infected cuttings, with many infected plants remaining asymptomatic. Certain genotypes, including CMR-89 and Ryong 11, were also susceptible to CMD [14]. The increased settling on CMR-89 likely facilitated more effective virus transmission, leading to higher disease severity. This relationship between whitefly behavior and SLCMV severity suggests that managing whitefly populations is crucial for controlling the disease. The findings from this study have significant implications for cassava breeding programs. Cultivars such as Kasetsart 50 and Huai Bong 80, which consistently showed lower whitefly numbers, should be prioritized for breeding and cultivation in areas prone to whitefly infestations and the SLCMV. The inherent resistance of these cultivars to whitefly settling could reduce the incidence of vector-borne diseases, thereby improving crop health and yield. Conversely, cultivars such as CMR-89 and Rayong 5, which attracted more whiteflies and exhibited higher SLCMV severity, may require additional management strategies or could be modified through breeding to incorporate resistance traits. Understanding the specific characteristics conferring resistance in Kasetsart 50 and Huai Bong 80 could guide the development of new, more resilient cassava varieties. The settling preference analysis further highlighted significant differences in whitefly settling behavior among the cassava cultivars. For instance, cultivars such as Rayong 9 and Huaybong 80 consistently exhibited fewer settling whiteflies than CMR 89 and Rayong 5. The observed variations in settling preference may be attributed to factors such as leaf trichome density and size. Indeed, the correlation analysis revealed intriguing relationships between whitefly settling preference, trichome density, and size. A negative correlation was observed between the number of settling whiteflies and trichome size, suggesting that larger trichomes may deter whitefly settling. Conversely, a positive correlation was found between settling preference duration and trichome density, particularly in cultivars such as CMR 89.

The SLCMV transmitted by the insect vector *B. tabaci* causes CMD in cassava plants [12]. Whiteflies select host plants upon landing, utilizing their stylets to create brief probes in the plant's epidermis for feeding and oviposition, a process not guided by olfactory cues [50]. This study exposed six cassava plants to field conditions infested with whiteflies. Interestingly, the initial leaf position of all six cassava cultivars showed lower whitefly infestation than the other leaf positions did. Variations in leaf position were found to impact the content of cyanide, amino acids, and crude protein. Specifically, cyanide and amino acid levels were higher in the upper leaf positions compared to the lower ones, while crude protein content exhibited the opposite trend, being higher in the lower leaves [51]. Cyanides and amino acids possess direct toxicity towards insect herbivores, with amino acids enabling nitrogen storage in a form inaccessible to them [52,53]. This inherent toxicity contributes to the aversion of whiteflies towards infesting the first leaf position more than others.

3.3. Trichome Density and Size

Trichomes affect insect behavior, influencing egg laying, shelter seeking, and feeding [54–57]. The results from the no-choice infestation test revealed differing whitefly infestation percentages among the six cassava cultivars. Notably, the Rayong 72 cultivar exhibited the lowest whitefly infestation compared to the Huaybong 80 and CMR-89 cultivars. The SEM of trichomes across all cassava cultivars revealed distinct differences. Rayong 72 displayed a smaller trichome size but higher trichome density. Trichome density has contributed to whitefly resistance in various crops, such as cotton, eggplant, tomato, chilies, soybean, and cucumber [58–62]. The non-glandular trichomes on the abaxial surface of all six cassava varieties serve as a defense mechanism against insect attacks, effectively restricting insect movement and as a barrier protecting the plant's epidermal layer from damage [57,63]. Consequently, the preference of whiteflies for egg laying, shelter seeking, or feeding on these plants is reduced.

The sparse distribution of trichomes or the thinness of the epidermis may have facilitated whitefly feeding, a crucial factor influencing the host plant preference of insect herbivores [64]. Whitefly feeding behavior on the host plant involves intracellular penetration by the insect's stylet, which is highly efficient for transmitting viruses to the host plant [65]. The SLCMV poses a significant threat to crop yield, leading to substantial economic losses and food insecurity [18]. This study investigated whitefly feeding behavior using EPG. Initially, the whiteflies traversed the abaxial surface of cassava leaves, generating an Np. Subsequently, the whitefly's stylet penetrated the epidermis and parenchyma tissue (waveform C). Upon reaching the phloem sieve tube, the insect released saliva (E1) and ingested phloem sap (E2). Across all cassava cultivars, the whitefly's stylet penetration of intracellular vascular tissue (Pd: potential drop) lasted only briefly. The Rayong 72 cultivar exhibited a short duration of phloem ingestion and a low number of ingestion events, potentially contributing to the limited spread of the SLCMV. Conversely, the Huaybong 80 cultivar showed an extended total duration of phloem salivation, increasing the likelihood of SLCMV transmission into the phloem sieve.

3.4. The Correlation Analysis

According to the findings in Table 2, a negative correlation was observed between the number of adult whiteflies settling on cassava cultivar leaves and the size of the trichomes. These findings are consistent with those of [66–68], who observed a negative correlation between trichome density on eggplant leaves and resistance to *B. tabaci*. Similar results were also reported by [69] in cucumbers, [70] in tomatoes, [71] in soybeans, and [72] in cotton plants. Furthermore, negative correlations were also identified between the settling preference of adult whiteflies and the number of penetration events, Np waveform, the waveform C duration per event per insect WDEI, and the Np waveform duration per insect. These results suggest that larger trichomes may deter adult whiteflies from settling, while certain EPG parameters indicative of feeding behavior were also negatively correlated with settling preference.

These findings underscore the complex interplay between trichome characteristics, EPG waveform parameters, and whitefly settling preference on cassava cultivars. Understanding these relationships is crucial for devising effective pest management strategies and breeding programs to enhance cassava resistance to whiteflies. Further research is warranted to elucidate the underlying mechanisms driving these correlations and their implications for cassava production and crop protection.

4. Materials and Methods

The experiment was conducted inside a laboratory and field at the Suranaree University of Technology (14°58′14.38″ N 102°06′7.06″ E), Nakorn Ratchasima, Thailand.

4.1. Source of Planting Materials

Six cassava cultivars (*Manihot esculenta* Crantz)—namely, Rayong 5, Rayong 9, Rayong 72, Kasetsart 50, Huaybong 80, and CMR-89—were sourced from Nakhon Ratchasima, Thailand, for the screening of antixenosis resistance to the *B. tabaci* whitefly. The DNA of each cassava plant was extracted from the auxiliary buds of both ends of the plants to confirm disease and non-disease status, using polymerase chain reaction (PCR) with specific

SLCMV primers [14]. The DNA extraction followed the protocols of [73,74], and [75], with quality and quantity assessed via spectrophotometry (Thermo Scientific™ NanoDrop™ 2000, Thermo Fisher Scientific, Waltham, MA, USA). The presence of the SLCMV was monitored by PCR, using AV1 gene-specific primers: AV1 forward (5′-GTT GAA GGT ACT TAT TCC C-3′) and AV1 reverse (5′-TAT TAA TAC GGT TGT AAA CGC-3′). PCR amplification involved a 25 μL reaction volume with 1XPCR buffer (PCR Biosystems, London, UK), 0.2 μM of each primer, and approximately 50 ng of the DNA template. The thermal cycling conditions were as follows: initial denaturation at 94 °C for 5 min, followed by 35 cycles of denaturation at 94 °C for 40 s, annealing at 55 °C for 40 s, extension at 72 °C for 40 s, and final elongation at 72 °C for 5 min. The PCR was performed using a Bio-Rad iCycler (Bio-Rad Laboratories, Hercules, CA, USA). DNA gel electrophoresis of PCR amplification from cassava samples (Figure S1). The non-disease cassavas were used for propagation and examined electropenetrography recording and whitefly *B. tabaci* settling experiment. The cassava stems were cut to 15–18 cm lengths, with 20 plants gathered in each cultivar. Six cassava plants were planted in 11 × 15 cm^2 pots containing a mix of sandy loam soil and manure (1:1 ratio). The plants were watered daily, and no insect- or disease-control chemicals were applied. When the plants reached six weeks of age, they were used for the settling experiments.

4.2. Whitefly Materials

B. tabaci colonies were collected from Nakhon Ratchasima Province, Thailand, cassava fields. These were maintained under laboratory conditions at 28 ± 2 °C, 60–65% RH, with a 12:12 (L/D) photoperiod. Susceptible CMR-89 cassava cultivars were used for mass rearing. Twenty cassava plants were potted in 5-inch diameter plastic pots and placed in BugDorm-6E620 Insect-Rearing Cages (W60 × D60 × H120 cm^3, MegaView Science Co., Ltd., Taichung, Taiwan). Two cages, each containing ten pots, were used, and plants were replaced every two weeks. Adult whiteflies within 24 to 48 h after emerging were used for experiments.

4.3. Electropenetrography Recording

Feeding-behavior experiments were conducted using a Giga-8 DC electropenetrography (EPG) system (EPG Systems, Wageningen, The Netherlands) with an input resistance of 10^9 Ω (1 GΩ) and adjustable plant voltage. The cassava plant and the insect with the EPG probe were placed within a Faraday cage (1.5 × 2 × 1.5 m^3) to block electrical noise. The system was installed in a temperature-controlled room at 26 ± 2 °C and 60 ± 5% RH. EPG signals were recorded by Stylet+ software (v01.34), adjusting the signal range from −5 to +5 V, and displayed on a computer [40]. Adult whiteflies were placed in a glass vial and immobilized on ice for 5–10 s before being connected to a gold wire electrode (2.5 cm long, 12 μm diameter) (Figure S2) (EPG Systems, Wageningen, The Netherlands), with water-based silver glue (Wageningen University) on the insect's pronotum. The other end of the wire was attached to a copper electrode (1 mm diameter, 2.5 cm long) (Sigmund Cohn Corp, Mt Vernon, NY, USA), which was connected to the EPG probe. The whitefly was placed on the abaxial leaf surface fixed with parafilm. Signal adjustments were made if needed, and data were recorded using Stylet+d software (v01.34). Each whitefly was observed for 3 h daily, with 30 whiteflies tested per cultivar in a completely randomized design. Annotated waveforms (non-probing, stylet pathway, phloem salivation, phloem ingestion, intracellular puncture, and xylem feeding) were analyzed using EPG Systems software and a modified Ebert 3.0 program in SAS Enterprise Guide 7.1, SAS 9.4 statistical software.

EPG data recorded by EPG Systems Stylet+d were manually annotated using EPG Systems Stylet+a software (v01.34). The annotated waveforms were non-probing (Np), stylet pathway (C), phloem salivation (E1), phloem ingestion (E2), intracellular puncture—potential drop (Pd), and xylem feeding (G). The waveform patterns were categorized by amplitude, relative voltage level, R/emf origin, frequency, and the waveform context as described in the previous EPG studies of *B. tabaci* [36–38,40,42–45]. Annotation files were then directly passed to a modified version of the Ebert 3.0 program in SAS Enterprise Guide 7.1, SAS 9.4 statistical software (SAS Institute, Cary, NC, USA) for further analysis, which produces the same parameters as the popular Sarria Excel workbook [76,77].

4.4. Whitefly *Bemisia tabaci* Settling Under Free Choice in Field Condition

The free choice test used six cassava plants (Rayong 5, Rayong 9, Rayong 72, Kasetsart 50, Huaybong 80, and CMR-89 cultivars). The experiment was conducted in a randomized complete block design with 40 × 35 m^2 pot dimensions, employing five replications per cultivar. Each treatment consisted of 30 cassava plants. Cassava stems were planted, measuring 15–18 cm in length with approximately four nodes each. The planting was conducted in mini-field conditions at the Suranaree Farm within the Suranaree University of Technology. A 1 × 0.8 m^2 planting distance was maintained between the cassava plants. The white net greenhouse covered the field (40 meshes, 1.3 mm/0.05″, aperture). To mitigate the impact of weeds and potential hosts of pests, manual weeding using hand hoes was performed. No plant protection measures (such as pesticide applications) were implemented throughout the trial. When the plants were eight weeks old, 250 adults of the whitefly species *B. tabaci* were introduced into the center of the white net greenhouse where the test plants were situated. Adult whitefly infestation levels were assessed by counting and recording the number of *B. tabaci* adults settling on each cassava plant.

For accurate findings, whitefly (*B. tabaci*) infestation was meticulously assessed. The number of whiteflies settled per treatment was divided by the total number of whiteflies settled per cassava plant within each replication (20 cassava plants). This calculation determined the proportion of whiteflies settled on each treatment (three times). Whitefly infestation on cassava plants was monitored weekly under field conditions for a period ranging from 1 to 9 weeks. Upon completion of the experimental period, cassava plants were maintained for up to 4 months. During this period, the plants were observed for symptoms indicative of SLCMV infection. The cassava plants that showed SLCMV symptoms were DNA extracted, and their disease status was determined using the PCR technique with specific SLCMV primers.

4.5. Trichome Size and Density

The morphology of the abaxial surface of the six cassava leaves (Rayong5, Rayong9, Rayong72, Kasetsart50, Huaybong80, and CMR-89) was analyzed for trichome density and size. Samples from the fourth leaf of 8-week-old plants were fixed in 37% formaldehyde overnight, washed in phosphate-buffered saline, and dehydrated in an ethanol series (70%, 80%, 95%, and 100%) [78,79]. Samples were critical-point dried and coated with gold-palladium [57,79]. The fixed surfaces of the leaves were then observed under a cutting-edge scanning electron microscope (SEM) (SEM™, FEI, Quanta450, Eindhoven, The Netherlands). Trichome density was counted manually per 100 and 50 μm^2, with ten biological replications and five images per leaf. Trichome sizes were measured for each treatment with 15 replications per cultivar.

4.6. Statistical Analysis

EPG waveforms related to feeding-behavior events were characterized: Np, C, E1, E2, Pd, and G; the variables total probing duration (TPD), total waveform duration (TWD), number of waveform events per insect (NWEI), waveform duration per event per insect (WDEI), and waveform duration per insect WDI) were also calculated (mean ± standard error), as described by [80]. The variables were compared using the Tukey–Kramer test at $p < 0.05$. Free choice test data were analyzed using the Mann–Whitney U-test at $p < 0.05$. All analyses were performed with SAS V9.4 software. The Tukey–Kramer test conducted mean comparisons for trichome density and size. Pearson correlation analysis was conducted on mean values of whitefly feeding behavior, insect settling, and trichome density and size using Origin Pro 2024 software, with significance set at $p < 0.05$ [81].

Data from the free choice in field condition, calculated as the mean member of whiteflies counted in the antixenosis resistance screening of six cassava cultivars at different time intervals (subjected to the Mann–Whitney U-test at $p < 0.05$ significance level), were computed using the SAS V9.4 software. All statistical analyses were calculated using Statistic Analysis System V9.4 software (SAS Institute, Inc., Cary, NC, USA). The mean was compared using the Tukey–Kramer test for trichome density and size.

For correlation analysis (Pearson correlation), conducted by [57,82,83] on the mean values of whitefly feeding behavior, number of settled insects, and trichome density and size to identify the relationships between parameters in this study. The differences were considered significant at a probability level of 5%, using Origin Pro 2024 software (OriginLab Corporation Northampton, MA, USA) [81].

5. Conclusions

In summary, the results indicate that trichome characteristics play a significant role in whitefly settling preference on cassava cultivars, with larger trichomes and higher trichome density correlated to reduced settling behavior. These findings have implications for pest management strategies, highlighting the potential of trichome-based resistance mechanisms in cassava breeding programs.

Supplementary Materials: The following supporting information can be downloaded at https://www.mdpi.com/article/10.3390/plants13223218/s1, Figure S1: PCR products of the AV1 gene using SLCMV specific primers. DNA gel electrophoresis of PCR amplification from cassava samples. Figure S2. The adult whitefly was connected to a gold wire electrode (2.5 cm long, 12 μm diameter).

Author Contributions: All authors contributed to the study's conception and design: conceptualization, J.R.; methodology, S.P. (Sudarat Pimkornburee), S.P. (Supawadee Pombud) and J.R.; software, S.P. (Sudarat Pimkornburee), S.P. (Supawadee Pombud) and J.R.; formal analysis, S.P. (Supawadee Pombud) and J.R.; resources, S.A.-o. and K.B.; data curation, S.P. (Sudarat Pimkornburee), S.P. (Supawadee Pombud) and J.R.; writing—original draft preparation, S.P. (Sudarat Pimkornburee) and J.R.; writing—review and editing, S.P. (Sudarat Pimkornburee), J.R. and W.N.; supervision, J.R.; project administration, J.R., K.B., W.N. and S.A.-o.; funding acquisition, J.R., K.B. and W.N. All authors have read and agreed to the published version of the manuscript.

Funding: The research was supported by the Thailand Science Research and Innovation (TSRI) and the Suranaree University of Technology (Grant no. FF3-302-65-12-100(22), 160342).

Data Availability Statement: Data can be provided upon request from the lead author.

Acknowledgments: The authors would like to thank the Center for Scientific and Technological Equipment, Suranaree University of Technology, for providing the equipment for this study. They would also like to thank the Synchrotron Light Research Institute for providing the chemicals and for observing the left surface during scanning electron microscopy.

Conflicts of Interest: The authors declare no conflicts of interest.

References

1. Ceballos, H.; Sánchez, T.; Chávez, A.L.; Iglesias, C.; Debouck, D.; Mafla, G.; Tohme, J. Variation in crude protein content in cassava (*Manihot esculenta* Crantz) roots. *J. Food Compos. Anal.* **2006**, *19*, 589–593. [CrossRef]
2. Legg, J.; Owor, B.; Sseruwagi, P.; Ndunguru, J. Cassava mosaic virus disease in East and Central Africa: Epidemiology and management of a regional pandemic. *Adv. Virus Res.* **2006**, *67*, 355–418. [CrossRef] [PubMed]
3. Legg, J.P.; Kumar, P.L.; Makeshkumar, T.; Tripathi, L.; Ferguson, M.; Kanju, E.; Cuellar, W. Cassava virus diseases: Biology, epidemiology, and management. *Adv. Virus Res.* **2015**, *91*, 85–142. [CrossRef] [PubMed]
4. Fargette, D.; Fauquet, C.; Thouvenel, J.C. Yield losses induced by African cassava mosaic virus in relation to the mode and date of infection. *Trop. Pest. Manag.* **1988**, *34*, 89–91. [CrossRef]
5. Rybicki, E.P. A Top Ten list for economically important plant viruses. *Arch. Virol.* **2015**, *160*, 17–20. [CrossRef]
6. Legg, J.P.; Fauquet, C.M. Cassava mosaic geminiviruses in Africa. *Plant Mol. Biol.* **2004**, *56*, 585–599. [CrossRef]
7. Legg, J.P.; Jeremiah, S.C.; Obiero, H.M.; Maruthi, M.N.; Ndyetabula, I.; Okao-Okuja, G.; Bouwmeester, H.; Bigirimana, S.; Tata-Hangy, W.; Gashaka, G.; et al. Comparing the regional epidemiology of the cassava mosaic and cassava brown streak; virus pandemics in Africa. *Virus Res.* **2011**, *159*, 161–170. [CrossRef]

8. Warburg, O. Die kulturpflanzen usambaras. *Mitt. Dtsch. Schutzgeb.* **1894**, *7*, 131.
9. Saunders, K.; Salim, N.; Mali, V.R.; Malathi, V.G.; Briddon, R.; Markham, P.G.; Stanley, J. Characterization of Sri Lankan cassava mosaic virus and Indian cassava mosaic virus: Evidence for acquisition of a DNA B component by a monopartite begomovirus. *Virology* **2002**, *293*, 63–74. [CrossRef]
10. Thresh, J.; Cooter, R. Strategies for controlling cassava mosaic virus disease in Africa. *Plant Pathol.* **2005**, *54*, 587–614. [CrossRef]
11. Maruthi, M.N.; Hillocks, R.J.; Mtunda, K.; Raya, M.D.; Muhanna, M.; Kiozia, H.; Rekha, A.R.; Colvin, J.; Thresh, J.M. Transmission of Cassava brown streak virus by *Bemisia tabaci* (Gennadius). *J. Phytopathol.* **2005**, *153*, 307–312. [CrossRef]
12. Kalyebi, A.; Macfadyen, S.; Parry, H.; Tay, W.T.; De Barro, P.; Colvin, J. African cassava whitefly, *Bemisia tabaci*, cassava colonization preferences and control implications. *PLoS ONE* **2018**, *13*, e0204862. [CrossRef] [PubMed]
13. Mrisho, L.M.; Maeda, D.G.; Ortiz, Z.M.; Ghanavi, H.R.; Legg, J.P.; Stensmyr, M.C. Influence of olfaction in host-selection behavior of the cassava whitefly *Bemisia tabaci*. *Front. Ecol. Evol.* **2021**, *9*, 775–778. [CrossRef]
14. Saokham, K.; Hemniam, N.; Roekwan, S.; Hunsawattanakul, S.; Thawinampan, J.; Siriwan, W. Survey and molecular detection of Sri Lankan cassava mosaic virus in Thailand. *PLoS ONE* **2021**, *6*, e0252846. [CrossRef]
15. Fiallo-Olivé, E.; Pan, L.L.; Liu, S.S.; Navas-Castillo, J. Transmission of begomoviruses and other whitefly-borne viruses: Dependence on the vector species. *Phytopathology* **2020**, *110*, 10–17. [CrossRef]
16. Horowitz, A.R. Population dynamics of *Bemisia tabaci* (Gennadius): With special emphasis on cotton fields. *Agric. Ecosyst. Environ.* **1986**, *17*, 37–47. [CrossRef]
17. Legg, J.P.; Thresh, J.M. Cassava virus diseases in Africa. In Proceedings of the Conference on Plant Virology in Sub Saharan Africa, Ibadan, Nigeria, 4–8 June 2001; IITA: Ibadan, Nigeria, 2003; pp. 517–552.
18. Ntui, V.O.; Kong, K.; Khan, R.S.; Igawa, T.; Janavi, G.J.; Rabindran, R. Resistance to Sri Lankan Cassava Mosaic Virus (SLCMV) in Genetically Engineered Cassava cv. KU50 through RNA Silencing. *PLoS ONE* **2015**, *10*, e0120551. [CrossRef]
19. Bellotti, A.C.; Smith, L.; Lapointe, S.L. Recent advances in cassava pest management. *Annu. Rev. Entomol.* **1999**, *44*, 343–370. [CrossRef]
20. Fereres, A.; Moreno, A. Integrated Control Measures against Viruses and their Vectors. In *Recent Advances in Plant Virology*; Caranta, C., Aranda, M.A., Tepfer, M., Lopez-Moya, J.J., Eds.; Caister Academic Press: Poole, UK, 2011; pp. 237–261.
21. Klinner, C.; König, C.M.; Werckenthin, A.; Daly, K.; Bisch-Knaden, S. Functional olfactory sensory neurons housed in olfactory sensilla on the ovipositor of the hawkmoth Manduca sexta. *Front. Ecol. Evol.* **2016**, *4*, 130. [CrossRef]
22. Rani, P.U.; Madhavendra, S.S. External morphology of antennal and rostral sensillae in four hemipteran insects and their possible role in host plant selection. *Int. J. Trop. Insect Sci.* **2005**, *25*, 198–207. [CrossRef]
23. Yang, K.; Gong, X.L.; Li, G.C.; Huang, L.Q.; Ning, C.; Wang, C.Z. A gustatory receptor tuned to the steroid plant hormone brassinolide in *Plutella xylostella* (Lepidoptera: Plutellidae). *eLife* **2020**, *9*, e64114. [CrossRef] [PubMed]
24. Avery, P.B.; Kumar, V.; Simmonds, M.S.; Faull, J. Influence of leaf trichome type and density on the host plant selection by the greenhouse whitefly, *Trialeurodes vaporariorum* (Hemiptera: Aleyrodidae). *Appl. Entomol. Zool.* **2015**, *50*, 79–87. [CrossRef]
25. Heinz, K.M.; Parella, M.P. Poinsettia (*Euphorbia pulcherrima* Willd. ex Koltz.) cultivar-mediated differences in performance of five natural enemies of *Bemisia argentifolii* Bellows and Perring, n. sp (Homoptera:Aleyrodidae). *Biol. Control* **1994**, *4*, 305–318. [CrossRef]
26. Mcauslane, H.J. Influence of leaf pubescence on ovipositional preference of *Bemisia argentifolii* (Homoptera: Aleyrodidae) on soybean. *Environ. Entomol.* **1996**, *25*, 834–841. [CrossRef]
27. Fereres, A.; Moreno, A. Behavioural aspects influencing plant virus transmission by homopteran insects. *Virus Res.* **2009**, *141*, 158–168. [CrossRef] [PubMed]
28. Calabrese, E.J.; Edwards, L.J. Light and gravity in leaf side selection by the green peach aphid, *Myzus persicae*. *Ann. Entomol. Soc. Am.* **1976**, *69*, 1145–1146. [CrossRef]
29. Simmons, M.A. Settling of crawlers of *Bemisia tabaci* (Homoptera: Aleyrodidae) on five vegetable hosts. *Ann. Entomol. Soc. Am.* **2002**, *95*, 464–468. [CrossRef]
30. McLean, D.L.; Kinsey, M.G. A technique for electronically recording aphid feeding and salivation. *Nature* **1964**, *202*, 1358–1359. [CrossRef]
31. Tjallingii, W.F. Electronic recording of penetration behavior by aphids. *Entomol. Exp. Appl.* **1978**, *24*, 721–730. [CrossRef]
32. Tjallingii, W.F. Continuous recording of stylet penetration activities by aphids. In *Aphid-Plant Genotype Interactions*; Campbell, R.K., Eikenbary, R.D., Eds.; Elsevier: Amsterdam, The Netherlands, 1990; pp. 89–99.
33. Janssen, J.A.M.; Tjallingii, W.F.; van Lenteren, J.C. Electrical recording and ultrastructure of stylet penetration by the greenhouse whitefly. *Entomol. Exp. Appl.* **1989**, *52*, 69–81. [CrossRef]
34. Lei, H.; Tjallingii, F.; van Lenteren, J.C. Probing and feeding characteristics of the greenhouse whitefly in association with host-plant acceptance and whitefly strains. *Entomol. Exp. Appl.* **1998**, *88*, 73–80. [CrossRef]
35. Lei, H.; van Lenteren, J.C.; Tjallingii, F. Analysis of resistance in tomato and sweet pepper against the greenhouse whitefly using electrically monitored and visually observed probing and feeding behaviour. *Entomol. Exp. Appl.* **1999**, *92*, 299–309. [CrossRef]
36. Johnson, D.D.; Waker, G.P. Intracellular punctures by the adult whitefly *Bemisia argentifoliion* DC and AC electronic feeding monitors. *Entomol. Exp. Appl.* **1999**, *92*, 257–270. [CrossRef]
37. Jiang, Y.X.; Lei, H.; Collar, J.L.; Martin, B.; Muñiz, M.; Fereres, A. Probing and feeding behaviour of two distinct biotypes of *Bemisia tabaci* (Homoptera:Aleyr-odidae) on tomato plants. *J. Econ. Entomol.* **1999**, *92*, 357–366. [CrossRef]

38. Jiang, Y.; De Blas, C.; Barrios, L.; Fereres, A. Correlation between whitefly (Homoptera: Aleyrodidae) feeding behavior and transmission of tomato yellow leaf curl virus. *Ann. Entomol. Soc. Am.* **2000**, *93*, 573–579. [CrossRef]
39. Rodríguez-López, M.J.; Garzo, E.; Bonani, J.P.; Fernández-Muñoz, R.; Moriones, E.; Fereres, A. Acylsucrose-Producing Tomato Plants Forces *Bemisia tabaci* to Shift Its Preferred Settling and Feeding Site. *PLoS ONE* **2012**, *7*, e33064. [CrossRef]
40. Milenovic, M.; Wosula, E.N.; Rapisarda, C.; Legg, J.P. Impact of host plant species and whitefly species on feeding behavior of *Bemisia tabaci*. *Front. Plant Sci.* **2019**, *10*, 1. [CrossRef]
41. Tjallingii, W.F. Aphid-plant interactions: What goes on in the depth of the tissues? *Proc. Exp. Appl. Entomol.* **1995**, *6*, 163–169.
42. Liu, B.; Yan, F.; Chu, D.; Pan, H.; Jiao, X.; Xie, W.; Wu, Q.; Wang, S.; Xu, B.; Zhou, X.; et al. Difference in feeding behaviors of two invasive whiteflies on host plants with different suitability: Implication for competitive displacement. *Int. J. Biol. Sci.* **2012**, *8*, 697–706. [CrossRef]
43. Yin, H.D.; Wang, X.Y.; Xue, K.; Huang, C.H.; Wang, R.J.; Yan, F.M.; Xu, C.R. Impacts of transgenic Bt cotton on the stylet penetration behaviors of *Bemisia tabaci* biotype B: Evidence from laboratory experiments. *Insect Sci.* **2010**, *17*, 344–352. [CrossRef]
44. Civolani, S.; Stefano, C.; Milvia, C.; Rison, J.L.; Andrea, B.; Juan, M.A.; Billy, I.A.; Giuseppe, P.; Massimo, G.; Elisa, A.F. An EPG study of the probing behavior of adult *Bemisia tabaci* biotype Q (Hemiptera: Aleyrodidae) following exposure to Cyantraniliprole. *J. Econ. Entomol.* **2013**, *107*, 910–919. [CrossRef] [PubMed]
45. Prado, M.N.K.; Garzo, E.; Moreno, A.; Navas-Castillo, J.; Fiallo-Olivé, E.; Spotti Lopes, J.R.; Fereres, A. Stylet penetration activities of the whitefly *Bemisia tabaci* associated with inoculation of the crinivirus tomato chlorosis virus. *J. Gen. Virol.* **2017**, *98*, 1515–1520. [CrossRef] [PubMed]
46. Jacobson, A.L.; Kennedy, G.G. Electrical penetration graph studies to investigate the effects of cyantraniliprole on feeding behavior of *Myzus persicae* (Hemiptera: Aphididae) on *Capsicum annuum*. *Pest. Manag. Sci.* **2014**, *70*, 836–840. [CrossRef] [PubMed]
47. Luo, X.; Yen, A.L.; Powell, K.S.; Wu, F.; Wang, Y.; Zeng, L.; Yang, Y.; Cen, Y. Feeding behavior of *Diaphorina citri* (Hemiptera: Liviidae) and its acquisition of '*Candidatus* Liberibacter asiaticus', on huanglongbing-infected *Citrus reticulata* leaves of several maturity stages. *Fla. Entomol.* **2015**, *98*, 186–192. [CrossRef]
48. Wu, T.; Luo, X.; Xu, C.; Wu, F.; Qureshi, J.A.; Cen, Y. Feeding behavior of *Diaphorina citri* and its transmission of '*Candidatus* Liberibacter asiaticus' to citrus. *Entomol. Exp. Appl.* **2016**, *161*, 104–111. [CrossRef]
49. Roddee, J.; Kobori, Y.; Hanboonsong, Y. Characteristics of sugarcane white leaf phytoplasma transmission by the leafhopper *Matsumuratettix hiroglyphicus* (Matsumura). *Entomol. Exp. Appl.* **2019**, *167*, 108–117. [CrossRef]
50. Van Lenteren, J.V.; Noldus, L. Whitefly-plant relationships: Behavioural and ecological aspects. *Whiteflies Their Bionomics Pest Status Manag.* **1990**, *47*, 49.
51. Chaiareekitwat, S.; Latif, S.; Mahayothee, B.; Khuwijitjaru, P.; Nagle, M.; Amawan, S.; Müller, J. Protein composition, chlorophyll, carotenoids, and cyanide content of cassava leaves (*Manihot esculenta* Crantz) as influenced by cultivar, plant age, and leaf position. *Food Chem.* **2022**, *372*, 131173. [CrossRef]
52. Huang, T.; Jander, G.; de Vos, M. Non-protein amino acids in plant defense against insect herbivores: Representative cases and opportunities for further functional analysis. *Phytochemistry* **2011**, *72*, 1531–1537. [CrossRef]
53. Van-Ohlen, M.; Herfurth, A.M.; Kerbstadt, H.; Wittstock, U. Cyanide detoxification in an insect herbivore: Molecular identification of β-cyanoalanine synthases from Pieris rapae. *Insect Biochem. Mol. Biol.* **2016**, *70*, 99–110. [CrossRef]
54. Heinz, K.M.; Zalom, F.G. Variation in trichome based resistance to *Bemisia argentifolii* (Homoptera: Aleyrodidae) oviposition on tomato. *J. Econ. Entomol.* **1995**, *88*, 1494–1502. [CrossRef]
55. Taggar, G.; Gill, R. Preference of whitefly, *Bemisia tabaci*, towards black gram genotypes: Role of morphological leaf characteristics. *Phytoparasitica* **2012**, *40*, 461–474. [CrossRef]
56. Xing, Z.; Liu, Y.; Cai, W.; Huang, X.; Wu, S.; Lei, Z. Efficiency of trichome-based plant defense in Phaseolus vulgaris depends on insect behavior, plant ontogeny, and structure. *Front. Plant Sci.* **2017**, *8*, 2006. [CrossRef] [PubMed]
57. Roddee, J.; Wangkeeree, J.; Hanboonsong, Y. Identification and evaluation of sugarcane cultivars for antixenosis resistance to the leafhopper *Yamatotettix flavovittatus* Matsumura (Hemiptera: Cicadellidae). *Plants* **2024**, *13*, 2299. [CrossRef]
58. Jindal, V.; Dhaliwal, G. Mechanisms of resistance in cotton to whitefly (*Bemisia tabaci*): Antixenosis. *Phytoparasitica* **2011**, *39*, 129–136. [CrossRef]
59. Hasanuzzaman, A.T.M.; Islam, M.N.; Zhang, Y.; Zhang, C.Y.; Liu, T.X. Leaf morphological characters can be a factor for intra-varietal preference of whitefly *Bemisia tabaci* (Hemiptera: Aleyrodidae) among eggplant varieties. *PLoS ONE* **2016**, *11*, e0153880. [CrossRef]
60. Andrade, M.C.; Da Silva, A.A.; Neiva, I.P.; Oliveira, I.R.C.; De Castro, E.M.; Francis, D.M.; Maluf, W.R. Inheritance of type IV glandular trichome density and its association with whitefly resistance from *Solanum galapagense* accession LA1401. *Euphytica* **2017**, *213*, 52. [CrossRef]
61. Lutfi, M.; Hidayat, P.; Maryana, N. Correlation between epidermis thickness, leaf trichome length and density with the whitefly *Bemisia tabaci* population on five local soybean cultivars. *J. Perlindungan Tanam. Indones.* **2019**, *23*, 23–31. [CrossRef]
62. Amini, S.N.; Golizadeh, A.; Tafaghodinia, B.; Razmjou, J.; Abbasipour, H. Interaction between host plant morphological characteristics with life history of the greenhouse whitefly *Trialeurodes vaporariorum*. *Arthropod-Plant Interact.* **2021**, *15*, 875–885. [CrossRef]

63. Kariyat, R.R.; Raya, C.E.; Chavana, J.; Cantu, J.; Guzman, G.; Sasidharan, L. Feeding on glandular and non-glandular leaf trichomes negatively affect growth and development in tobacco hornworm (*Manduca sexta*) caterpillars. *Arthropod-Plant Interact.* **2019**, *13*, 321–333. [CrossRef]
64. Hao, Z.P.; Zhan, H.X.; Gao, L.L.; Huang, F.; Zhu, L.N.; Hou, S.M. Possible effects of leaf tissue characteristics of oilseed rape *Brassica napus* on probing and feeding behaviors of cabbage aphids *Brevicoryne brassicae*. *Arthropod-Plant Interact.* **2020**, *14*, 733–744. [CrossRef]
65. Tjallingii, W.F.; Garzo, E.; Fereres, A. New structure in cell puncture activities by aphid stylets: A dual-mode EPG study. *Entomol. Exp. Appl.* **2010**, *135*, 193–207. [CrossRef]
66. Soundararajan, R.P.; Baskaran, P. Mechanisms of resistance in brinjal (*Solanum melongena* L.) to whitefly *Bemisia tabaci* (Gennadius). *Madras Agric. J.* **2001**, *88*, 657–659. [CrossRef]
67. Singh, D.; Jaglan, R.S.; Singh, R. Leaf morphological characteristics of brinjal in relation to whitefly incidence. *Haryana J. Hort. Sci.* **2002**, *31*, 289–291.
68. Ayyasamy, R.; Baskaran, P. Influence of certain leaf characters of brinjal accessions with incidence of *Bemisia tabaci*. *J. Food Agric. Environ.* **2005**, *3*, 333–334.
69. Shibuya, T.; Hirai, N.; Sakamoto, Y.; Komuro, J. Effects of morphological characteristics of *Cucumis sativus* seedlings grown at different vapor pressure deficits on initial colonization of *Bemisia tabaci* (Hemiptera: Aleyrodidae). *J. Econ. Entomol.* **2009**, *102*, 2265–2267. [CrossRef]
70. Oriani, G.M.A.; Vendramim, J.D.; Vasconcelos, C.J. Biology of *Bemisia tabaci* (Genn.) B biotype (Hemiptera: Aleyrodidae) on tomato genotypes. *Sci. Agric.* **2011**, *68*, 37–41. [CrossRef]
71. Silva, G.F.D.; Paulo, J.; Baldin, L.; Luiz, E.; Souza, S.D.; Lourencao, E.Y.; Lourenção, A.L. Assessing *Bemisia tabaci* (Genn.) biotype b resistance in soybean genotypes: Antixenosis and antibiosis. *Chilean J. Agric. Res.* **2012**, *72*, 516–522. [CrossRef]
72. Rustamani, M.A.; Khatri, I.; Leghari, M.H.; Sultana, R.; Mandokhail, A.S. Trichomes of cotton leaf as an aspect of resistance to sucking insect pests. *Sindh. Univ. Res. J. Sci. Ser.* **2014**, *46*, 351–356.
73. Kollar, A.; Seemüller, E.; Bonnet, F.; Saillard, C.; Bove, J. Isolation of the DNA of various plant pathogenic mycoplasmalike organisms from infected plants. *Phytopathology* **1990**, *80*, 233–237. [CrossRef]
74. Doyle, J.J.; Doyle, J.L. A rapid DNA isolation procedure for small quantities of fresh leaf tissue. *Phytochem. Bull.* **1987**, *19*, 11–15.
75. Nakashima, K.; Kato, S.; Iwanami, S.; Murata, N. DNA probes reveal relatedness of rice yellow dwarf mycoplas malike organisms (MLOs) and distinguish them from other MLOs. *Appl. Environ. Microbiol.* **1993**, *59*, 1206–1212. [CrossRef] [PubMed]
76. Sarria, E.; Cid, M.; Garzo, E.; Fereres, A. Excel workbook for automatic parameter calculation of EPG data. *Comput. Electron. Agr.* **2009**, *67*, 35–42. [CrossRef]
77. Ebert, T.A.; Backus, E.A.; Cid, M.; Fereres, A.; Rogers, M.E. A new SAS program for behavioral analysis of electrical penetration graph data. *Comput. Electron. Agric.* **2015**, *116*, 80–87. [CrossRef]
78. Roddee, J.; Backus, E.A.; Wangkeeree, J.; Hanboonsong, Y. Alteration in the Stylet Probing Behavior and Host Preference of the Vector *Matsumuratettix hiroglyphicus* (Hemiptera: Cicadellidae) After Infection with Sugarcane White Leaf Phytoplasma. *J. Econ. Entomol.* **2021**, *14*, 1081–1090. [CrossRef]
79. Roddee, J.; Wangkeeree, J.; Backus, E.A.; Hanboonsong, Y. Probing behavior of the leafhopper analyzed through DC electropenetrography and microscopy. *J. Insect Physiol.* **2023**, *151*, 104584. [CrossRef]
80. Backus, E.A.; Cline, A.R.; Ellerseick, M.R.; Serrano, M.S. *Lygus hesperus* (Hemiptera: Miridae) feeding on cotton: New methods and parameters for analysis of nonsequential electrical penetration graph data. *Ann. Entomol. Soc. Am.* **2007**, *100*, 296–310. [CrossRef]
81. Moberly, J.G.; Bernards, M.T.; Waynant, K.V. Key features and updates for Origin 2018. *J. Cheminform.* **2018**, *10*, 5. [CrossRef]
82. Ghaffer, A.B.M.B.; Pritchard, J.; Ford-Lloyd, B. Brow planthopper (*N. lugens* Stal) feeding behavior on rice germplasm as an indicator of resistance. *PLoS ONE* **2011**, *6*, e22137. [CrossRef]
83. Liu, T.; Liu, H.; Wang, Y.; Yang, Y. Climate change impacts on the potential distribution pattern of *Osphya* (Coleoptera: Melandryidae), an old but small beetle group distributed in the Northern Hemisphere. *Insects* **2023**, *14*, 476. [CrossRef]

plants

MDPI

Article

Conflicting Dynamics of Galling and Pollination: *Arastichus gallicola* (Hymenoptera, Eulophidae), a Specialized Parasitic Galler in Pistillate Flowers of *Thaumatophyllum bipinnatifidum* (Araceae)

Sergio Jansen-González [1,2,*], Simone P. Teixeira [3] and Rodrigo A. S. Pereira [4]

1 Programa de Pós-Graduação em Entomologia, Faculdade de Filosofia, Ciências e Letras de Ribeirão Preto, Universidade de São Paulo, Ribeirão Preto 14040-130, SP, Brazil
2 Escuela de Biología, Centro de Investigación en Biodiversidad y Ecología Tropical, Universidad de Costa Rica, San Pedro de Montes de Oca, San José 11501-2060, Costa Rica
3 Departamento de Ciências Farmacêuticas, Faculdade de Ciências Farmacêuticas de Ribeirão Preto, Universidade de São Paulo, Ribeirão Preto 14040-903, SP, Brazil; spadua@fcfrp.usp.br
4 Departamento de Biologia, Faculdade de Filosofia, Ciências e Letras de Ribeirão Preto, Universidade de São Paulo, Ribeirão Preto 14040-130, SP, Brazil; raspereira@usp.br
* Correspondence: sergio.jansen.gonzalez@gmail.com

Abstract: In the complex dynamics of plant–insect interactions, the specialized galling of reproductive structures presents unique evolutionary adaptations. This study investigates the parasitic relationship between *Arastichus gallicola* (Hymenoptera, Eulophidae), an ovary-galling wasp, and the inflorescences of *Thaumatophyllum bipinnatifidum* (Araceae). We employed field experiments and histological analyses to investigate the mechanisms driving this interaction. We reveal that ovule fertilization is not required for gall formation; however, pollination substantially enhances gall retention by reducing inflorescence abscission. Inflorescences exposed solely to galling presented a 64% abscission rate, whereas those with both galling and pollination experienced 33% abscission, underscoring pollination's role in mitigating inflorescence loss. Detailed observations of *A. gallicola* oviposition and larval development reveal the intricate gall formation process characterized by progressive tissue hypertrophy surrounding the larva. Galling and seed development were mutually exclusive, with only 9% of fruits containing both. This mutual exclusivity suggests a competitive interaction for developmental resources within the ovary. Our findings underscore the specialized larval biology of galling chalcid wasps, illustrating how interactions between gall formation and host reproductive strategies shape the evolution of gall induction in floral tissues. Our study advances the understanding of ovary-galling adaptations and the selective pressures shaping antagonistic and mutualistic interactions in plant reproductive structures.

Keywords: Chalcidoidea; gall-inducing insects; gall development; insect–plant interactions

1. Introduction

The interaction between plants and gall-inducing insects represents one of the most intricate and intimate plant–animal relationships. Gall formation, typically induced by holometabolous insects, begins with the female insect ovipositing into specific plant tissues, followed by the development of the larva within those tissues [1,2]. This process's success mainly depends on the female's ability to select the appropriate target tissue and initiate gall formation, which protects the egg and the first larval instar. This initiation is facilitated by tissue damage during oviposition and the injection of hormone-like substances or elicitors [3,4], potentially involving molecular or microbiological factors [5]. The selected tissue is usually in a physiologically reactive, undifferentiated stage, making it particularly susceptible to the gall-inducing factors introduced by the female and the subsequent larval secretions [6]. These factors induce cell division, tissue de-differentiation, hypertrophy, and the localized accumulation of nutrients and chemicals around the developing larva [7]. This intricate manipulation ensures that the gall remains attached to the host plant, promoting its continued growth, maintenance, and modification in favor of the galler.

In this study, we assessed the interaction between *Arastichus gallicola* (Zhang, Gates, Hanson & Jansen-González, 2022), a chalcid wasp (Eulophidae), and the inflorescences of *Thaumatophyllum bipinnatifidum* (Schott ex Endl.) Sakur. and Calazans & Mayo (Araceae). These species were selected as they represent a suitable system for investigating the relationships among pollination, inflorescence-abscission risk, and gall formation. The wasp was initially identified as belonging to the genus *Exurus*. However, a recent study reclassified it as a new genus, *Arastichus*, associated with the inflorescences of *T. solimoesense* (A.C. Smith) Sakur., Calazans & Mayo, and *T. bipinnatifidum* [8,9].

The difficulty of gall induction varies depending on the location and origin of the target tissue. Vegetative structures, which are generally less complex at the cellular and ontogenetic levels, are likely less challenging to gall than reproductive structures, where tissues of both gametophytic and sporophytic origins coexist within a single organ, such as flower ovaries or seeds [10–12]. The structural complexity of reproductive tissues not only complicates gall formation but also increases the risk of abscission, particularly when the retention of the organ hosting the gall depends on external processes like pollination, fertilization, and embryogenesis [10,13,14]. Consequently, it is unsurprising that surveys frequently report more insect galls in vegetative tissues than in reproductive tissues [15], a trend observed even in so-called "super host" plants [16]. These findings support the notion that galling reproductive structures is inherently more challenging, requiring additional adaptations to deal with complex ontological processes to mitigate abscission risk.

The mega-diverse superfamily Chalcidoidea (Hymenoptera) represents one of the few insect groups that have successfully evolved strategies to gall plant reproductive tissues. Despite most chalcid species being entomophagous parasitoids, gall-inducing habits have independently arisen in this group [17]. Phytophagous chalcid gallers exhibit a fascinating spectrum of interactions, ranging from highly refined mutualism to parasitism, each with distinct strategies to deal with the complexities of tissue ontogeny and the risk of abscission. The interaction between fig trees (genus *Ficus*) and fig wasps (Hymenoptera, Agaonidae) suggests that pollination and larval development strategies play a role in mitigating inflorescence abscission. In this obligatory nursery pollination mutualism, the wasps induce galls within ovules with just-fertilized embryo sacs, synchronizing their larval development with the early stages of plant embryogenesis. Inflorescence abscission is prevented by the pollination facilitated by the agaonid wasps, which specialize in galling the nucellus and endosperm. Thus, the wasp larvae rely strongly on the plant's embryological processes to complete their development [10].

Among fig wasps, certain non-pollinating fig wasps (NPFWs) from the families Epichrysomallidae and Pteromalidae also induce galls in fig ovaries. In NPFWs, synchronization between larval development and plant embryogenesis occurs; however, gall induction occurs in the nucellus, an ovular tissue whose formation does not depend on fertilization. This allows these wasps to adapt their development to ovule conditions irrespective of whether fertilization occurred [11,18,19]. The risk of abscission, in this case, is reduced either by ovipositing in pollinated fig flowers or, in the absence of pollination, through an as-yet-unknown mechanism where a higher number of galls appears to correlate with a lower likelihood of abscission [9]. Conversely, for chalcid seed predators, the strategy involves ovipositing in fruits and seeds at their early development stages. The larva then grows alongside the developing seed without notably disrupting embryogenesis and consumes the seed at a more advanced stage, when the risk of abscission is reduced [20,21]. The distinction between seed predation and gall-inducing feeding strategies is often blurred, as some species' larvae can induce hypertrophy in seed tissues, resembling the effects typically caused by true gallers [21].

The described adaptive strategies suggest no universal mechanisms govern inflorescence-abscission mitigation across different plant–insect systems. Further investigation into additional insect–plant interactions is necessary to identify the mechanisms shaping the dynamics of gall induction and inflorescence-abscission prevention. Therefore, we employed field experiments and developmental studies involving flower dissection and histological analysis to investigate the galling process and the relationship between pollination, inflorescence-abscission risk, and gall formation. We demonstrate that *A. gallicola* is a specialized ovary galler in *T. bipinnatifidum* plants. We also reveal that ovule fertilization is not required for gall formation, with pollination playing a crucial role only in preventing inflorescence abscission.

2. Results

2.1. Effects of Pollination on Inflorescence Retention

The abscission frequency significantly differed across the experimental treatments ($\chi^2 = 27.87$, degrees of freedom = 3, $p < 10^{-3}$). The treatment in which flowers were neither pollinated nor exposed to female wasps resulted in 100% inflorescence abscission. Flower pollination resulted in a 33% inflorescence-abscission rate, regardless of wasp exposure. In the treatment where flowers were exposed to wasps but not pollinated, the abscission rate was 64% (Table 1).

Table 1. Percentage of inflorescence abscission in *Thaumatophyllum bipinnatifidum* under different wasp oviposition/pollination treatments. The total number of inflorescences per treatment is shown in parentheses. Treatments sharing the same letter did not differ based on chi-squared pairwise comparisons with Bonferroni correction at the 5% significance level.

Treatment (Wasp/Pollination)	Percentage of Abscission
+/+	33.0 (15) [a]
+/−	64.3 (28) [a]
−/+	33.0 (24) [a]
−/−	100 (24) [b]

The mean number of fruits produced per inflorescence did not differ significantly across treatments, ranging from 231.2 to 313.5. In the treatments where flowers were pollinated, the number of fruits containing seeds was, on average, 33% higher in the inflorescences not exposed to the wasps. Regarding the wasp-exposed treatments, unpollinated inflorescences produced 40% more fruits exclusively containing galls. Few fruits (10.3%) produced a combination of galls and seeds in the treatment where flowers were pollinated and exposed to the wasps. The number of empty flowers, i.e., those containing neither seeds nor galls, did not differ significantly across treatments, ranging from 25 to 30 on average (Table 2).

2.2. Female Oviposition, Larval Development, and Gall Formation

Adult females inserted their ovipositors through the stigma or ovarian walls of the pistillate flowers (Figure 1a–c). The egg stage was observed from 1 to 5 days after oviposition. The eggs are elliptical (length: 0.218 ± 0.022 mm, width: 0.070 ± 0.007 mm, mean ± SD, $n = 20$ eggs) with a long pedicel several times longer than the egg (Figure 2a). Eggs are deposited individually or in groups of up to four within the locular space, each attached to the ovule funicle by its long pedicel (Figures 1d and 3a,b).

Table 2. Mean number (±standard deviation) of galls and seeds per inflorescence or infructescence in *Thaumatophyllum bipinnatifidum* across different treatment combinations. *n* is the number of examined inflorescences, and NA indicates the treatments for which, by definition, the data were not available.

Variable	Treatments (Wasp/Pollination)			ANOVA
	+/+	+/−	−/+	
Fruits (total)	313.5 ± 110.4	231.2 ± 60.7	265.3 ± 115.4	$F_{2,33} = 1.652, p = 0.207$
Fruits with only seeds	125.5 ± 112.7	NA	235.4 ± 99.3	$F_{1,24} = 6.8, p = 0.015$
Flowers with only galls	123.0 ± 109.7	206.3 ± 57.9	NA	$F_{1,18} = 4.51, p = 0.048$
Fruits with galls + seeds	32.3 ± 37.9	NA	NA	-
Empty flowers	32.7 ± 33.3	24.9 ± 19.4	29.9 ± 29.2	$F_{2,33} = 0.198, p = 0.821$
n *	10	10	16	-

* Due to inflorescence abscission, the treatment results were unbalanced across the experiments conducted on the nine *T. bipinnatifidum* individuals. Consequently, a mixed-effects model approach could not be applied to the data analysis. Therefore, *n* represents a combination of biological and technical replicates.

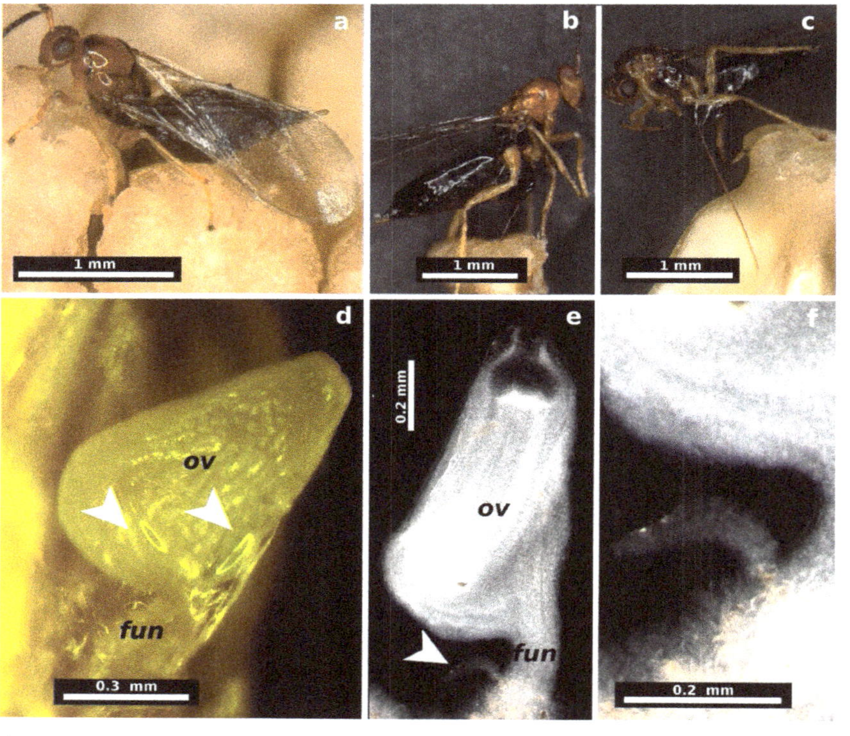

Figure 1. Oviposition in *Thaumatophyllum bipinnatifidum* flowers and early developmental stages of *Arastichus gallicola*. (**a**) Close-up of female wasp during oviposition; (**b**,**c**) ovipositor insertion sites: (**b**) through the flower stigma and (**c**) through the ovary wall; (**d**) detail of egg placement, with arrows indicating two eggs attached to the ovule base, each anchored to the funicle by a long peduncle; (**e**) position of the first larval instar (arrow) near the ovule funicle; (**f**) close-up of the first larval instar. Abbreviations: *fun* = ovule funicle, *ov* = plant ovule.

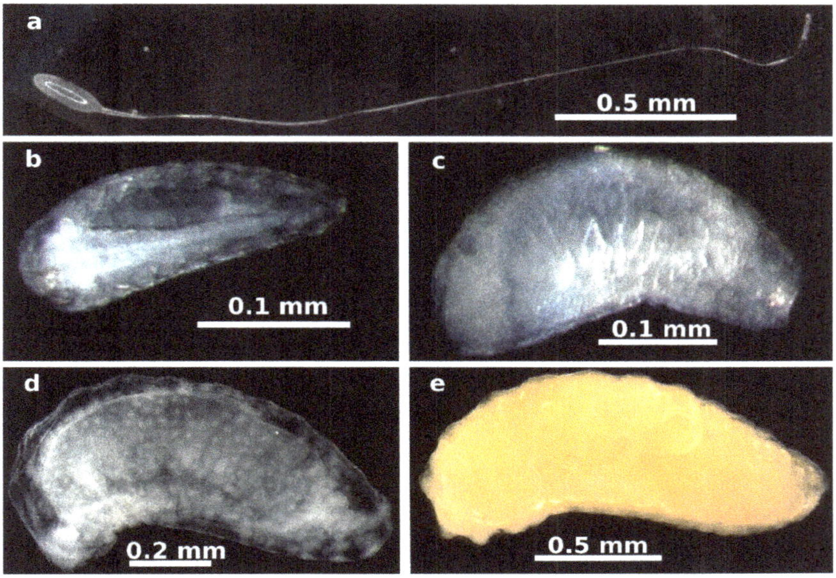

Figure 2. Immature stages of *Arastichus gallicola*. (**a**) Egg with peduncle; (**b**) first larval instar at 7 days; (**c**) first larval instar at 17 days; (**d**) second larval instar at 25 days; (**e**) third larval instar at 37 days.

We identified three larval instars by observing molts at two stages of larval development, where the cephalic capsule of the previous instar remained attached to the larva's body.

The first instar began five days after oviposition. At this stage, the larva was attached to the ovule funicle within the ovarian locule, likely feeding on the mucilaginous material filling the locule (Figures 1e,f and 3c–f). The larvae at this stage measured 0.312 ± 0.097 mm in length and 0.113 ± 0.050 mm in width (mean ± SD, n = 60 larvae, Figure 2b).

The second instar was observed starting 20 days after oviposition, with larvae measuring 1.193 ± 0.471 mm in length and 0.475 ± 0.172 mm in width (n = 40 larvae, Figure 2c). The gall began to form at this stage, characterized by hypertrophy of the plant ovule tissues to which the larva was attached (Figure 4). The increase in ovule volume was driven by disorganized cell division within the ovule's integuments and nucellus. Similar modifications were observed in the cells of the locular septa and the ovary walls adjacent to the larva (Figure 5a,b). These cells exhibited enlarged vacuoles and starch grains in the cytoplasm (Figure 4b,c). As the modified tissue grew, it enveloped the larva, forming gall tissue composed of hypertrophied cells from the flower ovule and adjacent tissues (Figure 5c).

The third instar was observed 30 days after oviposition, with larvae measuring 2.407 ± 0.562 mm in length and 0.791 ± 0.137 mm in width (n = 40 larvae, Figure 2d,e). By this stage, the gall was fully formed, and the larva continued to feed on the hypertrophied tissue within the gall (Figure 6a,b). Pupation occurs inside the gall, and adult wasps emerge when the spathe detaches from the ripe infructescence.

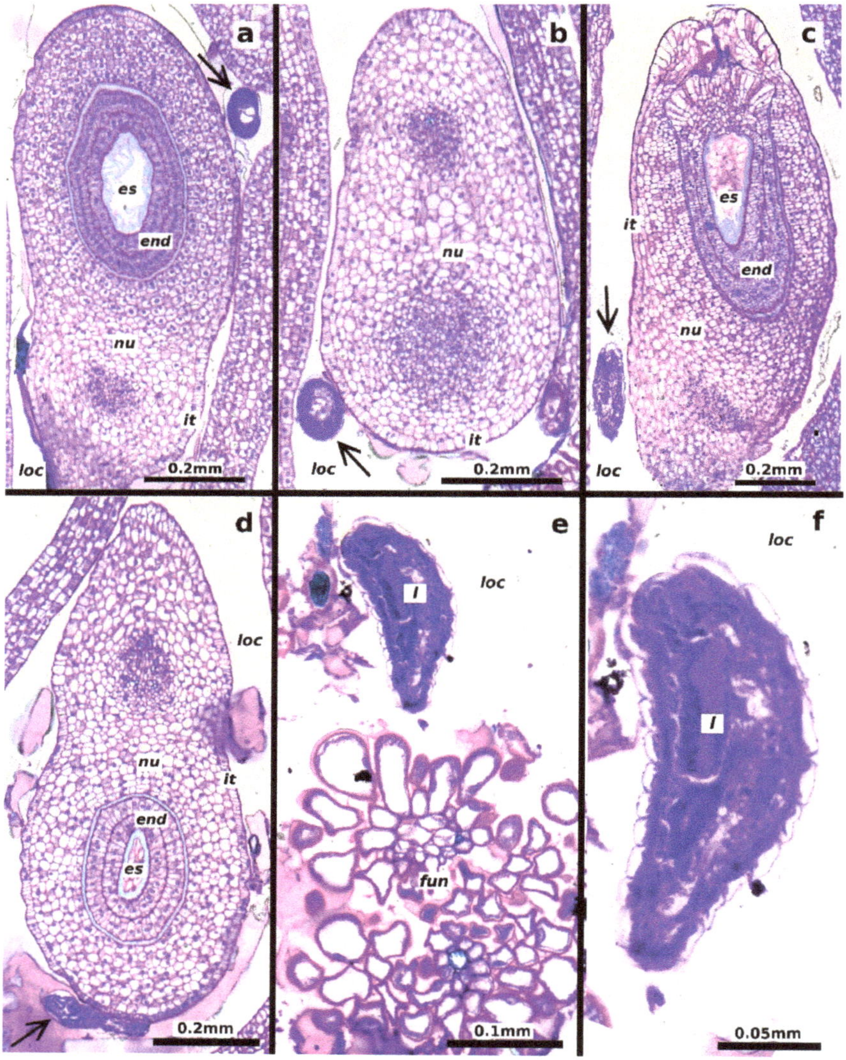

Figure 3. Longitudinal sections of *Thaumatophyllum bipinnatifidum* ovules showing early developmental stages of *Arastichus gallicola*. (**a**,**b**) Position of the wasp egg (arrows) relative to the ovule; (**c**–**f**) position of the first-instar larva (arrow) relative to the ovule; (**e**) detail of the larva near elements of the ovule funicle; (**f**) close-up of the larva. Abbreviations: *end* = endothelium, *es* = embryo sac, *fun* = funicle, *it* = inner integument, *l* = larva, *loc* = ovarian locule, *nu* = nucellus.

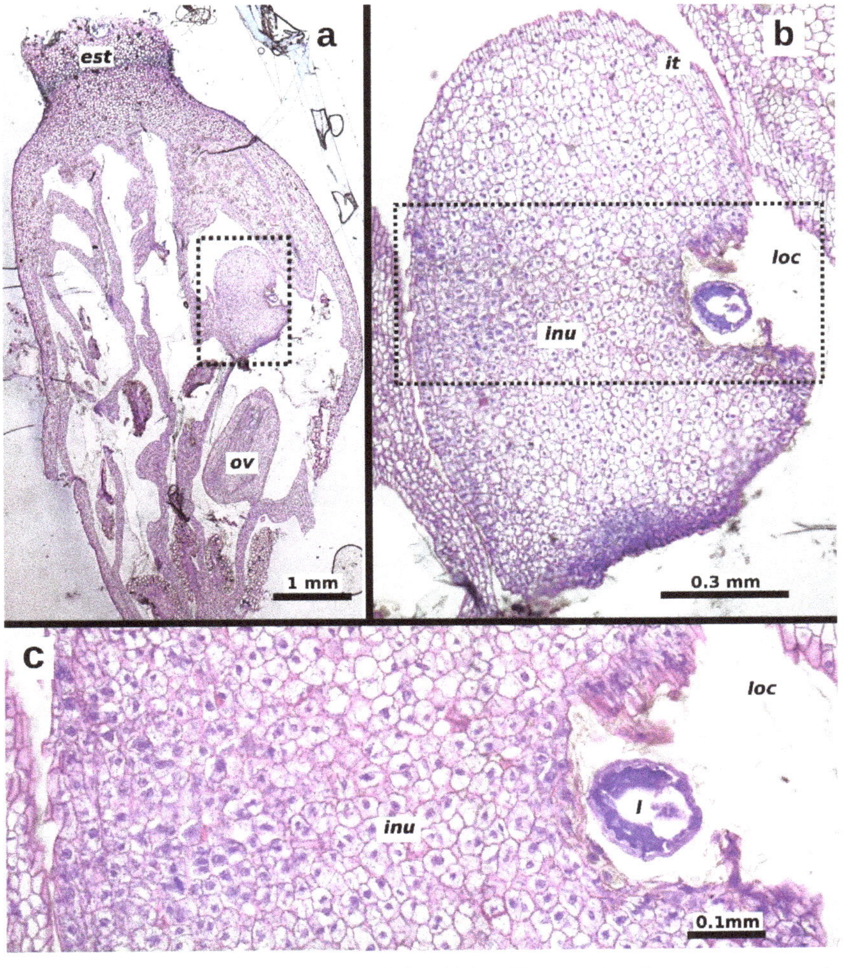

Figure 4. Longitudinal sections of *Thaumatophyllum bipinnatifidum* ovaries, showing an ovule undergoing gall formation and an unaffected ovule (ov), 20–25 days after oviposition. (**a**) Pistillate flower with affected (dotted rectangle) and unaffected ovules; (**b**) close-up of the affected ovule, showing the ovule elements; (**c**) close-up of the dotted area in (**b**), highlighting hypertrophied cells surrounding the larva, indicative of gall formation. Abbreviations: *inu* = induced nucellus, *it* = inner integument, *loc* = ovarian locule, *ov* = unaffected ovule.

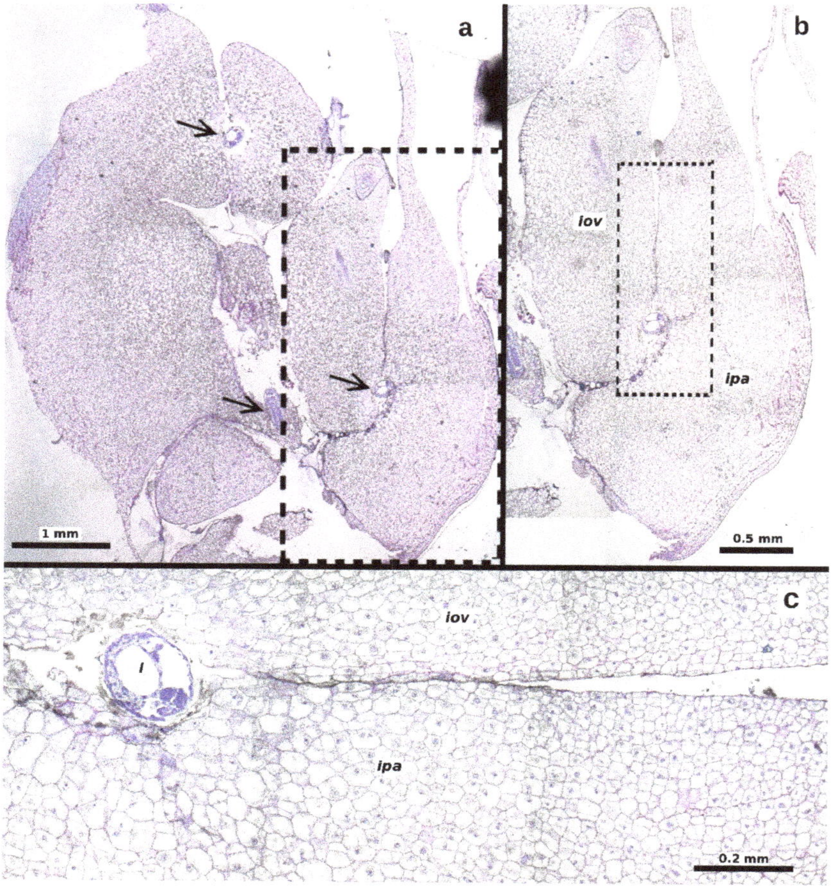

Figure 5. Longitudinal sections of a *Thaumatophyllum bipinnatifidum* ovary with galled ovules containing larvae (arrows). (**a**) Overview showing three larvae within ovule galls; (**b**) close-up of the dotted area in (**a**), detailing the affected ovule and parenchyma; (**c**) close-up of the dotted area in (**b**), illustrating hypertrophied cells throughout the tissue surrounding the larva. Abbreviations: *iov* = induced ovary tissue, *ipa* = induced parenchymal elements, *l* = larva.

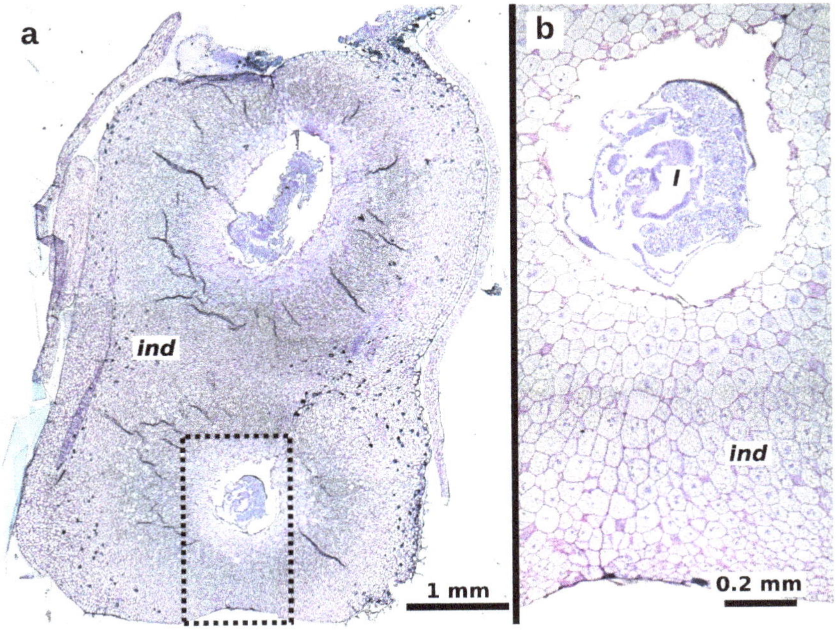

Figure 6. Longitudinal sections of fully developed galls of *Arastichus gallicola* in *Thaumatophyllum bipinnatifidum* ovary. (**a**) Section showing a galled ovary with two galls, each containing a larva; (**b**) close-up of the dotted area in (**a**), highlighting induced parenchymal elements. Abbreviations: *ind* = induced parenchymal elements, *l* = larva.

3. Discussion

Our study demonstrates that *Arastichus gallicola* is an ovary galler in the aroid *T. bipinnatifidum*. Insect eggs can be deposited in multiple ovules, with a single flower ovary capable of bearing several galls. This finding is consistent with previous observations that *A. gallicola* can induce multiple galls within a single ovary of *T. solimoesense* [9]. The females of *A. gallicola* oviposit within the ovaries of the pistillate flowers during anthesis, with the eggs being precisely attached to the ovule funicle. The precise egg deposition likely plays a crucial role for chalcid wasps specialized in galling plant ovarian structures. Similar sophisticated oviposition strategies have independently evolved in three chalcid families associated with fig trees, including the pollinating fig wasps (Agaonidae), the *Idarnes* species group *flavicollis* (Pteromalidae), and the genus *Sycobia* (Epichrysomallidae) [19,22]. In these lineages, regardless of whether the wasps oviposit from within the fig cavity or from the external fig surface, the ovipositor is inserted through the flower style, and the egg is deposited near the stylar canal entry [10,11]. These observations support the hypothesis that precise ovipositor insertion is essential to ensure undisturbed and successful larval development.

The larva of *A. gallicola* appears to be responsible for the gall-inducing process, as the formation of undifferentiated and hypertrophied gall tissue with a parenchymatic appearance begins when the larva changes to the second instar and comes into contact with the ovule and surrounding ovarian tissues, such as the septa and walls. Gall-inducing insects secrete effector molecules, including phytohormones, which stimulate surrounding tissue growth [3,4,23,24]. Borges [25] highlights that these insects produce auxins and cytokinins at levels manifold higher than those in ungalled plants, creating localized hotspots of tissue growth. A recent study suggests that gall induction by a cecidomyiid insect in cassava involves the integration of bacterial DNA vectorized by the insect into the plant's genome. The bacterial DNA was linked to genes influencing plant cell transformation and the ubiquitin–proteasome system, a regulatory mechanism involved in protein turnover and developmental processes [5]. In the case of *A. gallicola*, gall formation begins approximately 20 days after oviposition. The delay between oviposition and gall formation suggests that the mechanical stimulus from ovipositor insertion or any substances injected by the female during oviposition plays a limited role in initiating the gall. However, Elias et al. [26] demonstrated that venom glands in gall-inducing and non-galling fig wasp species have distinct peptide profiles, suggesting that the secretions injected during oviposition by gall-inducing species may play a role in modulating plant tissue and initiating gall formation.

Although the gall-inducing habit has evolved multiple times in six Chalcidoidea families [17], to the best of our knowledge, ovary chalcid gallers have been confirmed only in wasp species associated with fig trees and the aroid genus *Thaumatophyllum*, although ovule gall induced by the Ichneumonoidea *Allorhogas uberlandiensis* Joele & Zaldívar-Riverón, 2019 has been reported for *Miconia chamissois* Naudin (Melastomataceae) [12,27]. Additionally, seed predation without inducing abnormal tissue growth is known in genera such as *Megastigmus* (Torymidae: Megastigminae) and *Bephratelloides* (Eurytomidae) [20,21]. In the pollinating fig wasps and seed predators, which depend on embryo sac fertilization (in angiosperms) or megagametophyte development (in gymnosperms), the larvae employ a koinobiont strategy, allowing the seed to continue growing while feeding on its nutrient-rich tissues. This strategy likely evolved to avoid disrupting the proper embryonic development of the plant. In these instances, during the initial larval development, pronounced cellular modifications in the ovarian tissues in contact with the larva are not observed [10,28]. Conversely, in galling species that do not require pollination for larval development, such as *Idarnes* [11] and *Arastichus* (present study), the modification of the ovarian tissues surrounding the larvae is pronounced, resulting in aggressive gall development characterized by rapid and disorganized cell division.

Lack of pollination often leads to the abscission of flowers or inflorescences [14]. The risk of organ abscission may represent a substantial selective pressure on gallers associated with the reproductive structures of plants. This assertion is supported by the evidence that the capacity to avoid abscission has evolved in all chalcid galling lineages that do not depend on fertilized embryo sacs for larval development. Our experimental study confirmed that inflorescences with unpollinated flowers of *T. bipinnatifidum* abscise readily. However, our findings also demonstrate that unpollinated inflorescences containing galls of *A. gallicola* are less likely to abscise than solely unpollinated inflorescences, likely due to an unknown sink effect directed toward the galled organ [29].

A similar effect on preventing abscission is observed in Chalcidoidea and Ichneumonoidea wasps. Several *Megastigmus* species associated with Pinaceae species lay eggs inside the developing ovules of their host conifers before pollination occurs [20]. In *M. spermotrophus*, it has been demonstrated that the megagametophyte containing the insect larva continues to develop even without pollination [28]. Introduced *Ficus* species also provide evidence that ovary gallers can prevent inflorescence abscission. For instance, in Brazil, *F. benjamina* and *F. microcarpa* were introduced for ornamental purposes, and their associated wasps arrived decades after these introductions. The first species to recolonize these plants were ovary-galling NPFWs (*Sycobia hodites* and *Walkerella microcarpae* in *F. benjamina* and *F. macrocarpa*, respectively), which could prevent the abscission of unpollinated figs [18,30–32]. The ability to prevent fig abscission has also been experimentally demonstrated in a species of the *Idarnes* group *flavicollis* associated with the neotropical *F. citrifolia* [11,33]. Similarly, *Allorhogas* sp. (Ichneumonoidea), which induces galls in *Miconia calvescens* (Melastomataceae), also prevents fruit abscission, as higher rates of infestation were observed in plants with advanced phenology, where most fruits were mature [34].

We demonstrated that pollination influences the interaction between *T. bipinnatifidum* and *A. gallicola*. Although *A. gallicola* can induce galls in ovules with unfertilized embryo sacs, pollination enhances the insect's reproductive success by significantly reducing the risk of inflorescence abscission. However, when wasps were exposed to pollinated flowers, they succeeded in inducing galls in approximately half of the available flowers. Moreover, gall formation and seed development were mutually exclusive events, with only 9% of the fruits containing both galls and seeds. While our study design did not allow us to uncover the mechanism behind this pattern, it suggests that seed formation may interrupt gall development, or conversely, gall formation may inhibit seed development. Notably, the rate of fruit abortion (i.e., fruits not producing seeds or bearing galls) did not significantly differ across the experimental treatments, suggesting that the experimental conditions did not lead to differential larval mortality or seed abortion.

Our findings on the negative relationship between seed development and gall formation provide valuable insights into the selective pressures that may drive the evolution of the ovary-galling habit in insects and offer a plausible hypothesis for why this life history strategy has emerged in a limited number of biological systems. Within the framework of embryo sac fertilization negatively impacting gall development and vice versa, insects are expected to be individually selected to oviposit as early as possible in the flower ovaries before pollination occurs. Conversely, plants might be selected to limit exposure of their flower ovaries by investing in mechanical barriers, such as the aroid spathe, which restricts access to a brief window during anthesis, or by synchronizing flowering phenology at the population level to constrain the maintenance of insect populations year-round. While physical barriers may not universally prevent ovary-galling, as evidenced by the sophisticated oviposition behavior of chalcid wasps, facilitated by the complex structure of their ovipositors [35,36], phenological adjustments appear to be more effective. Notably, ovary-galling is observed predominantly in plant groups associated with host-specific pollinators and a year-round flowering pattern, such as *T. bipinnatifidum* (present study) and fig trees [37].

Under natural conditions, however, the prevalence of *A. gallicola* is likely limited by other insects interacting with *T. bipinnatifidum*. In first place, there is a physical interference with the pollinating dynastid beetles, *Erioscelis emarginata* (Mannerheim, 1828) [38], which gather around the pistillate flowers in large numbers during the same period that *A. gallicola* oviposits and disrupt and limit the wasps' ability to oviposit. The parasitoid *Prodecatoma philodendri* Ferrière, 1924 (Eurytomidae) can also exert control over *A. gallicola* populations [8,9]. Furthermore, closing the spathe and the subsequent filling of the cavity between the spathe and the spadix with fluid further restricts the oviposition opportunities for any remaining female wasps inside the inflorescence.

Our study advances the understanding of ovary-galling insects by highlighting the selective pressures on both plants and insects, which can lead to divergent evolutionary outcomes, such as mutualistic brood site pollination or antagonistic seed predation. The dependence on pollination seems to play an essential role in determining these pathways. For instance, it has been speculated that the agaonid *Ficus* pollinators evolved from a gall-inducing chalcid ancestor [39]. The urn-shaped inflorescence in the Moraceae clade formed by *Ficus* and Castilleae [40] likely drove the evolution of insect biology. This particular inflorescence shape may influence how floral resources are exploited and, in turn, shape the traits of the associated insects. If pollination enhances the larval success of ancestral gallers, pollination behavior could evolve as a beneficial trait. Conversely, in scenarios where pollination does not directly impact larval success, as in plant groups previously associated with specialized pollinators such as scarab beetles and aroid plants [41], the galler interaction may evolve into an antagonistic relationship. Our findings highlight the importance of considering the intricate ecological and evolutionary contexts when examining the relationships between ovary-galling insects and their host plants.

4. Materials and Methods

4.1. Study Site and Species

The experiments were conducted at the University of São Paulo campus in Ribeirão Preto (21°10′ S; 47°48′ W), Brazil, between 2009 and 2011. The campus features extensive gardens and lawns of various spontaneous and cultivated plant species. The local climate falls under the Aw category (Tropical Savanna) according to the Köppen classification, characterized by wet summers and dry winters. The lowest average monthly temperature is recorded in July (19.5 °C), while the highest occurs in October (24.8 °C). Annual rainfall averages 1384 mm, with January being the wettest month (mean of 256 mm) and July the driest (mean of 21 mm) [42]. The field study was conducted during the rainy months (i.e., September to March)

Thaumatophyllum bipinnatifidum plants grow naturally as scandents or hemiepiphytes, blooming year-round with individuals consistently producing inflorescences. The flowers are arranged on a condensed, finger-like spadix, surrounded by a spathe (bract). The pistillate flowers are in the lower portion of the spadix, while the staminate flowers are positioned in the upper portion. The spathe remains closed until anthesis, during which a strong, sweet scent is released, attracting both its beetle pollinators and female *A. gallicola* wasps [38,43]. The wasps arrive earlier than the beetles, entering through the first available opening, and reach the base of the inflorescence where the pistillate flowers are located to oviposit in them. The spathe closes approximately 24 h after anthesis, forcing the pollinators out as the staminate flowers release pollen. About 36 h after anthesis, the spathe closes completely, and the cavity between the spathe and the spadix fills with a fluid secreted by the plant. Months later, the spathe develops a dehiscence line at its base and detaches, revealing the ripe infructescence [8,9].

4.2. Effects of Pollination and Wasp Oviposition on Inflorescence/Infructescence Retention

To evaluate the effects of pollination and wasp oviposition on inflorescence and infructescence retention and fruit/gall development, we conducted experiments using controlled combinations of pollination and oviposition. Immature inflorescences were preventively enclosed in organdy bags. Once anthesis began, as indicated by the opening of the spathe, each inflorescence was subjected to one of the following treatments: (1) manual pollination + wasps, (2) manual pollination only (positive control), (3) wasps only (no pollination), and (4) no pollination and no wasps (negative control). Manual pollination was necessary because the organdy bags prevented natural pollination of the flowers. The 'pollination + wasps' treatment examined the interaction between pollination and galling. The 'pollination only' treatment evaluated the role of pollination in preventing inflorescence abscission. The 'wasps only' treatment investigated the galling process in the absence of pollination, while the 'no pollination and no wasps' treatment assessed the inherent rate of inflorescence abscission. Treatments were randomly assigned to each inflorescence by first attributing a unique number to each one and then using the 'sample' function in the R programming environment [44] to ensure randomization.

For the controlled pollination, pollen was collected 1–2 days in advance from different *T. bipinnatifidum* individuals and stored in 10 mL glass flasks at 4 °C until use. In the experimental inflorescences, a generous amount of pollen was applied to the pistillate flower's stigmata using a fine brush.

For the oviposition treatments, ripe infructescences containing galls with wasps nearing emergence were collected in advance from different *T. bipinnatifidum* individuals. The infructescences were enclosed in organdy bags under laboratory conditions until the adult wasps emerged. To keep the wasps alive, they were fed with a 10% sucrose solution on dampened cotton balls ad libitum, and mating was allowed within the bags. Only female wasps were selected for the experiment using an entomological aspirator. Approximately 30 female wasps were introduced into each bagged inflorescence for the oviposition treatments, where they were allowed to oviposit in the pistillate flowers. In treatments that combined pollination and wasp oviposition, manual pollination was performed first, followed by the introduction of the wasps. This experiment was repeated on nine different *T. bipinnatifidum* individuals, totaling 91 manipulated inflorescences.

The inflorescences were monitored every 4–5 days, from the day the treatments were applied until the dehiscence line at the base of the spathe became evident in the retained inflorescences. Inflorescence abscissions during the monitoring period were recorded. The ripe infructescences were collected and individually enclosed in organdy bags for wasp emergence. For all inflorescences and infructescences, we quantified the total number of flowers or fruits, the total number of flowers or fruits containing galls, the total number of fruits containing seeds, and the total number of fruits containing both seeds and galls.

The treatment results for each variable were compared by ANOVA using the R version 4.4.2 programming environment. We plotted the model residuals against the quantiles of the standardized normal distribution and the expected values. No substantial deviations from the ANOVA assumptions of residual normality and variance homogeneity were detected.

4.3. *Oviposition Observations*

To document the oviposition behavior of *A. gallicola*, we conducted field observations of female wasps interacting with receptive inflorescences of *T. bipinnatifidum*. Observations included both free-living wasps in natural settings and introduced wasps under experimental conditions. Photodocumentation was performed using a Canon S5 camera with a Raynox macro lens (Tokyo, Japan). To identify the pistillate flowers where eggs were deposited, we immediately killed any female observed during the oviposition process by applying a droplet of chloroform. In the laboratory, individual flowers containing the ovipositing wasp were carefully detached and dissected under a stereoscope at 40× magnification to locate the wasp egg within each flower. We dissected approximately 15 flower ovaries. The dissected flowers were photographed using a Leica MZ16 stereomicroscope coupled with a digital camera (Wetzlar, Germany).

4.4. *Larva and Gall Development*

The gall and larval development study was conducted on two inflorescences exposed to wasp oviposition without pollination. Immature inflorescences were preventively enclosed in organdy bags. At the beginning of the anthesis, 30 female wasps obtained from a different *T. bipinnatifidum* plant were introduced into each bag. Five developing fruits were sampled every five days using fine forceps, starting from the day the inflorescence closed until the larvae completed their development. To sample these galls without removing the entire inflorescence, a door-like 10 cm × 5 cm incision was made at the base of the spathe to gain access to the inflorescence. This incision was used for all subsequent sampling during gall development. After each sampling, the door-like cut was repositioned in the spathe and secured with a metal wire around it. The inflorescence remained enclosed in the organdy bag throughout the study to prevent insects and other small animals from accessing it.

The sampled fruits were fixed in FAA 50 (formalin: acetic acid: alcohol 50% [45]) for 24 h and then stored in 70% ethanol. Of the five fruits collected per sampling, two to three were dissected under a stereomicroscope to locate the immature stages of the wasps. Images of the dissected material and larvae were captured using a Leica MZ16 stereomicroscope coupled with a digital camera. Body length measurements of all extracted immature stages were taken using the Leica™ Application Suite version 3.3.0 software. The remaining sampled fruits were reserved for histological study. The material was processed following standard dehydration and softening protocols, embedded in Leica Historesin® [46], and sectioned into 5–6 μm slices using a Leica RM 2245 microtome. Serial sections were stained with 0.05% toluidine blue, pH 4.4 [47], and mounted on slides. Histological slides were photographed using a digital camera attached to a Leica DM 4500 microscope. All histological slides and wasp samples are held by R.A.S. Pereira as voucher material.

Author Contributions: Conceptualization, S.J.-G. and R.A.S.P.; methodology, S.J.-G., S.P.T. and R.A.S.P.; investigation, S.J.-G.; resources, S.P.T. and R.A.S.P.; data curation, S.J.-G.; writing—original draft preparation, S.J.-G. and R.A.S.P.; writing—review and editing, S.J.-G., S.P.T. and R.A.S.P.; funding acquisition, S.J.-G. and R.A.S.P. All authors have read and agreed to the published version of the manuscript.

Funding: This research was funded by Fundação de Amparo à Pesquisa do Estado de São Paulo (FAPESP), grant numbers 2009/10273-9 and 2022/11911-3; Conselho Nacional de Desenvolvimento Científico e Tecnológico (CNPq), grant numbers 303986/2023-9 and 304029/2023-8; and Coordenação de Aperfeiçoamento de Pessoal de Nível Superior-Brasil (CAPES), Finance Code 001.

Data Availability Statement: The data supporting the findings of this study are available from the corresponding authors upon reasonable request. The data are not publicly available because they form part of a larger ongoing research project, and public dissemination at this stage could compromise the ability to develop and publish further findings.

Acknowledgments: We thank Edmarcio S. Campos (FCFRP/USP) for the technical assistance.

Conflicts of Interest: The authors declare no conflicts of interest. The funders had no role in the study's design, data collection, analysis, interpretation, manuscript writing, or decision to publish the results.

References

1. Williams, M.A.J. *Plant Galls: Organisms, Interactions, Populations*; Oxford Science Publications; Published for the Systematics Association by Clarendon Press; Oxford University Press: Oxford, UK; New York, NY, USA, 1994; ISBN 978-0-19-857769-0.
2. Fernandes, G.; Carneiro, M.A.A.; Isaias, R. Gall-Inducing Insects: From Anatomy to Biodiversity. In *Insect Bioecology and Nutrition for Integrated Pest Management*; Panizzi, A.R., Parra, J.R.P., Eds.; CRC Press: Boca Raton, FL, USA, 2012; pp. 369–395, ISBN 978-1-4398-3708-5.
3. Elzen, G.W. Cytokinins and Insect Galls. *Comp. Biochem. Physiol. Part A Physiol.* **1983**, *76*, 17–19. [CrossRef]
4. Dorchin, N.; Hoffmann, J.H.; Stirk, W.A.; Novák, O.; Strnad, M.; Van Staden, J. Sexually Dimorphic Gall Structures Correspond to Differential Phytohormone Contents in Male and Female Wasp Larvae. *Physiol. Entomol.* **2009**, *34*, 359–369. [CrossRef]
5. Gätjens-Boniche, O.; Jiménez-Madrigal, J.P.; Whetten, R.W.; Valenzuela-Diaz, S.; Alemán-Gutiérrez, A.; Hanson, P.E.; Pinto-Tomás, A.A. Microbiome and Plant Cell Transformation Trigger Insect Gall Induction in Cassava. *Front. Plant Sci.* **2023**, *14*, 1237966. [CrossRef]
6. Weis, A.E.; Walton, R.; Crego, C.L. Reactive Plant Tissue Sites and the Population Biology of Gall Makers. *Annu. Rev. Entomol.* **1988**, *33*, 467–486. [CrossRef]
7. Oliveira, D.C.; Isaias, R.M.S.; Fernandes, G.W.; Ferreira, B.G.; Carneiro, R.G.S.; Fuzaro, L. Manipulation of Host Plant Cells and Tissues by Gall-Inducing Insects and Adaptive Strategies Used by Different Feeding Guilds. *J. Insect Physiol.* **2016**, *84*, 103–113. [CrossRef]
8. Zhang, Y.M.; Gates, M.W.; Hanson, P.E.; Jansen-González, S. Description of a Neotropical Gall Inducer on Araceae: Arastichus, Gen. Nov. (Hymenoptera, Eulophidae) and Two New Species. *J. Hymenopt. Res.* **2022**, *92*, 145–172. [CrossRef]
9. Gibernau, M.; Albre, J.; Dejean, A.; Barabé, D. Seed Predation in Philodendron Solimoesense (Araceae) by Chalcid Wasps (Hymenoptera). *Int. J. Plant Sci.* **2002**, *163*, 1017–1023. [CrossRef]
10. Jansen-Gonzalez, S.; Teixeira, S.P.; Pereira, R.A.S. Mutualism from the inside: Coordinated Development of Plant and Insect in an Active Pollinating Fig Wasp. *Arthropod-Plant Interact.* **2012**, *6*, 601–609. [CrossRef]
11. Jansen-Gonzalez, S.; Teixeira, S.P.; Kjellberg, F.; Pereira, R.A.S. Same but Different: Larval Development and Gall-Inducing Process of a Non-Pollinating Fig Wasp Compared to That of Pollinating Fig-Wasps. *Acta Oecologica* **2014**, *57*, 44–50. [CrossRef]
12. Bomfim, P.M.S.; Kuster, V.C.; Caetano, A.P.S.; Gonçalves, L.A.; Oliveira, D.C. Ovule-Gall Stimulating a Large Fake Fruit on *Miconia chamissois* Naudin (Melastomataceae): A Structural Overview. *Botany* **2021**, *99*, 115–126. [CrossRef]
13. Brody, A.K. Oviposition Choices by a Pre-Dispersal Seed Predator (*Hylemya* sp.): II. A Positive Association Between Female Choice and Fruit Set. *Oecologia* **1992**, *91*, 63–67. [CrossRef] [PubMed]
14. van Doorn, W.G.; Stead, A.D. Abscission of Flowers and Floral Parts. *J. Exp. Bot.* **1997**, *48*, 821–837. [CrossRef]
15. Isaias, R.M.S.; Carneiro, R.G.S.; Oliveira, D.C.; Santos, J.C. Illustrated and Annotated Checklist of Brazilian Gall Morphotypes. *Neotrop. Entomol.* **2013**, *42*, 230–239. [CrossRef]
16. Formiga, A.T.; Silveira, F.A.O.; Fernandes, G.W.; Isaias, R.M.S. Phenotypic Plasticity and Similarity Among Gall Morphotypes on a Superhost, *Baccharis reticularia* (Asteraceae). *Plant Biol.* **2015**, *17*, 512–521. [CrossRef]
17. La Salle, J. Biology of Gall Inducers and Evolution of Gall Induction in Chalcidoidea (Hymenoptera: Eulophidae, Eurytomidae, Pteromalidae, Tanaostigmatidae, Torymidae). In *Biology, Ecology and Evolution of Gall-Inducing Arthropods*; Raman, A., Schaefer, C.W., Withers, T.M., Eds.; Science Publishers, Inc.: Enfield, UK, 2005; pp. 507–537.
18. Farache, F.H.A.; Pereira, C.B.; Koschnitzke, C.; Barros, L.O.; de Castro Souza, E.M.; Felício, D.T.; Gatti, F.; Cardona, W.; Rasplus, J.Y.; Pereira, R.A.S. The Unknown Followers: Discovery of a New Species of Sycobia Walker (Hymenoptera: Epichrysomallinae) Associated with *Ficus benjamina* L. (Moraceae) in the Neotropical Region. *J. Hymenopt. Res.* **2018**, *67*, 85–102. [CrossRef]
19. Felício, D.T.; Pereira, R.A.S. Among Us: How to Find an Optimal Clutch Sex Ratio When the Clutch Sex Ratios of Co-Foundress Wasps Are Not Known Beforehand? *Acta Oecologica* **2023**, *121*, 103950. [CrossRef]
20. Rouault, G.; Turgeon, J.; Candau, J.N.; Roques, A.; Aderkas, P. Oviposition Strategies of Conifer Seed Chalcids in Relation to Host Phenology. *Naturwissenschaften* **2004**, *91*, 472–480. [CrossRef]

21. Jansen-González, S.; Teixeira, S.P.; Pereira, R.A.S. Larval Strategy of Two Species of Seed-Feeding Chalcidoidea Parallels That of Parasitoid Koinobionts. *Oecologia Aust.* **2020**, *24*, 903–916. [CrossRef]
22. Elias, L.G.; Teixeira, S.P.; Kjellberg, F.; Pereira, R.A.S. Diversification in the Use of Resources by Idarnes Species: Bypassing Functional Constraints in the Fig–Fig Wasp Interaction. *Biol. J. Linn. Soc.* **2012**, *106*, 114–122. [CrossRef]
23. Tooker, J.F.; Helms, A.M. Phytohormone Dynamics Associated with Gall Insects, and Their Potential Role in the Evolution of the Gall-Inducing Habit. *J. Chem. Ecol.* **2014**, *40*, 742–753. [CrossRef]
24. Takeda, S.; Hirano, T.; Ohshima, I.; Sato, M.H. Recent Progress Regarding the Molecular Aspects of Insect Gall Formation. *Int. J. Mol. Sci.* **2021**, *22*, 9424. [CrossRef] [PubMed]
25. Borges, R.M. Interactions Between Figs and Gall-Inducing Fig Wasps: Adaptations, Constraints, and Unanswered Questions. *Front. Ecol. Evol.* **2021**, *9*, 685542. [CrossRef]
26. Elias, L.G.; Silva, D.B.; Silva, R.; Peng, Y.-Q.; Yang, D.-R.; Lopes, N.P.; Pereira, R.A.S. A Comparative Venomic Fingerprinting Approach Reveals That Galling and Non-Galling Fig Wasp Species Have Different Venom Profiles. *PLoS ONE* **2018**, *13*, e0207051. [CrossRef] [PubMed]
27. Joele, F.R.; Rezende, U.C.; Samacá-Sáenz, E.; Cardoso, J.C.F.; Oliveira, D.C.; Zaldívar-Riverón, A. A New Species of *Allorhogas* (Hymenoptera: Braconidae: Doryctinae) Inducing Ovule Galls on *Miconia chamissois* Naudin, a Potentially Invasive Shrub in the Brazilian Cerrado. *J. Nat. Hist.* **2019**, *53*, 2073–2085. [CrossRef]
28. von Aderkas, P.; Rouault, G.; Wagner, R.; Rohr, R.; Roques, A. Seed Parasitism Redirects Ovule Development in Douglas Fir. *Proc. R. Soc. B-Biol. Sci.* **2005**, *272*, 1491–1496. [CrossRef]
29. Larson, K.C.; Whitham, T.G. Competition between Gall Aphids and Natural Plant Sinks: Plant Architecture Affects Resistance to Galling. *Oecologia* **1997**, *109*, 575–582. [CrossRef]
30. Figueiredo, R.A.; Motta Junior, J.C. Análise Da Relação Entre Ficus Microcarpa (Moraceae) e Um Microhimenóptero Hospedeiro Na Região de São Carlos, SP. In Proceedings of the Anais do VI Congresso da Sociedade Botânica de São Paulo, Campinas, Brazil, 22–26 September 1986; pp. 103–107.
31. Neves, L.J.; Isaias, R.M.S. Ocorrência de Agente Galhador Em Flores de *Ficus microcarpa* L. *Bradea* **1987**, *4*, 327–330.
32. Farache, F.H.A.; Pereira, R.A.S. New Occurrence of Non-Pollinating Fig Wasps (Hymenoptera: Chalcidoidea) in Ficus Microcarpa in Brazil. *Neotrop. Entomol.* **2009**, *38*, 683–685. [CrossRef]
33. Silva, P.C.A.; Pereira, R.A.S. How to Be a Good Non-Pollinating Fig Wasp: Galling Wasps (Idarnes Group Flavicollis) Do Not Interfere with the Floral Receptivity. *Oecologia Aust.* **2018**, *22*, 426–437. [CrossRef]
34. Badenes-Perez, F.R.; Tracy Johnson, M. Ecology and Impact of *Allorhogas* Sp. (Hymenoptera: Braconidae) and Apion Sp. (Coleoptera: Curculionoidea) on Fruits of Miconia Calvescens DC (Melastomataceae) in Brazil. *Biol. Control* **2007**, *43*, 317–322. [CrossRef]
35. Quicke, D.L.J.; Fitton, M.G.; Tunstead, J.R.; Ingram, S.N.; Gaitens, P.V. Ovipositor Structure and Relationships within the Hymenoptera, with Special Reference to the Ichneumonoidea. *J. Nat. Hist.* **1994**, *28*, 635–682. [CrossRef]
36. Elias, L.G.; Kjellberg, F.; Farache, F.H.A.; Almeida, E.; Rasplus, J.Y.; Cruaud, A.; Peng, Y.Q.; Yang, D.R.; Pereira, R.A.S. Ovipositor Morphology Correlates with Life History Evolution in Agaonid Fig Wasps. *Acta Oecologica* **2018**, *90*, 109–116. [CrossRef]
37. Cerezini, M.T.; Rattis, L.; Furini, P.R.; Pereira, R.A.S. Contrasting Phenological Patterns and Reproductive Strategies in Closely Related Monoecious Fig Tree Species. *Plants* **2024**, *13*, 1889. [CrossRef] [PubMed]
38. Gottsberger, G.; Silberbauer-Gottsberger, I. Olfactory and Visual Attraction of Erioscelis Emarginata (Cyclocephalini, Dynastinae) to the Inflorescences of Philodendron Selloum (Araceae). *Biotropica* **1991**, *23*, 23. [CrossRef]
39. Ramírez, W. Evolution of Blastophagy. *Brenesia* **1976**, *9*, 1–13.
40. Clement, W.L.; Bruun-Lund, S.; Cohen, A.; Kjellberg, F.; Weiblen, G.D.; Ronsted, N. Evolution and Classification of Figs (Ficus, Moraceae) and Their Close Relatives (Castilleae) United by Involucral Bracts. *Bot. J. Linn. Soc.* **2020**, *193*, 316–339. [CrossRef]
41. Schiestl, F.P.; Dötterl, S. The Evolution of Floral Scent and Olfactory Preferences in Pollinators: Coevolution or Pre-Existing Bias? *Evolution* **2012**, *66*, 2042–2055. [CrossRef]
42. CIIAGRO Portal Agrometeorológico e Hidrológico Do Estado de São Paulo. Available online: http://ciiagro.org.br/ (accessed on 4 December 2024).
43. Gottsberger, G.; Amaral, A., Jr. Pollination Strategies in Brazilian *Philodendron* Species. *Berichte Dtsch. Bot. Ges.* **1984**, *97*, 391–410. [CrossRef]
44. R Core Team R: A Language and Environment for Statistical Computing. Available online: http://www.R-project.org (accessed on 4 December 2024).
45. Johansen, D.A. *Plant Microtechnique*; McGraw-Hill Book Company Inc.: New York, NY, USA, 1940.
46. Gerrits, P.O. *The Application of Glycol Methacrylate in Histotechnology; Some Fundamental Principles*; Department of Anatomy and Embriology, State University Groningen: Groningen, The Netherlands, 1991.
47. O'Brien, T.P.; Feder, N.; Mccully, M.E. Polychromatic Staining of Plant Cell Walls by Toluidine Blue O. *Protoplasma* **1964**, *59*, 368–373. [CrossRef]

MDPI AG
Grosspeteranlage 5
4052 Basel
Switzerland
Tel.: +41 61 683 77 34

Plants Editorial Office
E-mail: plants@mdpi.com
www.mdpi.com/journal/plants

www.ingramcontent.com/pod-product-compliance
Lightning Source LLC
Chambersburg PA
CBHW061501210926
51850CB00008B/1783
* 9 7 8 3 7 2 5 8 4 5 7 3 6 *